Temporal Structures and Observer Perspectives

SIMULTANEITY

editors

Susie Vrobel

The Institute for Fractal Research, Germany

Otto E Rössler

University of Tübingen, Germany &
The Institute for Fractal Research, Germany

Terry Marks-Tarlow

The Institute for Fractal Research, Germany

Temporal Structures and Observer Perspectives

SIMULTANEITY

World Scientific

NEW JERSEY • LONDON • SINGAPORE • BEIJING • SHANGHAI • HONG KONG • TAIPEI • CHENNAI

Published by

World Scientific Publishing Co. Pte. Ltd.

5 Toh Tuck Link, Singapore 596224

USA office: 27 Warren Street, Suite 401-402, Hackensack, NJ 07601

UK office: 57 Shelton Street, Covent Garden, London WC2H 9HE

British Library Cataloguing-in-Publication Data
A catalogue record for this book is available from the British Library.

SIMULTANEITY
Temporal Structures and Observer Perspectives

ISBN-13 978-981-279-241-9
ISBN-10 981-279-241-4

Printed in Singapore by World Scientific Printers

FOREWORD

Simultaneity — The Next Revolution in Physics

Relativity theory (Einstein), quantum mechanics (Planck) and endophysics (Rössler) are the outstanding contributions of the 20th century to the science of space and time. What they have in common is a new conception of the role of the observer. These physical theories of the world have made us aware that the existence of the world is dependent on observation. We cannot prove the existence of the world without an observer. But even more: there is a relation between the world and the observer that defines each other mutually. The world is not a physical system with purely intrinsic qualities, just as the observer, the human subject, is not either. The observer is evidently part of the world, of the system, which he observes. The three theories we have mentioned could be defined in a common book with the title "Being and Observing" or "Existence and Observation", complementing the classics of Heidegger's "Being and Time" (1927) and Sartre's "Being and Nothingness" (1943).

The observer-dependency of the world declares an end to the classical dichotomies of subjectivity and objectivity, even in physics. Questions like whether time and space are subjective modes of perception (Kant) or absolute features of an objective reality are solved in a new way. George Berkeley said already in his "Treatise Concerning the Principles of Human Knowledge" (1710): "Esse est percipi" ["To be is to be perceived"]. He contended that individuals could only directly know sensations and ideas of objects, not abstractions such as "matter". His dictum "To be is to be perceived" means precisely "to be is to be observed". Not only is physics, the science and theory of the material world, evidently a construction of observers, but also the world itself is dependent on the perspectives and interactions of the observers.

Therefore we have to investigate not only space and time, but also spatial and temporal observer perspectives.

If we understand and accept that time and space are neither subjective nor objective notions, but products of the interaction between the world and the observer inhabiting the world, in the sense of systemic qualities of the interface which constitutes the relation between the world and the observer and are, therefore, neither qualities of the world itself nor the observer alone, we will be enabled to investigate and analyze in a new way the concepts of synchronicity, simultaneity versus succession, and sequence.

The phenomenon of entanglement is not only the quantum effect of twin photons, describing two particles in a single quantum state such that when one is observed to be spin-up, the other one will always be observed to be spin-down and vice versa – this despite the fact that it is impossible to predict, according to quantum mechanics, which set of measurements will be observed. As a result, measurements performed on one system seem to be instantaneously influencing other systems entangled with it. In this example, we see very clearly the decisive role of observable physical properties of systems and their correlations. We can describe the world as a system of correlations between observable physical properties. This description evidently includes the observer in the description of the world. Therefore the classical strong assumption of the objectivity of the world can perhaps be replaced by a weaker assumption: the objectivity of correlations. But the properties of the observed system depend on the observer. Because only between observable physical properties can correlations be established. These correlations predicted by quantum mechanics, which Einstein famously derided as "*spukhafte Fernwirkung*" or "spooky action at a distance", have been theoretically confirmed in 1964 by John Stewart Bell, in what is known as Bell's inequality. It dismantles the principle of locality, which states that information about the state of a system can only be mediated by interactions in its immediate surroundings. Results of subsequent experiments have supported Bell's non-locality.

The universe, up to now built on absolute space and time, has, under the influence of relativity theory, quantum mechanics and endophysics, become a system of correlations between observable particles in space and time. It is, therefore, a system that includes not only space but also non-locality, not only time but also non-temporality. In such a system, which is characterized by observer-dependent concepts of the world, simultaneity becomes a key concept in the description of the world. This is because simultaneity can only be stated by an observer, who observes different events and measures them in the parameter of space and time. In such a universe of correlations simultaneity can be seen as the principal structure of the world. After the concepts of relativity and observability, which revolutionized classical physics, the concept of simultaneity is the logical consequence and the next step towards another revolution in physics. Simultaneity clearly is an observer perspective which calls into question all our assumptions about space and time hitherto described as a succession of events. Simultaneity describes the world as the nucleus of the Now. It divides the world not only into observable and unobservable, into measurable and immeasurable, into compatible and incompatible elements, but also into spatial and non-spatial, temporal and non-temporal states. The Now can metaphorically be described as a church where all physical beliefs and concepts of spacetime cohabit. A philosophy of the Now, described as the physics of simultaneity, is the ultimate extinction of the standard model of physics describing the material world.

Space and time, described in the classical absolute way, are "the prison bars of reality" (O.E. Rössler). If the observer is part of the fabric of the world, the prison of space and time becomes also a mutual interaction between the observer and the prison. Simultaneity becomes the escape button of the prison of space and time. Symmetry and complementarity support the concept of simultaneity in destroying the concept of a linear succession. Symmetry described as self-similarity and connected to scale theory produces a theory of scale relativity, which is a fractal structure. Fractality of space and time rather than absolute space and time turns the universe into a Cantorian spacetime manifold, in the sense of Mohamed El Naschie. The universe described as the mechanics of time and space, as a time and space machine, was one side of the story, told today in digital philosophy as the universe as a quantum computer. The brain as a time and space machine is the other side of the story. With the concept of simultaneity we start a new story of the universe. It provides the key to open a new door, a new interface into the world of space and time and a new tool to liberate us from the prison of space and time.

Peter Weibel, Karlsruhe
15 November 2007

PREFACE

The Institute for Fractal Research was founded in 1997 in Kassel, Germany, as an independent and non-profit research organization. This festschrift celebrates its 10th anniversary with a selection of research results and ideas which tackle the notions of simultaneity and temporal observer perspectives. We hope that the interdisciplinary nature of this book will trigger a fresh discussion about the notion of temporal perspective between the humanities and the natural sciences. If it gives rise cross-insemination, our purpose has been achieved. Many thanks to our friends and colleagues for their fine contributions.

Many thanks also to my husband Barry Baddock for helping with editing tasks, to Franz-Günter Winkler for solving software problems and to Lakshmi Narayan for doing an excellent job as editor.

S.V.

CONTENTS

Part II: Identifying Temporal Observer Perspectives

Part III: Disentangling Temporal Simultaneous Contrasts

INTRODUCTION

Simultaneity is a fundamental concept which is implicit in all hypotheses, theories and models of the world. However, definitions vary greatly in the various disciplines and so it should not come as a complete surprise that there are numerous concepts in use which are not easily translatable into each other. In fact, some notions of simultaneity are incommensurate, as the underlying concepts of time and temporal structures relate to specific observer perspectives.

This book is an attempt to make visible implicit assumptions about temporal structures and observer perspectives in philosophy, psychology, physics, medicine, neuroscience, education, economics, logic and mathematics.

Part I deals with the epistemological background against which observer perspectives and the notion of simultaneity may be understood. **Susie Vrobel** introduces a phenomenological model of fractal time, which defines the structure of the Now as a nesting cascade of memory and anticipation. The resulting internal observer structure determines the extensions of the observer participant, whose boundaries are flexible and depend on where the interfacial cut is set between the observer participant and the rest of the world. Simultaneity arises both from a superposition of levels of description which are constrained by our perceptual apparatus and our internal nesting cascades, which together generate our Now. In order to explain our perceptions and the model we construct of our phenomenological selves, two mutually exclusive temporal dimensions of time must be assumed: Δt_{length} and Δt_{depth}: the length of time, which consists of incompatible events and generates succession, and the depth of time, which comprises compatible events and forms our simultaneity horizon.

Otto E. Rössler presents a deductive approach to brain function and temporality on the basis of the ordinary (classical) level of physics, to which he also assigns the level of Darwinian biology and brain physiology. Natural brains predictably possess two components: The force field generator, which has a preferred time direction, and the great simulator, which can, in principle, work in both directions of time. Mirror neurons are interpreted as a predictable manifestation of the great simulator. More incisive than the direction of time is the assignment of the Now. Only to people who are aware of the "double contingency" of both being at a certain moment in time and knowing why that double contingency is assigned at this very moment, does the Now have a special status. But people in love or caught in the gaze of a lizard know that time does not flow, know that it is frozen. But although we may experience frozen time, we always relapse in our daily lives and tend to forget to return afterwards. The structure of the Now may be pictured as a temporal self-similar fractal which bites itself into its own tail. Rössler introduces the notion of the "temporal fovea", which implies a self-similar structure of time over a certain range.

Mike Luetchford interprets the Japanese philosopher Dogen's concept of the Now as an indivisible whole in which we neither arrive nor leave: we are always here and now. Each Now is the universe and we are not able to escape from here-and-now reality. Remembering the past and anticipating the future also happens within our Now and it is in the Now that we construct a concept of self and the world. And when we come to accept that only the Now is real, we realize that everything is contained in the Now. There is no part of the universe we could name which is not part of the Now: apart from the present moment, it also contains both past and future. This view of time as being inseparable from the individual concrete events we experience describes the perspective of a participant rather than that of an observer.

A holistic viewpoint on systems and observers by **Franz-Günter Winkler** introduces the idea of observers and the interaction between observers and the world as two overlapping systems. Only a hypothetical outside observer has direct knowledge of the world. For the inside observer, knowledge is always about the dynamical interaction between the observer and the world. It is argued that the notion of overlapping systems, which is not inherent in the theory of self-organization, requires a holistic world view. The key assumptions of Winkler's spacetime holism are the containment of the whole's structure in the part and a duality relation for the structural properties of continuity and discontinuity in space and time.

In a systems-theoretical approach, **Nikolaus von Stillfried** suggests a Generalized Quantum Theory (GQT) to describe the phenomena of complementarity and non-locality. GQT suggests these phenomena are not the result of particular properties of elementary particles but arise from the systemic arrangement of these particles. Sufficient degrees of freedom in the subsystems and the conserved global property of the system as a whole are the most important systems-theoretical parameters for the description of non-local correlations. These are not linked by a causal interaction but may occur simultaneously. Against the background of these assumptions, synchronistic phenomena maybe interpreted as instances of system-inherent non-local correlations. The typical lack of reproducability is discussed as a consequence of experimental set-ups which do not take account of the temporal correlations which connect the whole, but rather focus on the notion of ideal isolated subsets.

Uri Fidelman's differentiation between brain time and physical time is based on Kant's idea that time and space are subjective modes of perception. If the concept of time changes according to the cerebral mechanism that deals with it, time is a subjective notion. The left hemisphere organizes the extracted data successively in time, while the right one organizes data adjacent to each other in space. The analytical nature of the left hemispheric perception of data implies the quantization of time, as that data is perceived like beads on a string and not as a continuous line. Thus, the Now is a discrete interval of time. However, the simultaneity generated by the right hemisphere causes it to perceive phenomena within continuous space. The inter-hemispheric interaction generates reality as we perceive it. In this process, both hemispheres are at work, but generate temporal structures which cannot be translated into each other but can only be mutually integrated. Based on the idea that there are physical incommensurate inputs (electromagnetic and gravitational forces) which are processed independently, Fidelman suggests that we have to assume a multidimensional right-hemispheric temporal "space", which comprises at least four independent and perpendicular axes.

In Part II of this volume, the generation and modification of temporal observer perspectives are identified. **Gerald L. Clore** looks at the impact of simultaneity on the way we construct momentary experience. Emotions are described as emergent states which exist only to the extent that multiple affective reactions to the same object occur at the same time. If lower-level affective reactions are re-iterated and re-interpreted, true emotions emerge in the shape of a meta-cognitive representation of one's embodied affective reactions to events. Clore describes how affective and emotional reactions act as information about the value of objects of judgement and of accessible cognitions and inclinations during tasks. They influence judgement and thought when they are experienced simultaneously with sensory data about the world. Affective influences thus depend on our inability to disentangle affective from descriptive perceptions. To the extent that affective reactions reflect different, incommensurate sources of value (e.g., utilitarian, moral, aesthetic), perceived persons or objects may be experienced as being transcendently good or evil. Experiments varying people's attributions for their affective experiences allow the separate roles of affective and descriptive information to be examined. However, it is the inability to parse everyday experience into its separate sources of evaluative and descriptive information that gives rise to a colourful and transcendent reality.

Hinderk M. Emrich, **Catharina Bonnemann** and **Detlef E. Dietrich** suggest envisaging the brain as containing aspects of a "time machine": The brain is a neurobiological physical system like other systems – existing in time – being an object of understanding by physics, by biophysics, especially of neurochemical oscillators and thus is subject to the concepts of chronobiology. The human brain is a "time machine", since it can memorize "stored regularities" and can recall them. It thus contains aspects of "time transcendence". Human beings are – in the sense of the phenomenology of Edmund Husserl – able to perform protension, i.e. the conceptualization of the future as well as retention, a reference to the past. We are also able to generate the time experience of the "now", of the "presence within time". This raises the question how to relate, how to synchronize the "outer", the so-called Aristotelian or "physical" time to the "time of inner experience", i.e. the time of Augustinus, and the Bergsonian time. Psychiatric disorders as well as extraordinary experiences of normal life tend to induce problems within this type of coordination, and depression represents a mood disorder, within which, also experimentally, a disturbance of the inner organization of time in relation to objective time can be demonstrated. The brain is a "time machine" in a Kantian understanding, insofar as the human mind can construe the universals of perception, especially the category of "time". The authors give an account of how emotionality of subjective time experience in depressed patients relates to the dominance of the past in relation to the spontaneity of the subject within a present moment of time. The role of forgetting to overcome such memory-related impairments is discussed within the context of possible functions of consciousness.

Martina Piefke and **Hans J. Markowitsch** describe memory formation and retrieval in episodic memory, which includes both autobiographic recollection and laboratory episodic memory. The episodic memory system operates on the level of explicit processing of information (conscious memory). Episodic memory is highly dependent on context, such as mood states, and is strongly interconnected with the ability to take a self-perspective and the formation of a self-concept. Vice versa, self-concepts can provide mental contexts against the background of which autobiographical events may be re-interpreted. While in studies of laboratory episodic memory, context parameters can be assessed by experimental manipulations, this proves difficult for autobiographical memory, for which context parameters are far more complex and cannot be easily controlled. Laboratory experiments suggest that contextual variables mainly affect prefrontal functions. In autobiographical memory, not only prefrontal, but also further medial temporal and posterior parietal regions, which mediate there-experience and emotional evaluation of personal memories, are highly susceptible to changing context variables of memory encoding and retrieval. The authors suggest that experimental and autobiographic episodic memory are influenced by both overlapping and differential context parameters.

Peter M. Allen looks at the hierarchical temporal organization in the evolution of systems and shows why investigation into temporal aspects of evolution needs to take into account multiple levels if description: Individuals are bundles of their internal components, the local community or organizations they form are bundles of these individual types, and ecosystems and larger structures they form are bundles of these local communities. The co-evolution of the successive layers of interacting elements is at work both horizontally and between levels, driven by the generation of microdiversity: diversity at the level below. His notion of "Evolutionary Drive" may be summed up by saying that evolution is driven by the noise to which it leads. Evolution and the creation of structures of seeming stability and large temporal spans are driven by the internal temporal hierarchies of processes. Allen denotes temporal islands of stability of emergent clusters of ideas, components, practices or forms "Structural Attractors": dynamical systems that emerge and persist for a time. Descriptions of seemingly stable structures may be useful in the short term, but will fail to provide a real indication of the potential for instability and structural change. Complexity teaches us, therefore, to permanently distrust our current knowledge and maintain constant doubt as to its truth.

Jerry LR. Chandler introduces a hierarchical nesting scheme to visualize the emergence of biological dynamics from predicate, sub-ordinate, ordinate and co-ordinate logics. The development of the capacity for scientific communication depends on creating synthetic symbol systems. Many such symbol systems co-exist, each justified by a localized interpretation of distributive and associative concepts. Therefore, communication among technical disciplines is difficult to realize in practice. Chandler introduces the concept of ur-symbols, which create a common basis for the ordinate logics. These, in turn, provide a common basis for integrating mathematical, physical and chemical reasoning. Multiple encodings and decodings of symbols play a crucial role in quantitative thinking. For example, the information encoded in atomic numbers is functionally separate and distinct from the information encoded in real numbers. The concept of ur-symbols facilitates multiple interpretations. Thus, correspondence between observations and scientific calculations depends on the choice of meaning for symbols used in the calculations.

Martin F. Gardiner discusses how skill of every kind must depend on brain processing that allows the individual to engage behaviourally to the extent of observed capability. For behavioural capability to improve the opportunities of the processing ensemble for brain engagement must improve as well. Gardiner discusses three types of context in which development of skill takes place, which can influence the extent and nature of capability achieved. These include context of use for the capability expected and intended by the learner, context of relationships of outside authority and personal experience of the capability, and context of relationship of processing components called upon by capability in one area of skill to similar processing components used to develop capability in other areas of skill. Gardiner proposes that learning connects arts skill to other skill learning may be especially useful to the skill learner by fostering incorporation of forms of processing especially favored by brain evolution. He proposes that the use of processing favored for other reasons by evolution can help to explain the appearance of the arts within human culture.

Terry Marks-Tarlow describes the advantages of perceiving time not as a linear, one-dimensional extension, but as a fractal structure, which reflects and resonates with the temporal nested structures within and around us. By adding the vertical axis of simultaneity to the horizontal axis of succession, time can be made arbitrarily dense. The nesting of temporal intervals of various lengths allows us to keep in mind the whole while being immersed in the moment. The simultaneity of fractal time is the opposite of multi-tasking, which leads to a fractured focus. In contrast, utilizing fractal time means focussing on the level of description of the current task or activity. Long-term projects embed projects which require shorter time spans. The art of utilizing fractal time consists of a level-hopping approach, in which total attention is given to a certain activity on one nested level of description, which fits the temporal extension available. Activities linked with longer or shorter time spans, which embed and are nested in the currently focussed level, form a background against which the momentary activity is connected to the whole fractal temporal nesting cascade. As our bodies are fractal timepieces, tuning into and becoming aware of temporal fractal patterns increases our ability to self-reflect on our unconscious knowledge.

Part III of this volume takes a closer look at temporal simultaneous contrasts and how they can be disentangled. **Laurent Nottale** and **Pierre Timar** present the theory of scale relativity, which is an extension of the principle of relativity to scale transformations of the reference system, in a fractal geometry framework where co-ordinates become explicitly dependent on resolutions. Applied to an observer perspective, this means that the scales of length and of time, usually attributed to the observed object as being intrinsic to it, have actually no existence by themselves, since only the ratio between an external scale and an internal scale, which serves as unit, is meaningful. Oliver Sacks' observations on patients suffering from temporal and spatial distortions in Parkinson's disease and encephalitis lethargica disease offer a particularly relevant field of application for such a scale-relativistic view.

Bill Seaman unpacks simultaneity for differing observer perspectives and qualities of environment. He writes about differing modes of observation including direct human observation; human observation that is augmented by differing tools and/or machines; computational models functioning as machinic observers, in which humans become super-observers; chemical observers where particular changes register particular qualities; and potential future observers: neosentient machines. For each of these different kinds of observer, Seaman discusses how simultaneity might potentially be seen to be problematic. In particular, he points out that notions of time are different on the micro and macro scale. He also discusses the fact that humans have been observed to include up to a one half-second time buffer before awareness of an event is registered, adjusting for the body's distributed sensing functionality and, potentially, its embodied relation to memory and past thought.

Dieter De Grave talks about the link between the theory of complexity (philosophy) and Freudian/Lacanian thinking (psychoanalysis). Using two key concepts – sexuality and death – he talks about the castration complex as a way to apply complex thinking to the human mind and human existential suffering. In this enterprise, the overlap and the differences between these two fields are brought to the fore. Both fields stand in a synergetic relationship and, with a little effort towards translating certain insights from one approach to the other, we can come to novel ideas and advances in each field. But in the synergetic approach we always have to keep in mind that certain concepts are specific to the field itself and if we simply equate them with other ideas from the other field important losses should be considered. The idea of self in self-organization is used as an example of these bridging difficulties.

Serge F. Timashev and **Yuriy S. Polyakov** present a review of Flicker-Noise Spectroscopy (FNS), a time-series analysis which is based on the idea of measuring chaotic signals simultaneously on nested levels of description. Correlation links between irregularities and discontinuities in derivatives of various orders on all levels of the spatio-temporal hierarchy of a system are treated as the main information carriers. Apart from determining the characteristic parameters of open complex systems, FNS, identifies precursors to abrupt changes in the state of open dissipative systems and determines flow dynamics in distributed systems on the basis of correlations in stochastic signals which are measured simultaneously. FNS has been applied in the analysis of the structure and dynamics of physiochemical, electrochemical, biological, medical, geophysical and astrophysical processes. An overview of FNS is presented and exemplified by an analysis of EEG signals.

Raymond Pavloski's approach to conscious observer perspectives portrays the emergence of phenomenal experience as a kind of hidden pattern formation. His neuron models support the hypothesis that some recurrent networks can generate hidden patterns which may be interpreted as hidden perspectives. In these models, clusters of neurons represent meta-states as whole entities, some of which are capable of generating emerging properties. A circular causality emerges as a result: model neurons use indices of cluster states to determine their own states, which are in turn parts of cluster states. The states of cluster-cluster interactions determine their influence on the states of other clusters. Descriptions of subjective perspectives may fit self-organized patterns embodied in the dynamics of the neural network. These patterns are revealed only by considering each neuron's perspective on this network.

Michail Zak defines common sense as feedback from the self-image and proposes a physical model of common sense: Human behavior is governed by feedback from the external world, a process which can be simulated by control systems. However, if the external world does not provide sufficient information, human beings turn for "advice" to their experience, which is associated with common sense. As many natural and social phenomena exhibit some degree of regularity only on a higher level of abstraction, i.e. in terms of some invariants, within the mathematical formalism of the model of mental dynamics, this means that only variables of the highest level of abstraction are capable of classifying changes in the external world. As follows from this model, the difference between behaviours of living and non-living systems is fundamental: The evolution of living systems is progressive in the sense that it is directed to the highest levels of complexity if complexity is measured by an irreducible number of different parts that interact in a well-regulated fashion. This property is not consistent with the behaviour of closed Newtonian systems, which cannot increase their complexity without external forces.

Part IV of this volume focusses on the notion of synchronization. **Fabian Ramseyer** and **Wolfgang Tschacher** approach the phenomenon of synchrony from a framework of embodiment and embodied cognition. A review of the evidence of nonverbal synchrony in human communication reveals the complexity of synchrony in interactive systems. Synchronized nonverbal movement of patient and therapist in dyadic psychotherapy is a readily accessible example of synchrony and an observable (embodied) manifestation of relationship formation. The authors introduce a new empirical approach for the study of nonverbal synchrony in psychotherapies under field conditions that is based on an automated video-analysis algorithm. This new technique allows them to explore the connection between nonverbal synchrony and the evolution of the quality of the therapeutic bond during the course of psychotherapeutic treatment. Their findings indicate that synchrony as a global concept of interrelatedness has high integrative value and its application can be broadened to human interaction in general.

Gerold Baier and **Thomas Hermann** describe the advantages of multi-scale auditory inspection of experimental data over approaches which are based on visual signals only in the diagnosis of dynamical diseases. Sonification provides a means of highlighting the temporal aspects of a time series rather than static ones as are associated with data in a visual representation. As bodily rhythms occur on many nested time scales, from milliseconds for neural firing to minutes for intracellular messenger signaling to hours for episodic hormone release, an integrated understanding of a dynamical disease requires that many or all parallel rhythms of the body must be studied. Sonification allows the almost real-time study of multirhythmic processes, such as those associated with dynamical diseases. Our ability to detect highly differentiated parallel temporal patterns with our ears creates a new temporal perspective in the scientific observer.

Christian Tschinkel introduces the concept of spatial composition which results from the perception of simultaneous auditive contents. Simultaneity is created by dichotic listening, which means that each ear is fed a different content. Dichotic listening was originally employed in lateralization experiments on brain hemispheres. This new kind of hearing emerges as a result of the superposition of auditive contents. The degree of complexity associated with dichotic hearing correlates with the degree of attention given by the listener and his appreciation of the auditive content. Thus, the internal complexity of the listener decisively determines the soundscape of the auditory reality he generates.

Gus Koehler shows that two-dimensional simulation programs do not take account of heterochrony and the time-ecology of nested temporal systems, which are based on Frazer's five hierarchically nested temporal levels in nature (nootemporality, biotemporality, eotemporality, prototemporality and atemporality). As multiple layers of temporal levels are interconnected both locally and, via pipes, also to other local portions of the time-ecology, causal relationships are difficult to identify. Therefore, Koehler calls for a program which takes account of varying nested hierarchies of time, heterochrony and limitations imposed by velocity cones. A virtual time-ecology landscape could produce emergent higher-level agents as a result of evolutionary selection of, among other parameters, temporal boundaries, such as velocity cones, and heterochronic flows and rates.

Artur P. Schmidt and **Otto E. Rössler** offer an endo-view on economics. A new interface is described that systematically analyses hidden patterns in complex systems. Economic temporal structures evolve through parallel processes that are correlated through feedback loops. It is only possible to reach an endo-economic-perspective and make hidden structures visible when they are able to design a simultaneous interface that is capable of looking behind the curtain of attractors and forces. The authors' first article proposes a new damping method for the rational exploitation of flaring-type economic phenomena. When systems flare or get out of control, they can show irrational exaggeration. The so-called Livermore Indicator (calculated through the software I-Matrix) can be used for anti-flaring steering procedures, for example, by the Federal Reserve Bank. The second article shows the existence of a new attractor in economics which can be made visible through matrix analysis. This software-based observing method reveals the existence of an organizing force vector in complex dissipative systems like the economy. The third article proves that the cybernetic analysis of inflation can be used as an x-ray tool to show hidden patterns in the economy. Any boom-and-bust cycle can be identified if one uses the spectacles of a detached observer in an approach called endonomics.

Uwe Kämpf investigates the possibility of accounting for cortical synchronization as being the result of the Wheeler-Feynman absorber effect. In quantum brain dynamics, this interpretation would provide an anticipative resonance coupling model for aspects of cortical synchronization and recurrent visual action control. This model describes registered activation patterns of neuronal assemblies in synfire chains not as resulting from retarded communication in processing but, rather, as the surface of a system of standing waves generated by visual processing. An analogy is drawn between the resonant transactional control mechnisms of visual space-time representation and spatio-temporal non-locality as observed on the quantum scale. In his time-loop model, Kämpf suggests that findings about mirror neurons may be at least partially associated with temporal rather than spatial mirror functions of visual processing, similar to the phase conjugate adaptive resonance coupling of nonlinear optics. On a speculative note, Kämpf suggests that the reported findings, which support the hypothesis of the existence of macroscopic, mesoscopic and microscopic absorber effects, may provide us with an analogy between the resonant transactional control mechanisms of visual space-time representation and the spatio-temporal non-locality effects which have been found on a quantum-mechanical scale and, possibly, also on a neurophysiological and even on a cosmological scale.

Otto van Nieuwenhuijze describes how time is always connected to our involvement with/in our context. Time is therefore always related to the functioning of our body. In our body, our experience of time is very complex: there are many different cells, each of which has its own time base. All of these cell processes are interrelated, because all of our body cells have a common origin in the zygote, the first cell of our body. This approach offers a basis for unifying our understanding of time, and for understanding the experience of time and timing in our body and provides a fundamental notion for the understanding of health. All cells synchronise their actions (for this purpose, each cell has a Gap Phase for adjusting the timing), thus diseases of timing may arise in our bodies. Van Nieuwenhuijze concludes that Vrobel's Time Fractal offers a useful research tool for integrating all of these aspects.

The notions of simultaneity and temporal observer perspectives presented in the various chapters differ with respect to their presuppositions and logical distinctions. They comprise hierachical and non-hierarchical models, nested and overlapping structures as well as local and non-local perspectives.

It is the aim of this book to trigger a fresh discourse between disciplines on the nature of time, temporal perspectives and the structure of our Now: our interface and only access to the world.

Part I

Observer Perspectives: Epistemological Background

CHAPTER 1

FRACTAL TIME: EXTENDED OBSERVER PERSPECTIVES

Susie Vrobel

The Institute for Fractal Research
Goethestrasse 66, 34119 Kassel, Germany
E-mail: Susanne.Vrobel@t-online.de

In order to describe a temporal observer perspective, we have to assume two mutually independent temporal dimensions: the depth and the length of time, which respectively correspond to simultaneity and succession. These define the degree of complexity a temporal observer perspective displays. Observer types may be distinguished in terms of their ability to create simultaneous contrasts or to resist contextualization. The setting of the interfacial cut between the observer participant and the rest of the world determines the structure of the fractal temporal observer perspective by assigning rhythms to the observer or to the rest of the world. Misattributions occur when the observer perspective becomes hard-wired and immune to external perturbations, i.e., if only the observer perspective but no external rhythms determine his interface reality.

1. Introduction: One Dimension is not Enough

If we wish to describe the nature and structure of time, we face a Gödel limit. As observer participants, we are already embedded in the subject matter we wish to define: time. We cannot describe temporal structures from an exo-perspective (from an outside vantage point of view), as this perspective is reserved for idealized constructs, such as Laplace's demon, which find no counterpart in the real world. We are limited to an endo-perspective (the view from within), as we are part of the system we wish to describe. The notion of the endo-perspective was introduced by

Otto E. Rössler in his seminal work *Endophysics*[1]. In a nutshell, this perspective is determined by the observer's internal organizaton, i.e., his internal microscopic movements. As Ernst Pöppel pointed out, also our modes of perception act as constraints to the way we generate reality: our models and theories are based on our primary experiences of time, namely, succession, simultaneity, duration and the Now[2]. These primary experiences are based on embodied cognition. On a cognitive level of description (hereafter denoted as LOD), our endo-perspective is also structured by metaphors which arise from the limitations set by the observer's sensory-motor system[3].

As observer participants, we have no access to the *noumenon* and face a limit to conscious introspection of the prerequisites which constrain our perceptions. Given these limitations, a phenomenological approach which focusses on the world-observer interface is a promising starting point. The ideas and assumptions presented in this chapter therefore refer to interface reality – the boundary which embraces both the observer participant and the environment he interacts with. Particular emphasis is given to the notion of the observer's Now, an extended and structured interval which defines the observer's event horizon. As Rössler put it, Nowness is pure interface[4].

The way we generate succession, simultaneity and our Now is not a trivial matter. Our perceptional constraints determine whether we experience a visual, auditory or tactile sensation as one event or as a number of successive ones. An auditory event spans a shorter interval than a tactile or a visual one. So we experience several modes of perception in parallel, as a temporal nested pattern, which forms our Now. The Now is therefore not a point which divides the past from the future, but an extended field, which hosts a structure of temporally nested events – a fractal.

The first person to describe a nested structure of the Now was the German phenomenologist Edmund Husserl[5]. He could not refer to the term fractal, as his treatise appeared 80 years before Mandelbrot's seminal book[6]. Husserl asked the simple question as to why we are able to perceive a tune, as opposed to a series of unrelated notes. He concluded that this must be so because we remember the note we have just heard (retention) and, assuming that this was not all, anticipate the next note (protention), in our consciousness of the present, the Now. Every Now embeds the preceding one, including its differentiated structure of retentions and protentions, i.e., the memories of the notes that have been played and those we anticipate to follow. This leads to a nesting cascade of Nows, each hosting both retentions and protentions. The current Now forms the framework time into which all previous and expected events are embedded, thus generating a fractal structure. If we were only able to hear successive, uncorrelated notes, no embedding could take place and our Now would not be extended by simultaneous nested structures, i.e., we would not be able to hear a tune or perceive any other time series as a meaningful entity. However, as we *are* able to perceive a tune and not just a succession of isolated notes, we must assume the Now to be extended and provide for *both* succession and simultaneity.

Below, I shall introduce a fractal model of time by means of which we can describe nested temporal structures in an extended Now by differentiating between succession and simultaneity as two mutually exclusive temporal dimensions.

2. Fractal Time

My Theory of Fractal Time adopts Husserl's notion of a nested Now[7,8,9]. It provides a means of quantifying the internal structure of the observer's interface by differentiating between the length of time, $\Delta t_{length,}$ the depth of time, Δt_{depth}, and the density of time, $\Delta t_{density}$. These concepts allow us to describe our temporal interfaces as nested temporal perspectives. Δt_{length} is measured in the number of incompatible events in a time series, i.e., events which cannot be expressed in terms of during-relations (simultaneity). It defines the length of time, the temporal dimension of succession for individual LODs. Δt_{depth} is measured in the number of compatible events in a time series, i.e., events which can be expressed in terms of during-relations. It defines the temporal dimension of simultaneity for two or more LODs. $\Delta t_{density}$ is the fractal dimension of a time series. It measures the density of time, i.e., the relation between successive and simultaneous events.

Δt_{depth} provides the framework times which allow us to structure events in Δt_{length} on individual LODs. Therefore, Δt_{depth} logically precedes Δt_{length}: there is no succession without simultaneity. Δt_{length} and Δt_{depth} are mutually exclusive, i.e., an observer participant cannot generate both at the same time. In retrospect, however, an interval of a time series may be measured in both dimensions and the resulting fractal dimension, which measures the density of time in that interval.

Any multi-layered signal may be considered as an example of a fractal time series. For instance, the least complex frequency ratios of musical notes which are played simultaneously is 2:1. This ratio defines the interval between the notes as an octave. The next overtone of a frequency of, say, 440 Hz (the note A played on an oboe) would have a frequency of 880 Hz, and so on[10]. As the overtones are integer multiples of the fundamental frequency, the overtones generate a nesting cascade whose LODs can be translated into each other. If a temporal pattern displays a self-similar structure, i.e., if the embedded LODs host structures which are identical to those of the embedding LODs, albeit of different extension in Δt_{length}, it is always possible to translate between nestings. In the musical example of nested overtones, this leads to consonance created by overlapping frequencies which are easily translatable into each other in terms of Δt_{length} and Δt_{depth}.

In general, we can say that both temporal dimensions (Δt_{length} and Δt_{depth}) must be assumed in order to explain our perception of a multi-layered signal. $\Delta t_{density}$ is a useful measure for comparing time series. To define a temporal observer perspective, however, the notions of Δt_{length} and Δt_{depth} suffice.

Fractal time is a generalization, as the Newtonian metric of time may be defined as a special case of fractal time metrics. A fractal clock can be imagined as an infinite number of pointers running on the perimeter of the triadic Koch island (a snowflake-like structure with an infinitely complex boundary: see Figure 1), with all pointers ticking away simultaneously[11,12].

So while pointer no. 1 ticks, say, only three times (per lap), pointer no. 2 is ticking 12 times, pointer no. 3 is ticking 48 times, and so on, ad infinitum. If the infinitely nested structure of the triadic Koch curve is projected onto a one-dimensional straight line, it forms a continuum, and thereby, a Newtonian metric: the set of points generated in this way is the set of rational numbers. Thus, in terms of fractal time, the Newtonian metric may be defined as Δt_{length} of the nesting level ∞, i.e., $\Delta t_{depth} = \infty$.

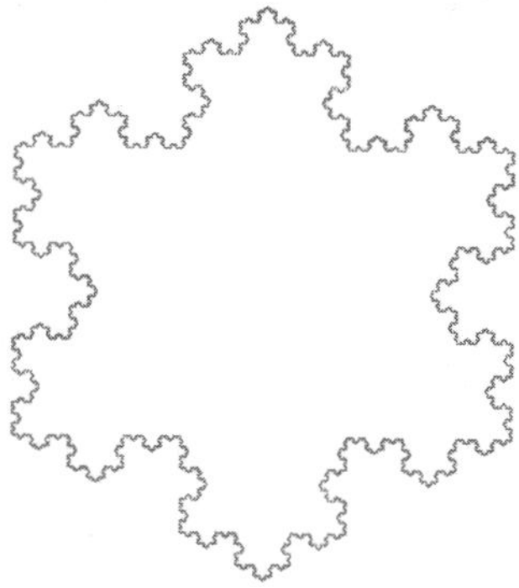

Fig. 1: A fractal clock

Fig. 2: The Prague Orloj – a fractal clock

The oldest fractal clock is the astronomical clock in Prague (Orloj: see Figure 2), dated around 1410. It shows several nested timescales: minutes, hours, days and months, as well as the night sky throughout the year, the ascent and setting of the stars, the lengths of the days and nights throughout the year, the moon phases and the festive days. It also shows Bohemian time (from sunrise to sunset) and the signs of the zodiac. Underneath this clock is another, divided into 24 sections. It depicts the seasonal changes and corresponding scenes of rural life.

3. The Nested Now: Simultaneity Horizons

The nested structure of the Now is the result of fractal temporal structures within the observer and in his environment. Outside the observer, for instance, tidal rhythms are embedded into seasonal rhythms, which again, are embedded in astronomical ones. Within the observer, for example, neural oscillations are embedded into much slower metabolic ones[13]. Also our sensory perception is fractal: Pöppel[2] showed in his neuro-scientific approach how our perceptual constraints create a simultaneity horizon, at which we perceive audio and visual signals as occuring simultaneously. We perceive audio signals as non-simultaneous if they are separated by an interval of roughly 6 milliseconds. If that separating interval is shorter, we perceive audio signals as being simultaneous. Visual impressions which are separated by an interval of 20 to 30 milliseconds are experienced as non-simultaneous. Below this threshold, impressions are perceived as simultaneous. The simultaneity threshold for tactile impressions lays at approximately 10 milliseconds. Thus we perceive tactile, visual and audio events as temporally nested, since the interval which defines them as one event rather than two successive ones varies for each sensory realm.

According to Pöppel, this temporally nested structure is perceived as one Now if we have a clustering of perception-related experiences which are based on meaningfulness, i.e., perceptual gestalts which are constructed by our brains. He does not refer to a nested or fractal structure, but his examples implicitly describe a fractal perspective. Thus, the extention of the Now depends on the mental capability of the person who experiences an event. The more differentiated and richer the language, the more complex are the perceptual gestalts a person may construct and the more extended is the Now. An extended Now requires both the identification and integration of perceptual gestalts[2] which, in terms of fractal time, may be denoted as a nesting of meaningful events, and memory, the nesting cascade of past Nows within the current one. The measuring rod for the Now is thus situated within ourselves: the temporal extensions of events, which may be turned into perceptual gestalts by means of integration. Therefore, we may say that a complex observer perceives a complex environment in a decomplexified manner[14,15,16]. The observer participant's degree of fractality determines the temporal complexity of the world he generates.

These observations also support the view that the more complex the observer participant's internal differentiation is, the higher his empathic skills will be, as he is able to simulate and contextualize what happens in the external world.

Contextualization is both an innate and an acquired skill. There are innate correlations such as the female cycle and moon phases, mating seasons etc. On the other hand, there is acquired entrainment, which comes in the shape of adjusting to artificial timetables and other external constraints. However, the ability to contextualize may be impaired for certain observer participants, so these individuals generate simultaneity contrasts which differ from those of observers whose nesting capacity is not compromised.

An example of a simultaneous contrast, i.e., an illusion which some observer types fall for and others resist, is Dakin's experiment, in which individuals with symptoms of schizophrenia and a control group were exposed to a simultaneous contrast[17]. Either an isolated disc (see Figure 3, left margin) or a disc with a high-contrast surrounding disc was presented. Subjects were then asked to report which disc appeared to be of higher contrast. Control group subjects fell for the visual illusion induced by the simultaneous contrast. To them, the central disc appeared to be of lower contrast, as the embedding, larger disc of higher contrast influenced their perception of the embedded disc. Schizophrenics did not contextualize the nested disc, and thus did not fall for the visual illusion because the outer high-contrast disc did not have any impact on their perception of the embedded one.

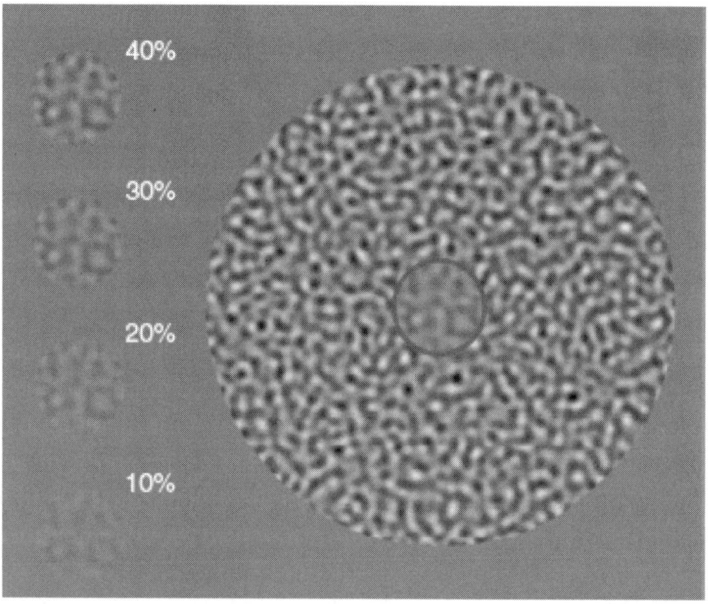

Fig. 3[17] [Reprinted from S. Dakin, et al.] Weak suppression of visual context in chronic schizophrenia. *Current Biology* 15, p. R823 (2005) with permission from Elsevier.

Unconscious nesting performances, such as the one described above, shape the structure of our interface by means of inducing or inhibiting simultaneous contrasts. As Dakin's experiment shows, sometimes an apparent handicap turns out to be a blessing in disguise, when failure to contextualize results in the correct matching of external stimuli on one LOD. Although matching discs is not in itself a useful skill, the ability to resist contextualization in various sensory and cognitive realms can be of advantage when we try to rid ourselves of misattributions. In Chapter 7 of this volume, Gerald L. Clore gives an example of how a misattribution disappears when the true sources of a feeling can be disentangled from a merged cluster of experiences.[18]

4. The Extended Observer

Hitherto I have referred to the observer participant in an unspecific manner, without defining the boundary which limits him. But the question as to where we may set the interfacial cut between the observer and the rest of the world is crucial for our reality generation game[19,20]. But deciding what we assign as being part of the observer and what as being part of the rest of the world is a tricky matter.

Misattributions are common and often not easily identified as such by introspection or reasoning. Simple conditioning effects can turn individuals into extended observers. The famous example of the rubber hand illusion illustrates this nicely. In 1998, Botvinick and Cohen conducted an experiment in which a person is sitting at a table, with his left arm resting on it[21]. His other hand is screened off from view and a rubber hand is placed in front of him. Next, he is asked to look only at the rubber hand while both the rubber hand and his own hand which is hidden from view are stroked with paintbrushes simultaneously. The visual feedback linked the observer's visual and tactile perceptions, so that after a while, he reported that he felt his own hand was being stroked when only the rubber hand was touched by the paintbrush. The intermodal matching sufficed for self-attribution. This conditioning effect (neurons that fire together wire together) changed his perspective by shifting the location of the interfacial cut between the observer and the rest of the world.

There are also examples of diminished observer extentions. Ramachandran and Blackeslee describe a patient who refuses to believe that his leg belongs to him, although it is obviously attached to his body. Rather than accepting it as being part of himself, he comes up with elaborate explanations why the leg does not belong to him[22]. This condition, which is well-documented under the name "neglect", shifts the observer-world interface towards the observer and sets an interfacial cut within his body.

The other extreme, shifting the interface further and further into the world, well beyond the observer's body, has been described by Metzinger, who reports the case of a patient who spends all day looking out of the window, making the sun move across the sky[23]. The patient is convinced that the apparent motion of the sun is the result of his volitional act. This may be an extreme case, but then many managers and politicians find themselves in a similar situation.

The position of the interfacial cut determines which temporal rhythms are attributed to the observer and which to the outside world. Positive feedback results in hard-wired expectations which give rise to self-delusions. These are next to impossible to dissolve by means of reasoning, let alone lecturing[24]. Our temporal observer perspectives are defined by their fractal structures generated by nestings of rhythms or oscillations. And, as our brain's output controls its inputs, rather than the other way around[13], our fractal temporal interfaces are fairly immune to external perturbations.

5. Conclusion

In order to explain the way we perceive meaningful entities such as a tune or any other time series as a gestalt, we must assume two mutually independent dimensions of time: Δt_{length} and Δt_{depth}. The ability to fall for or resist simultaneous contrasts shapes our temporal fractal interfaces: contextualization generates Δt_{depth}. Thus, our observer perspectives can be defined in terms of the number of nestings at our disposal. The resulting interfacial cut generates our very personal realities.

This cut can lead to advantages or disadvantages. If my interfacial cut is valid for one observer frame only, namely mine, I may, as the individual who has generated this perspective, well be diagnosed as suffering from a cognitive disorder. The same is true for a group of individuals who are embedded in a larger context. They may be branded as suffering from a collective hallucination if their perspective does not match that of the individuals who make up the embedding context. Scientific paradigms are also subjected to and constrained by the interfacial cut set by the scientific community. Scientists who adhere to a scientific paradigm may be falling for a collective hallucination, which becomes hard-wired as a result of positive feedback and results in educated incapacity.

There is no such thing as a pure observer, an exo-observer who does not engage in participation. Neither is there a pure participant, as this state would not allow for an observer perspective. Individuals who approach a state of pure participation would find themselves suffering from oceanic boundary loss or dreadful self-dissolution.

Our knowledge of the world is limited by perceptual and cognitive constraints. We cannot rule out that there may be observer participant extentions we cannot yet assign to a physical or mental cause. Private observer extentions cannot easily be pinned down by external observers, even when we think we can discern an unusual interfacial cut. I shall spare you a long excursion into anecdotal evidence of such phenomena. But, on a lighter note, here's an example of an extended observer participant whose extention we may observe but cannot explain: "... He stands like a statue, becomes part of the machine, feeling all the bumpers, always playing clean, plays by intuition, the digit counters fall, that deaf and dumb and blind kid sure plays a mean pinball ..."[25].

References

1. O.E. Rössler, *Endophysics.* World Scientific, Singapore (1998).
2. E. Pöppel, *Grenzen des Bewußtseins – Wie kommen wir zur Zeit, und wie entsteht die Wirklichkeit?* Insel Taschenbuch, Frankfurt (2000).
3. G. Lakoff and R. E. Núñez: *Where Mathematics Comes From: How the Embodied Mind Brings Mathematics into Being*. Basic Books (2000).
4. O.E. Rössler, Intra-Observer Chaos: Hidden Root of Quantum Mechanics? In: *Quantum Mechanics, Diffusion and Chaotic Fractals*. Edited by Mohammed S. el Naschie, Otto E. Rössler & Ilya Prigogine. Pergamon, Elsevier Science, pp. 105-112 (1995).
5. E. Husserl, *Vorlesungen zur Phänomenologie des inneren Zeitbewußtseins.* First published in 1928. Niemeyer (1928/1980).
6. B.B. Mandelbrot, *The Fractal Geometry of Nature.* W.H. Freeman, San Francisco (1982).
7. S. Vrobel, *Fractal Time*. Houston: The Institute for Advanced Interdisciplinary Research, Houston (1998).
8. S. Vrobel, Fractal Time and the Gift of Natural Constraints. *Tempos in Science and Nature: Structures, Relations, Complexity*. Annals of the New York Academy of Sciences, Volume 879, pp. 172-179 (1999).
9. S. Vrobel, How to Make Nature Blush: On the Construction of a Fractal Temporal Interface. In: *Stochastics and Chaotic Dynamics in the Lakes: STOCHAOS*. Edited by D.S. Broomhead, E.A. Luchinskaya, P.V.E. McClintock and T. Mullin. New York: AIP (American Institute of Physics), pp. 557-561 (2000).
10. J. Fauvel, J. et al, *Music and Mathematics – From Pythagoras to Fractals.* Oxford University Press, p. 27 (2003).

11. S. Vrobel, Fractal Time and Nested Detectors. In *Proceedings of the First IMA Conference on Fractal Geometry: Mathematical Techniques, Algorithms and Applications*. DeMontfort University, Leicester, pp. 173-188 (2004).
12. S. Vrobel, Fractal Time in Cognitive Processes. Paper presented at the 2nd *International Nonlinear Science Conference*, Heraklion (2006).
13. G. Buszáki, *Rhythms of the Brain.* Oxford University Press, Oxford, U.K., p. 134 (2006).
14. S. Vrobel, Simultaneity and Contextualization: The Now's Fractal Event Horizon. Talk held at *13th Herbstakademie: Cognition and Embodiment.* Ascona, Switzerland (2006).
15. S. Vrobel, Nesting Performances Generate Simultaneity: Towards a Definition of Interface Complexity. In: *Cybernetics and Systems* Vol 2 (Edited by Robert Trappel). Austrian Society for Cybernetic Studies, Vienna, pp. 375-380 (2006).
16. S. Vrobel, Reality Generation. In: *Complexity in the living – a problem-oriented approach.* Edited by R. Benigni, A. Colosimo, A Giuliani, P. Sirabella, J.P. Zbilut. Rome: Rapporti Istisan, pp. 60-77 (2005).
17. S. Dakin et al, Weak suppression of visual context in chronic schizophrenia. In: *Current Biology* 15, pp. R822-R824 (2005).
18. G. L. Clore, Simultaneity in Emotional Moments. Chapter 7 of this volume, §4: Mood Influences on Judgment.
19. S. Vrobel, Temporal Observer Perspectives. In: *SCTPLS Newsletter*, Vol. 14, No. 1 Society for Chaos Theory in Psychology & Life Sciences, October 2006.
20. S. Vrobel, Ice Cubes And Hot Water Bottles. In *Fractals. An Interdisciplinary Journal on the Complex Geometry of Nature.* Vol. 5 No. 1. World Scientific, Singapore, pp. 145-151 (1997).
21. M. Botvinick and J. Cohen, Rubber Hands 'Feel' Touch That Eyes See. In: *Nature*, Vol. 391, p. 756 (1998).
22. V.S. Ramachandran and S. Blakeslee: *Phantoms of the Brain*. Fourth Estate, London (1999).
23. Metzinger, T. *Being No One: Consciousness, the Phenomenal Self and the First-Person Perspective*. Foerster Lectures on the Immortality of the Soul. The UC Berkeley Graduate Council (2006). http://video.google.com/videoplay?docid=-3658963188758918426&q=neuroscience
24. S. Vrobel, The Extended Observer Revisited. *Systems Research and Cognition.* Vol. 1. G.E. Lasker M. Malatesta (Eds.) The International Institute for Advanced Studies in Systems Research and Cybernetics, Tecumseh, Canada 2007.
25. P. Townshend, Pinball Wizard, *Tommy, A Rock Opera*. Track, Polydor, UK (1969).

CHAPTER 2

MIRROR NEURONS: EVIDENCE FOR THE GREAT SIMULATOR AND VROBELISM

Otto E. Rössler

Division of Theoretical Chemistry, University of Tübingen
Auf der Morgenstelle 8, 72076 Tübingen, F.R.G.

The deductive biology of brain function is reviewed and brought in contact with recent neurobiological findings. Natural brains predictably possess two components: the force field generator (brain equation) and the Great Simulator (virtual-reality machine). The force field generator contains a preferred time direction, while the Great Simulator can in principle work in both directions of time. The "mirror neurons" discovered empirically by Rizzolatti are interpreted as a predictable manifestation of the Great Simulator. Artificial intelligences that share with us many aspects of the experience of time are possible. The notion of a "temporal fovea" is introduced. It implies a self-similar structure of time over a certain range. This finding confirms a proposal made by Susie Vrobel.

1. Introduction

A juxtaposition between a deductive approach to the brain on the one hand, and the modern neurobiological finding of mirror neurons on the other, is attempted. The motivation for the new synthesis stems from fractal-time theory.

2. Deductive Biology

The experimental discovery by Giacomo Rizzolatti of "mirror neurons" in the cerebral cortex[1] is a bounty for a branch of theoretical biology called deductive biology[2]. In 1973, ethologist Konrad Lorenz produced the following definition: "thinking is acting in imagined space" (*Denken ist Handeln im vorgestellten Raum*), in his pre-Nobel book "Behind the Mirror"[3]. Now, an empirical substrate for this quip appears to have been found in the prefrontal motor cortex by Rizzolatti. To make the connection, deductive theoretical biology[4] can be invoked.

After many decades of non-perturbative observation of animal behavior, Lorenz formulated an intuitive approach to the "necessary properties" of behavior-controlled animals[3]. In long conversations we had in 1966, he mentioned Charles Darwin, author of "The expression of Emotions in Man and Animals"[5], as ultimately responsible for his daring new notions. They included: innate releasing mechanisms, little-child schema, appetitive behavior, fixed motor patterns, taxis component, action-specific drives, endogenous mood pressure, relaxed field, bonding drive, imprinting. This almost inexhaustible catalog of observation-based constructs was rooted in equal parts in observation and an interpretation of Darwin's thinking. The greylag geese, brood-caring fishes and dog and wolf were his main inspirations. Like Darwin himself, he was a deeply humble and kind person (and protector of "non-Aryan" family members in the dark age).

But were all these constructs not just artful fantasy? Unexpectedly, the legitimacy of his results could be confirmed by a deductive mathematical approach to position-in-space dependent survival[2]. This deductive approach to "positional adaptation" is complementary to Darwin's deductive approach to "metabolic adaptation". Although formally a subbranch of Darwin's theory, the spatial branch is much more specific in its predictions because it reduces to a well-posed optimality problem in the sense of mathematics.

Specifically, adaptive behavior in space solves a "traveling-salesman problem"[2]. Any system whose survival depends on position in space and time in a lawful manner has to solve a well-posed mathematical problem. The "travelling-salesman-with-alarmclocks problem"[2] consists of the following task: to solve (up to a small finite error bound allowed by the procreative replacement of lost individuals) the task of always visiting "just in time" a source of the right kind from among about ten types of sources randomly scattered in space – so that a maximum time interval between two visits of the same type (think of water replacement) is never exceeded[2]. The other source types could be shelter for the night, food or mates; "negative sources" that invariably need to be avoided or fled from can be incorporated without appreciably changing the mathematics. A pertinent sufficiently optimal solution *in space* cannot be afforded "not" to be followed by nature – just as Darwin's environmental precondition of a random variation of survival-relevant properties *in time* made an "endogenous heritable variability" predictably unaffordable "not" to be implemented by nature.

The obtained *predictive theoretical biology of behavior* implies the existence of a "brain equation". The latter – to be found in Equation (13) in Rössler[2] – describes a time-dependent field of directional forces, each attached to an environment-based point-shaped "source" of a specific type. This momentary-position centered, time-dependent force field yields only a local ("direction optimizing") solution rather than an (as eventually needed) supra-local or "path-optimizing" solution to the travelling salesman problem. Nonetheless, the former approximative solution is already irrefutable as a "fall-back option" if rapidly changing environmental situations can occur as natural. The brain equation is hence a predictable part of the unavoidable path-optimizer. The required combination of fast local (direction) optimization and slow supra-local (path) optimization can be achieved by simply complementing the brain equation with a universal flight simulator[2] (or, synonymously, "cognitive-map system"[6] or "virtual reality machine" – VR for short). The VR is needed anyhow in case the travelling system in question is equipped with non-panoramic sensors ("focusing-type eyes")[2], as is the case with vertebrates and molluscs in terrestrial biology. At the same time, many details of the endogenous force field of the brain equation (its time dependencies, its autcatalytic features, its attractive and repulsive components[2]) turn out to be utterly familiar – not only from ethology's descriptive terms, but also from everyday subjective experience.

The combined system (force field generator plus VR) supports "path optimization" rather than mere "direction optimization" as we have seen. Moreover, whereas the unique optimal solution to the travelling salesman problem (with zero error bound) would require an infinitely large and complex path-optimizing computer[2], a good suboptimal solution as needed in biology can be found in a short time by the combined force-field generator and simulator. Hence the latter's existence is predictable both for terrestrial and in extraterrestrial biology (soon to be discovered on the moon Europa in a watery environment, and inside Jupiter itself under non-watery conditions with B-N- rather then C-C-backbone molecules – perhaps).

Specifically, the combination of force-field generator and VR allows for many merely "simulated" runs of local optimization, before the convergent supra-local result is finally carried out in reality. "Thinking is acting in imagined space" holds true for the combined system.

The described two-level method of path optimization is, by the way, unavoidable only in "non-eusocial" (more precisely, "single-brain type") animal species. A deductive biology of eusocial animals (like ants or mole rats) is currently out of reach.

3. Parallel Picture Processing

An artificial design of the described two-level path-optimizing machine is possible[7]. While the force-field generator ("great motivator") is implementable in many ways, the picture-processing part ("Great Simulator") is rather constrained for an analog-type parallel implementation – for example, for a multicellular system with contact-type and bath-type chemical and electrical interactions, as was chosen in terrestrial biology.

Under this "modular" constraint, both frequency-space calculations and sequential programming methods familiar from the von Neumann computer are *un*available. In the parallel picture-processing that is required, several levels of spatial resolution of the surrounding environment have to be combined[8]. Since the picture hereby necessarily gets shifted across different resolution levels, such a machine automatically generates "Platonic ideation" in the sense of Heinz von

Foerster[9]. Specifically, many (up to 15) different levels of spatial resolution of the environment, each differing by a factor of 2 in linear extension from the next, are to be combined in an additive (if not quite linear) fashion. More specifically, a cut-off of all "aliasing"-induced negative values (at the edges) and a multiplicative rather than linear superposition at the edges, when sharper-resolution subcomponents are added to lower-resolution ones, are predictable. Hence so-called "tolerance attractors"[8] form under iteration. That is, *invariant objects* are being generated: of constant length in 1D, of constant radius or with sharp angles of 90 and 60 degrees in 2D, and spheres, tubes, tori, cubes and pyramids in 3D.

And just as *space* is represented by many parallel resolution levels, superposed, so *time* involves many levels of resolution of the spatially represented pictures, too. Older environmental scenes and events and planned actions remain retrievable over a certain length of time – with a decaying resolution as far as both space and time are concerned. Thus, there necessarily exists a *temporal grading* in the Great Simulator or, synonymously, "inner eye".

The described machine has apparently never been built, but there is little doubt that it *can* be built any time soon[10]. In our present context, it is not so much the Artificial Intelligence aspect of deductive biology that is at stake, but rather the usefulness of the latter for a better understanding of the elements of our own subjective experience as far as *time* is concerned.

4. Mirror Neurons

While it is possible to prove that nature "cannot not" produce an implementation of the above picture-processing scheme (Great Simulator), it is currently open how *exactly* nature solved this problem hardware-wise in its vertebrate and mollusc brains. Recent progress in imaging technology (like positron emission tomography and functional magnetic resonance imaging) greatly facilitates the study of ongoing processes in the brain.

Rizzolatti's discovery of mirror neurons[1] belongs to this context; cf. the recent book edited by Arbib[11]. The empirically observed "locking-on" onto an observed sub-scene (the "taking act" of a watched hand that belongs to another individual[1]) *is* a predictable activity of the Great Simulator. Hence the Great Simulator of deductive biology has apparently been identified empirically for the first time. It hereby turned out that the *same* neuronal machinery is used over and over in the same context[1]. It can be hoped that the location and further features of the Great Simulator in the hardware of the vertebrate and human brain will be mapped out in great detail soon.

An empirical question is hereby still open: Is the observed functional coupling between neural assemblies transitory and wave-like, that is, based on phase- or frequency-binding[12], or is it each time identically repeated and thus anatomically fixed[13]? I remember in this context a talk given by Jack Cowan at the Center for Nonlinear Studies in Los Alamos in 1983, in which he showed that many different levels of spatial resolution co-exist in the visual cortex in a layered fashion in close proximity to each other.

In computer science, nonlinear wavelet-like filters are in most cases implemented in frequency space. The real-space image analysis[8] can be realized in many ways in parallel-processing computers. In technology, the shifting of images would still not be done physically: address re-labelling remains the method of choice. In biology's slow neural machinery, in contrast, this elegant way of sparing the task of having to shift pictures across many units in an analogous physical manner (with the consequence of unavoidable distortions and idealizations) is clearly unavailable. But how else could one possibly shift pictures around freely on the Big Screen of the Great Simulator as necessary? This quandary remained unsolved for a long time.[14] Now – on remembering Jack Cowan's talk – the solution suddenly is obvious: *multiple representation* at different resolutions. That is, "dormant" portions present in many places can be combined with a special coupling-in signal that determines where the picture is "active" momentarily. An activating signal ("an awakening kiss") therefore predictably exists. The latter may take the form of a synchronizing wave. If so, he two theories – the constant-site and the wave theory[13], which up until now have competed with each other – would *both* be vindicated for once. As a case in point, Roger Shepard's discovery of "analogical" picture processing in human beings under the mental rotation of a chair[15] would become amenable to a first explanation, hardware-wise.

5. Pandaka

In the present neurobiological context, a proposal made some time ago – to set up an international "Pandaka pygmaea Institute" devoted to the study of one of the smallest vertebrate brains[16] – comes to mind again. This particular, very small (only 0.9 cm long) fish has the asset that it possesses a close relative (Gobius niger) of some 20 cm in length whose intimately related brain is about a thousand times bigger – so the bridge toward the human brain (again a factor of a thousand) would be covered "half-way" as it were. Werner Reichardt's early decision (made at age 27) to devote his life to the housefly's brain – as he told me when he was twice as old – has yet to be followed up by a similar vow, pledged to the not very much larger *vertebrate* brain of Pandaka, by an equally dedicated young scientist. Modern methods (including the knocking-out of genes) would greatly facilitate the task. In this way, both "deductive" and "inductive" brain theory could meet halfway for once.

There is a nice story told by Lorenz in 1966, of a mouth-breeding little fish that inadvertently had put both many offspring and a similarly shaped small piece of food into its mouth: what would happen? The animal was visibly stunned, remaining motionless for minutes on end. Then, suddenly, it spat out the whole contents of its mouth, ate the food and carefully recollected the offspring – while the human eye-witness could not help applauding.

6. Time on the Big Screen

Our topic proper – to come to it at last – is a specific subquestion: How is the Great Simulator wired so as to support *temporal self-similarity*? The spatial resolution of bygone environmental inputs – like where one was a moment ago – decreases exponentially in time, as we have seen. This is unavoidable because new inputs are bound to take the places of the old ones. The formal alternative – complete preservation followed by abrupt cessation – is clearly dysfunctional since it arbitrarily excludes 50 percent of all time slices in an intermittent fashion. Sneaking out exponentially as far as the occupation of harware is concerned is thus predictable. Of course, long-term passive memory can always be employed to re-load an old scene, but this cannot provide for an unbroken functional chain into the past as we saw. Because of the exponential decay, the active-memory solution is automatically confined to a short time window.

It follows that a "temporal fovea" that is moving across objective time is bound to exist in the Great Simulator of active memory. It can at first sight only exist in the form of a "half fovea", since the future cannot be included by definition. However, since in simulation future situations are anticipated as if real (as we have seen: "thinking is ... "), symmetry is in fact restored. Hence a "red line" that is continually moving across objective time as the center of a time-symmetric "nowness fovea" is bound to exist in the Great Simulator as the seat of the highest possible resolution. Ichiro Tsuda once posed the question of whether nature has succeeded in solving the problem of "perfect sensitivity" of an excitable cell, since anything less would be counterproductive.[17] We know from the ear and the eye that their sensitivities do reach the boundaries of the physically possible. By way of analogy, it is possible to ask at this point whether a "perfect Now" can get formed in the image-processing multi-neuron pseudo-simultaneous machine of the Big Screen. How does such an optimal machine have to be wired? This question is open. The "smell of the Now" (or, synonymously, the "gong of the Now") would be a predictable implication of such a machine. The term "gong of the Now," by the way, came up in conversation with Art Winfree in 1980. How sharp is the tip of this temporal iceberg, the *fovea of existence*?

While there is hope that the Great Simulator will be understood better in its anatomical location and physiological features any time soon in the wake of the motion programmer detected by Rizzolatti, the associated question of the programmer's temporal capabilities, and especially the features of the temporal fovea, is wide open. A further empirical inroad has apparently been opened up by the "error signal" (in the case of nonconvergent simulation) discovered in 1990 by Michael Falkenstein (see Jordanova et al)[18].

On the deductive side, an "overlap buffer" got incorporated into the scheme of the Great Simulator in 1981[6]. It allows for several levels of iteration to be implemented simultaneously. The apparent existence of this buffer in mammals was first glimpsed when watching a red panda (the animal was actually grey and lacked a bushy tail) in Stuttgart's "Wilhelma" Zoo in 1966: The cat-sized animal, whilst walking straight on the narrow concrete wall bounding its territory, would abruptly stop in mid-gait as if frozen – with one or the other paw still up in the air – in order to just as abruptly resume its gait a second or two later. It looked as if the animal could not (through a mutation?) simultaneously "move and think" – that is, prepare the next step while moving. The overlap buffer enables a smooth superposition of both activities so that the ongoing internal simulation is invisible on the outside.

7. Self-Similarity

An implication of the arrived-at organization of the Great Simulator is that a "temporal self-similarity" becomes accessible to the autonomous optimizer-plus-VR, in both natural and artificial brains. While more and more "idealized" spatial objects survive as remnants from each scene under the passage of time, as we have seen (with irregularly shaped "potatoes" turning into perfect Platonic spheres, on receding into older temporal layers until eventually fading from the screen), the overlap buffer – by its not being a part of the recycling machinery proper – allows for several different temporal patterns to be superimposed simultaneously. Mozart said of himself that he could "see" a whole planned composition laid out in front of him simultaneously as an acoustic painting as it were. In music, a superposition of self-similar waves of differing lengths may be artificially achievable (and even sound pleasant). A maximally simple example is the famous, always increasing but never really increasing in pitch, sequence of musical chords. It tells us something about the structure of our own brain physiology in the present context.[19] On listening, one almost feels reminded of a rat's wheel type of structure – endless progress without progress.

One could call the underlying principle a "temporal rectifier"[20]: Half of the past is maintained, the other half suppressed – much as a water mill's wheel is extracting one particular direction of rotation out of a linearly flowing watery stream. The continually propelled wheel is not a miracle to watch for us post-Archimedeans since the opposite symmetric motion, inducible by the same watery stream on the other side of the wheel, does not make itself felt to us as "visibly missing" in this mechanical rectifier. In its ingenuity, the trick behind the simple machine no longer meets the eye. In the same vein, one is in the case of *time* tempted to suspect that we may be continually "rectifying" a primarily nonflowing stream into a flowing one.

While the preceding simile is undoubtedly overstretched, the "extraction" of one particular direction of temporal progression could prove to be a nontrivial thing to accomplish in a machine called the "brain" – as a feature that needs to be specially wired-in. The brain clearly possesses this feature of "temporal rectification." In the same vein, Susie Vrobel[21] refused to attribute a flow to her "fractal time"[22]. This casually made comment is hard to understand at first sight, since to us modern post-hunter-gatherer humans the fact that time is "flowing" appears totally self-evident in the Western culture. But, of course, time does not flow in reality, as pre-Socratic philosopher Parmenides first proved. Karl Popper[23] discussed this problem at length with Einstein in 1949. Some American-Indian cultures for whom there exists no division between the sacred and the profane, make similar claims[24].

8. Toward Applications

If time does not flow, there also is no "gong of the Now" since the Now is always there. Only to people consciously aware of the "double contingency" of both being at a certain moment in time and knowing why that double contingency is assigned at this very moment, does the Now have a special status. If there is a moment of being called by one's name and being talked to (compare Johnny Cash's song "Five more minutes to go"), the gong does exist. On the other hand, the lizard that captivates you by holding your gaze for seconds or hours in a row (you cannot tell afterwards) knows that time is frozen. The Buddha knows. A person in love knows. The poet knows. Peter Landsberg[25] knows. The Shechina knows (the "time of advent" in the West). But we always relapse in our daily lives – perhaps legitimately so, because other persons need us. We just tend to forget to return afterwards.

In principle, an *experiment* is possible, in which every second time segment (of ten seconds duration, say) is obliterated by acoustic and optical noise while the other (non-obliterated) segment in between is recorded and played back inverted in time to the experimental subject, in an interactive fashion under a combined mechanical, acoustical and optical coupling while the subject is standing on a running belt with fixed handrails.[26] Will the first volunteer for this experiment find it to be a perfect analog to the famous "inverted-spectacles" experiment of earlier times? The hypothesis was that it may be possible to "adapt" to such an inversion of past and future in time, no less successfully than to an inversion of top and bottom in space in the inverted-spectacles example. The dissipative arrow of time in physics[27] would prove to have nothing to do with the subjective arrow of time if the latter could in principle be constructed by the brain the other way round. This new type of causal wiring would, if successful, of course have disastrous consequences when implemented outside the laboratory.

While it is highly probable that the idea is misleading[28], it shows that it can be great fun to mentally picture oneself as a demodulating rectifier in the giant radio set of the universe. The real scandal hereby, of course, remains the assignment of the Now, which is even more incisive than the direction of time.

9. Conclusions

No one seems to have thought more about the question of the structure of living time than Susie Vrobel[22]. I therefore propose to use the name "Vrobelism" for the deep study of the physiological arrow of time and its potentially self-similar elements. Kurt Vonnegut's story "Slaughterhouse Five" fictitiously addresses a subelement of the problem (the free choice of the Now).

To conclude, a deductive approach to brain function and temporality was presented on the basis of the ordinary (classical) level of physics. Darwinian biology and brain physiology belong to this level. For a recent workshop on time, devoted to all levels including the quantum level (with its Bell-Everett micro time reversals), see Ungerer and Smolin[29]. Even on the trivial classical level already, the "abyss of the Now" adopts the structure of a selfsimilar fractal that bites itself into its own tail – as the most elegant "dragon" in the sense of Benoit Mandelbrot. When I was discussing with him the question of a "circular self-similar fractal" (in which the macro end feeds into the micro end) with him, at Okan Gurel's chaos conference of the New York Academy of Sciences in 1977, the question of *time* making this impossible dragon possible had not yet occurred to us. Nor did the fractal half of the same question arise in a later discussion with Peter Landsberg about his circular "Ouroboros of time" (the snail that bites itself into its own tail[25]), at John Casti's 1987 Abisko workshop on Artificial Minds. Only now, in the aftermath of Vrobel's fractal-time theory, does it appear that the *combination* – the self-similar whirling abyss of the eternal Now – is beginning to take shape in science and technology.

Acknowledgments

I thank Dieter Fröhlich, Mohamed ElNaschie, Siegfried Zielinski, Anthony Moore, Gerold Baier, Hans Diebner, Bill Seaman, Peter Weibel, Masaya Yamaguti, Ilya Prigogine, Wolfgang Engelmann, George Lasker, John Casti, Kunihiko Kaneko, Ichiro Tsuda, Maria Haase, Erdmute Sommerfeld, Werner Ebeling, Günther Palm, Valentin Braitenberg, Niels Birbaumer, Gottfried Mayer, Maria Haase, Okan Gurel, Jeanette Fischer, Yrish Sefla, Rosi Paschon, Vladimir Gontar, Roger Shepard, Piet Hut, Ulrich and Roland Wais, Andy Hilgartner, Artur Schmidt, Boris Shapiro, Peter Plath, Gerard Caris, Thomas Feuerstein, Michael Langer, Artur Schmidt, Albrecht Arendale, Jonathan Kemp, Wilfried Hou je Beck and Martin Howse for discussions. For J.O.R.

References

1. L. Fadiga, L. Fogassi, G. Pavesi and G. Rizzolatti, Motor facilitation during action observation: a magnetic stimulation study. *J. Neurophys.* 73, 2608-2611 (1995).

2. O.E. Rossler, Adequate locomotion strategies for an abstract organism in an abstract environment: A relational approach to brain function. *Lect. Notes Biomath.* 4, 342-369 (1974).
3. K. Lorenz, *Behind the Mirror.* London: Methuen, 175 (p. 128 of German edn.) (1977).
4. O.E. Rossler, Deductive biolgy: Some cautious steps. *Bull. Math. Biol.* 40, 45-58 (1978).
5. C. Darwin, *The Expression of Emotions in Man and Animals.* New York: Appleton (1899).
6. O.E. Rossler, An Artificial Cognitive-map System. *BioSystems* 13, 203-209 (1981).
7. O.E. Rossler, An Artificial Cognitive-plus-motivational System. *Progr. Theoret. Biol.*13, 203-209 (1981).
8. O.E. Rossler and B. Uehleke, Tolerance Attractors: Attracting Patterns in Reverberatory Image Processing. *ACM Computer Science Conference* (J.W. Hamblem, ed.), Abstracts, 34. St. Louis: Association for Computing Machinery Publications (1981).
9. H. von Foerster, Circuitry of clues on Platonic ideation. In: *Aspects of the Theory of Artificial Intelligence* (C.A. Muses, ed.), 43-82. New York (1962).
10. Bill Seaman, personal communication (2005).
11. M. Arbib, ed., *Action to Language via the Mirror Neuron System.* Cambridge: Cambridge University Press (2006).
12. Péter Erdi, personal communication (1987).
13. C. Koch and S. Greenfield, How does consciousness happen? *Scientific American* 297, Oct., 50-57 (2007).
14. Michael Klein, personal communication (1984).
15. R. Shepard and J. Metzler, Mental rotation of three-dimensional objects. *Science* 171, 701-703 (1971).
16. O.E. Rossler, G. Kampis, W. Nadler, W. Musterle, B. Shapiro and P. Urban, Highly parallel implementation of an autonomous direction optimizer with cognition (abstract). *Biophys. J.* 57, 194a (1990).
17. Ichiro Tsuda, personal communication (1994).
18. J. Jordanova, M. Falkenstein, J. Hohnsbein and V. Kolev, Parallel systems of error processing in the brain. *Neuroimage* 22, 590-602 (2004).
19. Manfred Euler, personal communication (1987).
20. Serge Desmangles, personal communication (1964).
21. Susie Vrobel, personal communication (2007).
22. S. Vrobel, *Fractal Time.* IAIR, Houston (1998).
23. K. Popper, in: *The Philosophy of Karl Popper*, ed. by P.A. Schilpp, Open Court, LaSalle, Vol. 1, p. 102 (1974).
24. Andy Hilgartner, personal communication (2005).
25. P.T. Landsberg, *The Enigma of Time.* Bristol: Adam Hilger (1982).

26. Dieter Fröhlich, personal communication (1999).
27. H. Price, *Time's Arrow and Archimedes' Point.* Cambridge University Press (1996).
28. Dieter Fröhlich, personal communication (2007).
29. R.M. Ungerer and L. Smolin, An Interdisciplinary Conference on Time Including Topos Theory, www.perimeterinstitute.ca/en/Events/Evolving_Laws/Evolving_Laws_Overview/

CHAPTER 3

THE CONCEPT OF NOW IN DOGEN'S PHILOSOPHY

Michael Eido Luetchford

Dogen Sangha, 21 Melbourne Road
Bristol BS7 8LA, UK
E-mail: mjl@gol.com

Belief in the totality of the present moment forms one of the fundamental tenets of Buddhist philosophy. According to Buddhist thought, the past is not real, because it no longer exists, and the future is not real because it has not yet come into existence. Only the present moment is real existence. Early Buddhist philosophers developed a theory of existence as a point-instant, and this view finds parallels in the poetry of T.S. Eliot. The 13th century monk/philosopher Dogen wrote extensively about our concept of time set against our experience of time. In his masterwork, *Shobogenzo*, he expounds on the nature of the Now in a stimulating essay rich in imagery and metaphor.

1. Introduction

Belief in the totality of the present moment forms one of the fundamental tenets of Buddhist philosophy. According to Buddhist thought, the past is not real, because it no longer exists, and the future is not real because it has not yet come into existence. Only the present moment is real existence. But any attempt at analysis of the very moment in which we exist leaves us running in circles. How then, can we approach a description of the structure of the Now, this most difficult of philosophical problems, when, as Wittgenstein points out, the end of this sentence is in a different present moment to its beginning? How can we reconcile our received linear view of time with a view that makes the instantaneous present the only reality? If Now is the only time that is real, where is our past, and where will our future be? Does the present contain the past and future? How is that possible? What is the structure of a Now that contains all? [a]

2. The Theory of Instantaneous Being

This problem was addressed by two great Indian Buddhist philosophers in the 6th and 7th centuries – Dignaga and Dharmakirti – under the title "Ksanika-vada" (The Theory of Instantaneous Being). An excellent exposition of their work can be found in the book *Buddhist Logic* by Stcherbatsky, who summarised their ideas as follows:

"Such is the leading idea of Buddhism – there is no other ultimate reality than separate, instantaneous bits of existence. Not only eternal entities, be it God or Matter, are denied reality, because they are assumed to be enduring and eternal, but even the simple stability of empirical objects is something constructed by our imagination. Ultimate reality is instantaneous."[1]

The Buddhist philosopher and teacher Nagarjuna had already addressed the problem in the 2nd century in his work *Mulamadhyamakakarika*, in a chapter entitled "Kala-pariksa" (An Examination of Time).

Surely if the present and the future exist relative to the past,
They will both be contained in time past.
But if the present and the future are not both contained in the past,
How can they be related to it?
Again, it is impossible to establish the existence of either the present or the future without their being related to the past.
Thus present and future time do not exist as such.
Using the same process, it can be clarified whether present time – or future time –

[a] I am using the phrase "the Now" in the sense of "all that is here and now", not just to represent time at this moment, but embracing the present time and present place as one indivisible whole. In this sense, it represents the oneness of space-time-existence, and has the same meaning as Dogen's "Uji" – Existence-Time or Time-Present.

Comes before, after, or between the other two, or whether they are all a unity.
The nature of time is instantaneous and so it cannot be grasped.
Static time, which can be grasped, does not exist.
Since time is impossible to grasp, how can we understand it?
If time depends on things, how can time exist when there are no things?
And as there are no things anywhere, how can there be time?[2]

In the last lines of this chapter, Nagarjuna points out that our concept of time depends on our belief in the existence of "things". By this, he is referring to the naming of things, to our conceptualisation of the world. To Nagarjuna, reality exists in front of us in a preconceptual, uninterpreted state, that he calls "empty" (sunya) in the sense that it is empty of concepts. He states that our concepts of reality are always different from reality itself.

The poet T. S. Eliot, who is known to have studied Buddhist thought, was also much interested in the concept of instantaneous time, which he alludes to in the opening stanza of the first of his *Four Quartets*, "Burnt Norton":

Time present and time past
Are both perhaps present in time future
And time future contained in time past.
If all time is eternally present
All time is unredeemable.
What might have been is an abstraction
Remaining a perpetual possibility
Only in a world of speculation...

Time past and time future
What might have been and what has been
Point to one end, which is always present.

3. The Concept of Now in Dogen's Philosophy

The 13th century Japanese monk/philosopher, Eihei Dogen, expounds on the nature of time and existence in his lifework *Shobogenzo*, in a chapter entitled "Uji". Composed of the two Chinese characters "U" and "Ji", this is often translated as "Being-Time". However, I much prefer the translation "Time-Present", since this really does focus us on the immediate. This short chapter is full of imagery and opposing viewpoints and its contents provide us with a treasure chest full of inspiration for investigating the nature of the Now itself. Here is my modern interpretation[b] of the full text of the chapter, which is based on the translation by Nishijima & Cross.[3] The chapter begins with a poem that describes the independence of each moment of existence. Dogen then comments on the poem from different viewpoints. He then quotes a koan story to illustrate the conduct of someone who has realised the nature of the Now, and adds his commentary. This is followed by another poem that illustrates natural behaviour in the here and now. Dogen then concludes with his own version of the poem. I have inserted lines in italic after each section giving a summary of the main points he is making.

A buddha of old said:

Time-present is standing on the mountain heights.
Time-present is sinking to the depths of the ocean.
Time-present is an angry demon, time-present is a buddha.
Time-present is a formal ceremony, time-present is the temple compound.
Time-present is an everyday individual, time-present is pervading the whole Universe.

[b] Dogen's original essay is written in 13th century Japanese, and the stories and poems are quoted in Chinese. The Nishijima-Cross translation, although an excellent translation, is strictly literal and often quite difficult to understand, since it follows the original grammatical constructions quite closely. In my interpretive version, I have tried to present the text in easier-to-understand phrases.

> Time-present means that time is the present and the present is time. The state of buddha is always made real at a time, and because of this, the state is illuminated by the light of the present.

[Dogen quotes this poem by Chinese Zen Master Yueshan Weiyan (745-828) as an illustration of time identified with concrete events in the Now, rather than time as a uniform flow. His comments identify time and present existence.]

> But we must also study time as 24-hour time this day. Anger has its time, and so it is 24-hour time. 24-hour time is relative; although we can't be sure if one period of 24-hours is longer or shorter, or faster or slower, than another, we still say that a day lasts for 24 hours.

[Time can be seen as a continuous process. But although time is experienced as a subjective and relative experience, it is usually thought of as an objective measurable period.]

> And although we never doubt our ability to trace clearly the passage of time, we cannot be sure that we know clearly what time is. As ordinary people we have many doubts about many things, but our doubts are not always the same; they are always changing. But although this is true, every one of those changing doubts is time-present.

[But although we don't doubt our received notion of time as a linear trace, do we really know what time is? Although we doubt many things, our doubting always takes place in the Now.]

> We understand that the Universe consists of all the things contained in this world. We must also understand that all the people and things in the Universe have their time-present. Just as individual things are independent of each other, so individual time-presents are independent of each other. This means that many people can want to find the truth at the same time, and there can be many times at which a single person wants to find the truth.

[The Now is both independent and inclusive. Each person and all things in the Universe experience the Now independently. The Now includes an array of personal Nows; a person experiences a series of Nows.]

This is true, not only of wanting to find the truth, but also of actually realizing the truth, and actually practicing a Buddhist life. This person who I think of as myself is a "person" that I put together at one time-present. We can apply this thinking to everything in the Universe. This kind of intellectual analysis is the starting point of Buddhist practice.

[To "realise the truth" is to realise what is real at each concrete Now, This is to practice a Buddhist life. The Self is a construct, put together at each moment; in the same way, our interpretation or conceptualization of the Universe is an instantaneous construct.]

But a person who has clarified their real state sees only each thing, each thing, each thing, and lets go of understanding the nature of each thing. And at that moment, time-present contains the whole of time, and that time contains all things. Thus the whole of existence, the whole Universe, is present at each moment of time. Have a quick look to see if you can find any part of the Universe that has escaped from this present moment.

[Someone living fully in the Now is not interpreting the world, but experiencing the world. Realising the instantaneous nature of reality means realising that each Now contains all. Nothing is left out.]

Until we study the Buddhist view, it is normal to think that "time-present" means at one particular time we become angry, and at another particular time we become a buddha. We imagine events as parts of a journey, as though we were crossing a river or walking over a mountain pass. And although we feel fairly sure that the mountain pass or the river are still present back there where we crossed them, we have already passed them and moved on to be illuminated by the present, leaving them behind in the far distance.

[We construct a line of time by joining events into a "journey" with events passing from the present backwards into the past; we model the present as an illuminated band that sweeps along our constructed journey]

> But this is not the only way to think about it. At the time when we are crossing the mountain pass, or crossing the river, we are present there, and so time is present there. Time cannot elude the present. Accepting that time does not appear and disappear, the time when we are crossing the mountain pass is also a real time-present. And even if time were to appear and disappear, the time when time is actually present is also time-present. Then how can the time when we are crossing the mountain pass or crossing the river not be swallowed up in the time when we are illuminated by the light of the present. How can we say that the time when we are crossing the mountain pass or crossing the river does not spew out this present brightness?

[Time exists only at the present. Although we construct present Nows and past Nows, all time-presents are enfolded here in the eternal Now.]

> Although the angry person is an event in the past, and the buddha is here today, we can also think of time as if going up into the mountains and looking out over thousands of peaks, rather than seeing time as passing by us. The time-peak when we become angry is then a time-present, even though it seems to have receded into the past. And the time-peak when we become buddha is also a time-present, even though it seems to be back there. Then a pine tree is a time-present, and a bamboo thicket is a time-present.

[The Now is compared to a spatial array of time-presents. All presents are arrayed in the Now; the past and future are arrayed in the Now. The Now is inseparable from concrete objects (pine trees, bamboo thickets) existing at the present moment.]

> Seeing time simply as flowing away is not enough. Thinking that the only property time has is the ability to flow is not enough. If we

think of time only as flowing away, then there must be gaps between the instants of time-present as they pass. Ordinary people only see time as something that flows away, and this is why they do not experience time-present, and have not heard it explained. In actual fact, all the things in the whole Universe are time-presents that are both continuous and separate. Real time is always time-present, and so it is always this time-present.

[Our concept of time as a linear trace does not fully capture the nature of time. If moments flow, then they must be discrete and separated by gaps, as represented by the metaphor of time as consecutive frames in a film. To see time only as flowing is to lose the experience of being fully in the Now. Time can be modelled as both continuous and discrete; it can also be experienced as neither (or both) of those. The Now is eternally present.]

But time-present can also be thought of as a process. Time proceeds from today to tomorrow. Today proceeds back to yesterday. Yesterday proceeds on to today. Today proceeds on through today. Tomorrow proceeds on through tomorrow. Although time can be seen as a process like this, times arriving do not pile up on top of times past, neither do they extend out in a continuous line. And for this reason we see that Master Seigen's life was made up of moments of time-present, and so were the lives of Master Obaku, Master Baso, and Master Sekito. Both I and the external world exist in time-present, and so practice and realization of the truth exist at time-present. All our daily struggles are also at time-present. People have their own views today, and they think these views are produced by their own causes and circumstances. Although they think that their views form the basis of, and are based on, their experiences, they are not the basis of their reality. They themselves are only reality presenting its results at this time-present as causes and circumstances.

[Time is a marker for process. But instants of time have no existence per se; they do not accumulate or extend. Buddhist masters of the past lived

in the Now. The self and the external world are conceptualisations of an undivided Now. Our life takes place in the Now. People mistake their views of reality formed from their past experience for reality itself, but reality is not based on past experience. It exists Now.]

> People do not see reality itself as being this time-present – what exists at this moment – and so they feel that the state of a buddha is not time-present, but a state that they must make efforts to attain. But even the moment in which they make their efforts and reject the state of buddha because they cannot see that it is their state, is also time-present – it is the time-present of a person who has not realized who they are. It is time-present that causes the hour of 12 o'clock and the hour of 2 o'clock arranged as they are today as they rise up and drop away at their place in the Universe. Two o'clock is a time-present; 6 o'clock is a time-present; living beings are time-present; buddhas are time-present. The concrete time-present of being angry makes the whole Universe real. The concrete time-present of being a buddha makes the whole Universe real. The time-present in which the Universe makes itself real is called perfectly real. The state of buddha which wants to search for the truth, practice Zazen, attain the truth, and enter the balanced state to make the state of buddha real, exists at time-present. It is making perfectly real that the whole of time is the whole of existence, with nothing left over. But since something left over is just left over, even making the whole of time-present only half real is making half of time-present wholly real. Even the times when we seem to be blundering along heedlessly are at time-present. To leave those heedlessly blundering time-presents as they are is to see that they have their own places as time-present. The vivid state of being in our own place in the Universe is time-present. We should neither confuse it with not being present, nor force it into being present.

[Because people do not see that only Now is real, they project their aims into a future "present" and make efforts to achieve them. People mistakenly think that the state of a buddha is something to be attained in the

future by efforts in the present. They do not see that to be present in the Now is the state they are trying to attain.[c] *But even their misguided efforts take place in the Now. Times are markers for concrete Nows; a concrete Now is the Universe itself. Time-present is the only reality and is indivisible; the Universe is created at this moment; we cannot escape from the Now; acceptance of this reality at this moment is neither to avoid being in the Now nor to try to be in the Now.]*

> We suppose that time just passes away, and do not understand that time never arrives. Although our understanding itself is time-present, our understanding of time does not influence circumstances. People see time as coming and going, and there are few people who see it clearly as time-present here and now. Then how can it be easy for them to get the time to attain the truth? Even among those who see time as what is present now in the Universe, who can express clearly what the ineffable something is that they have already got? And even among those who have long been asserting that they have it already, none of them have stopped groping around trying to find out who they already are. But leaving these people to their time-present, even attaining the truth and the balanced state is just time-present that appears to come and go.

[The Now neither arrives nor passes; it is eternally present. Our attempts to understand the Now themselves take place in the Now, but our understanding does not influence the character of the Now. It is difficult to grasp the nature of the Now intellectually, and even more difficult to express what is already present.]

> In short, time-present is always made real without ridding ourselves of hindrances and delusions. The benevolent forces that appear to the right and left of us are the time-present of our wholehearted efforts in this moment. All the living beings of the lands and oceans are also the time-present of our wholehearted efforts in this moment.

[c] See Wittgenstein, *Culture & Value*; 7e: *"The place I really have to get to is a place I must already be at now."*

All the beings in this world and other worlds are also the time-present of our moment-by-moment wholehearted efforts. Our practice teaches us that without our own moment-by-moment effort in time-present, nothing could ever be made real, or continue from one moment to the next.

[It is not a question of getting rid of things that prevent us being in the Now; whenever we make wholehearted efforts we are in the Now. Our wholehearted effort at this moment not only brings us into the Now, it creates the Now at each moment.]

But we should not understand that this momentary continuance is like the wind or rain sweeping from east to west. The whole Universe is progressing from one moment to the next; not static, but also not a continuous process. An example of this momentary progression is Spring. It has many different aspects, the passing of which we call the progress of Spring. But our practice teaches us that the passing of time from moment to moment involves no external object.

[There is a difference between our concept of a succession of Nows and the experience itself; the progression is neither continuous nor discrete. There is nothing external that regulates the passing of time from moment to moment.]

For example, momentary Spring passes through a process that we call Spring. But the passing is not Spring itself; we just think of Spring as the movement of momentary Spring through the process Spring. But Spring is made real at every time-present in the process of Spring. We need to come back to this again and again. In discussing the momentary passing of time, if we think that there are only individual things that change on the outside, but there is some real unchanging entity that passes from moment to moment moving through all worlds and all times, we are becoming lax in studying Buddhism in our practice.

[Concepts are not reality; what we call "Spring" is only a process we conceptualize by connecting together a succession of momentary Springs. Reality is not something unchanging that underlies our conceptual world]

One day, at the suggestion of Master Sekito Kisen, Master Yakusan Igen asked Master Baso Doitsu a question: *"I have more or less understood the meaning of Buddhist theory and its classifications. But I want to ask you, what was Bodhidharma's intention in coming from India to China?"*

On hearing this question, Master Baso Doitsu replied, *"At one time-present he was moved to raise his eyebrow and wink, at another time-present he was not moved to raise his eyebrow and wink. At one time-present, to raise his eyebrow and wink was right, and at another time-present, to raise his eyebrow and wink was wrong."*

When Master Yakusan heard these words, the whole of Buddhism became clear to him, and he said to Master Baso, *"When I studied with Master Sekito, I was just a mosquito sitting on the back of an iron ox!"*

[Dogen quotes this record of a conversation between two Chinese Zen masters to illustrate natural behaviour in the Now. Bodhidharma was the third son of a southern Indian king and the successor of Master Prajnatara. He sailed to India in the sixth century and became the first Buddhist patriarch in China. The question "What was Master Bodhidharma's intention in coming from India to China?" is really a question about how to live as a Buddhist; about whether Bodhidharma intentionally planned where he was going in his life. The answer affirms that Bodhidharma acted naturally at each moment, following the circumstances of his life. His actions were not limited by ideas of right and wrong. These independent and momentary actions in each Now formed Bodhidharma's real life.]

No-one else could have uttered these words of Master Baso's. The eyebrow Master Bodhidharma raised or the eye he winked may indicate a mountain or an ocean, because the mountains and oceans and his eyebrows and eyes are one whole. Being moved to raise an eyebrow he sees the mountain. Being moved to wink, he rules the ocean. Being right has become part of him and pulls him in. Wrong is not always not done, and not doing is not always wrong. All these are time-present. The mountains are time-present and the oceans are time-present. Without time-present the mountains and the oceans would not exist. We cannot deny that time-present *is* the mountains and the oceans here and now. If time-present decays, so do the mountains and the oceans. If time does not decay, neither do the mountains and oceans. It is from this truth that the bright star appeared to Gautama Buddha, the state of buddha appeared, his ability to see the truth appeared, and his acknowledgment of Mahakasyapa appeared. All *are* time-present. Without time-present nothing could be here and now.

Master Kisho from the Sho region is a descendent of Master Rinzai and the successor to Master Shuzan Shonen. On one occasion he preaches to the assembly:

> *At one time-present will is present but words are absent.*
> *At one time-present words are present but will is absent.*
> *At one time-present both will and words are present.*
> *At one time-present both will and words are absent.*

[Dogen comments that these simple actions that formed Bodhidharma's life were not separate from the circumstances of each moment. Each Now is complete in itself; our conceptualisation of the Now produces time and space, right and wrong, people and circumstances, observers and observed, out of one indivisible whole. Dogen then quotes a tenth century Chinese Zen master describing in a very realistic way the experience of preaching to his students. Sometimes we want to say something, but the words don't come. Sometimes we have the words, but not the impetus to speak.]

Both will and words are time-present. Both presence and absence are time-present. Before the moment of absence has ended, the moment of presence has come. Before the will has left, the words arrive. To be present does not mean that something has arrived. To be absent does not mean that something has left. Time-present is like this. Presence is just being present, it is not being absent. Absence is just being absent, it is not being present. The word "will" describes will itself, and depicts will as an object. The word "words" describes words themselves, and depicts words as an object. The word "description" describes description itself, and depicts description as an object in time-present. Descriptions describe things, but a description of a thing is never the thing itself. When I meet a person, a "person" meets another "person." When I meet myself, a manifestation meets a manifestation. If there were no time-present none of this could happen. In summary, will is one time-present that makes the Universe real. Words are one time-present in the balanced state. Being present is just the time when everything is here and now. Being absent is just the time when a fact is not here and now. This is how we should understand and make ourselves real.

[The Now is beyond conceptual explanation; all actions take place in the Now, the eternal present. Language misrepresents what is being described. We describe events in time as arriving and leaving, but in reality, there is only the instantaneous Now. Now it is present. Now it is absent. Now we have the will. Now we have the words. Now we have both. Both the will and the words are just facts at the moment of the present. Both presence and absence are facts in the here and now.]

Although this poem is how one master of the past expressed himself, I feel moved to express my own understanding:

Will and words being half present is also at time-present.
Will and words being half absent is also at time-present.

We can study the situation like this:

He was moved to raise his eyebrow and wink in half of time-present.
He was moved to raise his eyebrow and wink in the jumble that is time-present.
He was moved not to raise his eyebrow and wink in half of time-present.
He was moved not to raise his eyebrow and wink in the jumble that is time-present.

To experience arriving and leaving like this is to experience time-present. To experience being present and being absent like this is to experience time-present.

[Dogen presents his own version of the poem, in which he replaces time-present by half of time-present. In doing this, he deconstructs our conceptualisation of the Now, which is indivisible. He suggests that our experience of the Now is an instantaneous experience; Now, Now, and Now…]

4. Time and Concrete Existence are Indivisible

Bringing together the summaries of the main points of each paragraph, we obtain the following overall view of what Dogen is saying in this chapter.

Time is always identified with individual concrete events, and time does not exist separately from them. We can view time as a continuous process, both subjective and relative, and objective and measurable.

Although we do not usually doubt that time is continuous, a linear progression, we cannot grasp the true nature of time. Although our belief in the concept of linear time is virtually complete, any doubt we have about this will also occur in the Now. The Now is experienced independently by each individual, but the Now is also inclusive. (The Now is both a series and an array.)

We construct our concept of self in each Now, and we construct (perceive and conceptualise) the world we live in at each Now. To be present

Now is to experience the world, not to interpret or conceptualise the world. To realise that only Now is real is to realise that everything is contained in the Now. We see the progression of events as a journey in time, with the present as an illuminated band sweeping along our constructed journey. But we can also view past, present, and future events as all arrayed in the Now.

Our linear concept of time is not sufficient to describe reality. Although time can be seen as continuous (analogue?) or discrete (digital?), time is actually inseparable from the Now. The Now is the only reality, and this time and this place are inseparable. Each Now is the Universe created and we can never escape from this reality. We do not arrive in the Now and we never leave the Now; we are always *here*. But no matter how hard we try, we cannot grasp this fact intellectually. It does not require an effort to be present, but our action in this moment itself *is* the Now.

Our concept of the Now and our actual experience are different; in reality, the Now is neither continuous nor discrete. Concepts are not reality, but neither is reality something unchanging that underlies our conceptual world. When we are present in the Now, our behaviour is natural. And natural behaviour is complete in itself. The Now is beyond conceptual explanation and complete in itself, unmoved by our inability to express it. But although we cannot express it, we can experience it.

5. Conclusions

In today's world, more than 800 years after Dogen wrote about Time-Present, the linear view of time still rules, and other personal experiences of time we may encounter are labelled as "subjective". When, immersed in a task, we find that "the hours have flown past", we have no hesitation in accepting that our experience is just a feeling, and that in fact, the seconds have ticked past exactly as dictated by the clock. That is our received notion of reality. Our life moves on as the hands of the clock revolve. The river of life flows onwards.

Einstein conceived that time is always relative to the observer. He showed that it is not always linear and not the immovable standard that we had come to accept. At the frontiers of modern physics, situations are emerging in which events cannot always be meaningfully ordered in time. We may need to fundamentally review our notion of time itself.

Dogen uses a three-layered theory of time, in which each of the three layers is represented by a different and mutually exclusive viewpoint.

- Layer 1 (Continuous): The traditional view of time; time is an objective linear stream in which the events in our lives occur. Time is quantifiable and the same in all locations.
- Layer 2 (Discrete): The conclusion reached by Buddhist philosophers such as Dignaga and Dharmakirti: The theory of instantaneous existence – "Every duration in time consists of point-instants following one another [like the frames in a film]. Every extension in space consists of point-instants arising in contiguity and simultaneously, every motion consist of these point-instants arising in contiguity and in succession. Therefore there is no time, no space, and no motion over and above the point-instants of which these imagined entities are constructed by our imagination."[4]
- Layer 3 (Eternal present): The one irrefutable fact is that we are always here in this place at this time. The present is thus *always* present. It is *always* Now. Both past and future are contained in the Now. The Now is eternal. It never starts and never ends; it has no definable duration. The present reality does not come from somewhere and then pass on to somewhere else; everything is always part of the Now. How can we imagine this kind of Now? How is it possible for the moment of the present to embed all that was, is, and will be? What kind of mathematical structure is needed to model something with no duration?

References

1. Th. Stcherbatsky, *Buddhist Logic* Vol. 1, P. 80, Dover Publications, 1962.
2. M. Luetchford, *Between Heaven & Earth – A Translation and Interpretation of the Mulamadhayamakakarika,* Chap. 19, Windbell Publications, 2002.
3. Nishijima & Cross, Master Dogen's Shobogenzo Book 1, Chap. 6, Windbell Publications, 1994.
4. Th. Stcherbatsky, *Buddhist Logic* Vol. 1, P. 84, Dover Publications, 1962.

CHAPTER 4

SYSTEMS AND OBSERVERS FROM A HOLISTIC VIEWPOINT

Franz-Günter Winkler

Stumpergasse 63/11, 1060 Vienna, Austria
E-mail: fg.winkler@aon.at

The observer problem is tackled from a systems-theoretical viewpoint: The observer and the interaction game between observer and world are regarded as two overlapping systems. This construction requires a foundation of systems theory in a holistic framework.

1. Introduction

Knowledge is not just about something, it is always someone's knowledge. If we believe that the holder of any kind of knowledge and the process of knowledge creation cannot be abstracted away, we have to deal with the problem of observation. The domain of the observer problem, i.e. the field in which we suppose the contextuality of knowledge to be relevant, cannot be reduced to psychology and sociology. The status of the observer has to be addressed in the philosophy of science and even in theoretical physics.

In this article, we tackle the observer problem from a holistic perspective: more precisely, from the perspective of spacetime holism. After outlining the observer problem on the basis of a distinction between inside and outside concepts of observation, we will give an overview of spacetime holism. In its framework, the notion of a (self-organizing) system will be re-introduced or, rather, dissolved. The main argument concerning the observer problem, which is a systems-theoretical one, will be discussed in the context of the so-called representation problem of cognitive science.

2. The Observer Problem

When talking about observation, what we have in mind is a scenario consisting of the world and an observer holding a model of the world. This scenario is the result of an analytical cut (the Heisenberg cut). It is connected to three critical questions: *(i)* What is the world? *(ii)* What is the observer? *(iii)* What is the role of the interactions between the observer and the world?

If these questions could be analyzed separately, the observer problem would be an easy one, at least conceptually. The questions, however, are complexly intermingled. For example, the world is not directly accessible - it is given to us as a modeled and therefore observed world. The observer, of course, is an integrated part of the world – modeling the world successfully means, therefore, to model the observer as well. The same holds for the interactions between the observer and the world.

It is thus clear that the Heisenberg cut has a limited analytical value. A useful account of the observer problem should address the shortcomings that are produced by the Heisenberg cut without losing its benefits. The suggested holistic worldview provides the basis for such an undertaking. As a result of this holistic move, we will shed some new light on the third of the above questions dealing with the relation between observer and world. Only by answering this question can the status of an observer's knowledge of the world be clarified.

2.1. Inside versus Outside Observation

We start our analysis of the observer problem by introducing a distinction between inside and outside observation.

By outside view, we mean the ideal of natural science, namely, the construction of a God-like perspective of the world, where observation is no longer a physical, interactive process. From the outside, we see things "as they are". However, all our observations are inside operations, as we ourselves are part of the physical universe, and each observation, in the final analysis, is an interaction between an observer and the thing observed.

As a first example, think of a simple length measurement of an object by the use of a meter stick. The statement "this object is one meter long" says something about the object, the meter stick and the measuring procedure, and is therefore the result of an inside observation (using a "meter" stick that is twice as long would yield half a meter as a result of the length measurement). Of course, by normalizing measuring instruments and by defining measurement procedures, we can construct a perfectly useful outside perspective, which allows us to treat the result of a measurement as an objective property of the measured object. This is much less clear in the second example: It is hard to think of a statement like "this person is nice" as an outside observer's statement capturing an objective property of a person.

2.2. Observation as a System

As has just been argued, every observation is an inside observation in the final analysis: To observe means to interact. This statement does not only refer to the fundamental limitation to "passive" physical observation posed by quantum theory: it holds on virtually all conceptual levels. Whenever we observe, we bring our expectations, prejudices and models into contact with the raw perceptual data. These data are not just there: they are perceived by us, because we direct our attention to something and because we force the world to answer to our actions.

Every cognitive concept and even every measurement result is therefore the result of an interaction game between observer and world. We will regard this interaction game as a dynamical, self-organizing system, which leads to a new understanding of knowledge. An observer's knowledge is no longer about some part of the world, but is about the interaction game between observer and world. In order to make the views of dynamical systems theory applicable, the connection between holism and systems theory has to be clarified.

2.3. The Representation Problem and Two Aspects of Observation

In cognitive science, the observer problem forms the core of the so-called representation problem.[a] To formulate it cautiously, the representation problem addresses the question in which way the knowledge held by a cognitive system is related to the outside world. Representations are regarded as identifiable states of a cognitive system, which *(i)* refer to states of the system's environment, and *(ii)* are processed by the cognitive system.

The first aspect focuses on the relation between observer and world. A representation is related to the world and therefore meaningful. The second aspect addresses the computational dimension of a representation. A good model allows us to compute future states of the world on the basis of the representations it holds.

Elsewhere,[1] we have argued that these aspects of a representation form a duality relation which does not allow the full realization of both aspects at the same time. The more meaningful a concept is (i.e. the more relations it has), the less computable will it be and vice versa. On this basis, a very general duality relation for soft and hard sciences can be formulated. Only well defined concepts as postulated by the hard sciences allow the computation of future states of a system. The concepts used by the soft sciences are much more associative and, for exactly this reason, more meaningful. In the final analysis, a full realization of both meaning and computation cannot be expected according to the duality relation.

In the light of these considerations, the Heisenberg cut appears as a necessary precondition for computable models while at the same time limiting the potential meaning of the models. This analysis is far from new. What other analyses often lack, though, is the perception of the incompatibility of the duality relation with the still prevailing reductionist world view. In the following, we briefly introduce the basic

---

[a] Other aspects of the representation problem deal with the coding of information in biological and artificial cognitive systems.

assumptions of spacetime holism, which takes the duality relation as fundamental.

3. Spacetime Holism

To hold a holistic world view means to believe that the analytical method of extracting parts (both physical and conceptual) and of modeling their relations cannot, in principle, produce a full description of the world. This is, of course, a rather poor, because negative, definition which seems to open the door for complete arbitrariness of models. In this section, we give a brief introduction of *spacetime holism*,[b] which adds some positive features to the negative definition.

3.1. Ontology

To a certain degree, spacetime holism accords well with David Bohm's concept of "undivided wholeness" (Bohm[2]). What are usually regarded as physical objects existing independently of each other are understood as necessarily imperfect patterns that an observer cuts out of an irreducible and infinitely complex whole. If it is at all useful to talk of ontology, only the whole deserves an ontological status. Any attempt to describe the world as a collection of interacting parts (that exist as such) is therefore doomed to failure.

Once the impossibility of a complete and consistent ontological order of the world is accepted, there is a pragmatic way of dealing with ontology. If we do not demand that the entities of our most different models and descriptions (from physics to sociology) can be fully defined and reduced one to the other, we can work (or better: go on working) with a virtually infinite ontology. In this heterogeneous picture there is no priority given to some entities or some perspective.

Beside the incompleteness of any description, there is yet another limitation to scientific knowledge of the material world for spacetime holism, namely, the impossibility of achieving a perfect outside view. As

---

[b] More on spacetime holism and its application to philosophical problems can be found in Winkler[1,3,4].

inside observers, we can only *construct* more or less consistent outside views.

3.2. The Universe as one Spacetime Whole

The idea that space and time form some kind of unity is part of the (standard) interpretation of the theory of relativity. However, there is no common treatment of space and time in other disciplines. The reason for this lies mostly in the big difference between our everyday perceptions of space and of time (time "flows", space does not), but may also have something to do with the way relativity theory is understood.[c] Though agreeing with known holistic approaches in their treatment of entities and relations in space, we would like to criticize holism for its treatment of the time dimension. Most holistic approaches are far away from integrating space and time.[d] Instead, time is seen as something special, and different concepts are used for the description of relations in time (e.g. Bohm's "flowing movement", non-determinism in quantum theory, unpredictable emergence in the theory of self-organization). In our approach, we go one step further by fully integrating the time dimension: Everything in spacetime must be regarded as one indivisible whole. As a consequence, the same conceptual tools can be applied to both space and time.

3.3. Infinite Complexity and Self-Containment

To assume the fundamental incompleteness of any description of the whole in terms of perfectly definable parts is tantamount to assuming that the spacetime whole is infinitely complex. This allows us to postulate an additional feature of the whole without running into contradiction, namely, full containment of the structure of the whole in each "part". This, of course, does not imply that the structure of the whole can be studied equally well in each part.

---

[c] An alternative interpretation of relativity theory has been suggested by the author in Winkler[5,6].

[d] Though usually not seen as a holistic theory, general relativity can be regarded as a prominent exception.

3.4. Spacetime Structure and the Duality Relation

The assumption of infinite complexity of the spacetime whole does not mean that no understanding or description of the structure of spacetime is possible. For our investigation, we suggest a terminology that, on the one hand, is related to concepts in mathematics, physics, and systems theory. On the other hand, the terminology is soft enough to be applicable to a wide range of less precise domains, e.g. to psychology and sociology.

3.4.1. *Continuity and Discontinuity*

We talk about *continuity* along one dimension of a description when several entities differing in position along that dimension have something in common. While *continuity* forms the basis for the recognition of relations between entities, *discontinuity* may lead us to make distinctions and to draw boundaries, it is thus a precondition for the concept of an object.

3.4.2. *Levels of Description*

The necessity for a distinction between levels of description can be illustrated using the distinction between continuity and discontinuity. What appears as a discontinuity on a lower level of description may be described as a continuous sequence of discontinuities on a more abstracted level. An object can be seen as a discontinuity in space - yet there can be a continuous arrangement of objects in space. An event can be seen as a discontinuity in time - yet there can be a continuous sequence of events.

3.4.3. *Non-locality and Locality*

The distinction between non-locality and locality is an abstraction of the distinction between continuity and discontinuity. The ideal object has a location in space and is unaffected by its environment. Beside its spatial discontinuity, the object is extended in time. Its existence in time is thus not bound to a short interval, as is the case for its spatial existence. In this sense we call objects non-local in time. When reversing space and time, we come across phenomena that are non-local in space, but local in time. The ideal wave[e] is identical for all (spatial) points. On the time axis, there is discontinuity: The behavior of a (spatial) point in a wave is not primarily correlated with prior or later states of that point, but much more to other spatial points (at the same time). Like the object marks a point in space, the wave marks a point in time.

3.4.4. *The Duality Relation*

As a generalization of the last paragraph, we assume a very general linkage between structural properties along space and time dimensions:

> *Continuity (non-locality) along one dimension is correlated with discontinuity (locality) along the other dimension.*

Only when discontinuity concentrates on the spatial axes is a description as an object justified. In the same way, it is the discontinuity along the time axis that characterizes the wave. For us, both sides of the duality are necessarily incomplete – in general, there is continuity and discontinuity on all levels in space and time. In the following, we will talk about a duality between particle and wave aspects that correspond with the pairs *(discontinuity in space, continuity in time)* and *(continuity in space, discontinuity in time).*

The classical scientific world view exemplified by Laplace´s demon appears as the belief in the sufficiency of the particle aspect: If a demon knew the positions and velocities of all particles of the universe at some point in time, he could calculate every state of the universe at any time! From this viewpoint, time is a purely continuous dimension - there should be a "right" level of description that makes the passage of time a transformation that neither brings anything new nor loses anything.

---

[e] The metaphorical use of the term "wave" requires some clarification. What we have in mind is a standing wave rather than a propagating one. Associations with quantum theory are intended – the so-called collapse of the wave function shows the postulated properties of spatial non-locality (a simultaneous link between distant parts) and temporal locality (non-predictability, at least from local parameters).

4. Spacetime Holism and Systems

Our argument on the interaction between observer and world is a systems-theoretical one. It is therefore necessary to connect the principles of spacetime holism with systems-theoretical concepts.

4.1. Self-Organization and Holism

The theory of self-organization is usually introduced in the form of a critique of the still prevailing reductionist world view. In the following, we will take a closer look at the holistic aspects addressed by the theory of self-organization.

The whole is more than the sum of its parts. This well-known statement reflects that the theory of self-organization is about collective phenomena of systems (which thereby constitute as wholes) that cannot be reduced to the behavior of their parts, at least not in the traditional straightforward way.

A key concept for self-organizing systems is *emergence.* The collective behavior of the whole is said to emerge from the interaction of its parts. Depending on the interpretation of emergence used, there is sometimes also an influence from the emergent level on the level of the parts assumed.

Following from this, there is a lot of confusion about the notion of *causality*. There is talk about circular causality, efficient versus final cause, bottom-up versus top-down causation, weak and strong causality, and non-causality (chance). The critique of the traditional understanding of causality which stands behind these discussions may have well contributed to the popularization of the theory of self-organization, as it addresses the uncomfortable feelings we have with a completely deterministic universe leaving no room for chance and free will. However, scientifically speaking, the messiness of the treatment of causality is a major weakness of the theory of self-organization.

From the perspective of spacetime holism, the theory of self-organization lacks an important aspect of part-whole relationship, namely, the full containment of the structure of the whole in each part.

Another important point of critique from a holistic perspective is what could be called the re-introduction of a typical reductionist aspect, namely, the restriction to systems hierarchies. Even those approaches, which quite frankly assume top-down causation in addition to bottom-up causation, draw strict hierarchical pictures of system relations (e.g. physical – biological – cognitive – social).

4.2. The System From the Viewpoint of Spacetime Holism

The notion of a system is both central and hard to define, not only in the theory of self-organization. From the perspective of spacetime holism, all that can be done is to describe certain aspects (and their relations) which are relevant for treating something as a system.

(i) **External particle aspect.** A system shows some independence from its environment and some stability.

(ii) **External wave aspect.** A system is open to the environment: it interacts and thus takes part in multiple meta-systems (hierarchical or overlapping[f]). By this, it is possibly subject to changes.

(iii) **Internal particle aspect.** A system has some internal structure, which can be described as constituent parts in interaction.

(iv) **Internal wave aspect.** A system maintains relations between its parts. ("The whole is more than the sum of its parts").

(v) **Representational aspect.** The part represents the system; the system represents the meta-system. This aspect has to do with the postulated feature of part-whole containment.

These aspects compete with each other and cannot be fully present at the same time. The more some aspect dominates, the less we are inclined to talk of a self-organizing system. For example, mere causal systems that can be reduced to the deterministic behavior of their well-definable parts (particle aspect) cannot represent, which presupposes a strong wave aspect.

The aspects (i) to (iv) are illustrated in Figure 1. The illustration is based on the geometrical metaphor of a dominating axis of continuity that may have different spacetime orientations. In the orthogonal

[f] cf. section 4.3.

direction, discontinuity dominates (this reflects the duality relation linking continuity and discontinuity in space and time). For the ideal wave, the axis of continuity is spatial, for the object the axis of continuity is temporal. For the wave, there are relations in space; for the object, there are relations in time.

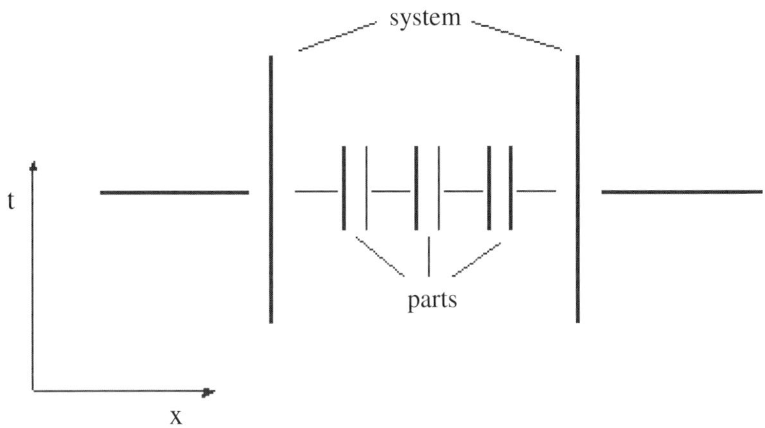

Fig. 1. The axes of continuity for the particle and wave aspects of a system.

4.3. Overlapping Systems

As has already become clear from the discussion so far, spacetime holism gives a secondary status to the concept of a system. The system must be understood as a mixture of aspects that, if taken as absolute, contradict each other. In addition to this, being a system is not an objective property. Instead, the question what to treat as a system highly depends on the observer and the level of abstraction. From different views, there may appear different systems, which do not at all have to form a hierarchy.

Figure 2 shows schematically what we mean by overlapping systems. Each circle represents a system as it appears to some observer from her or his perspective. For reductionism in particular, but also for the usual understanding of the theory of self-organization, overlapping systems are forbidden.

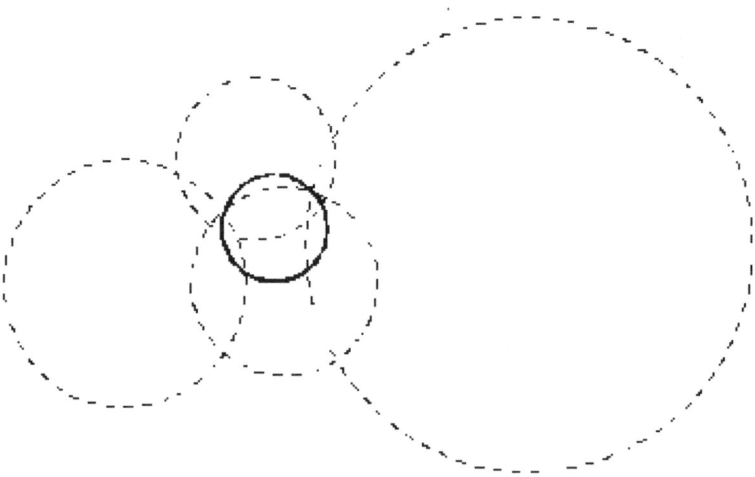

Fig. 2. Overlapping systems.

However, it is easy to argue for the necessity to accept overlapping systems. Maybe the best example is the highly interdisciplinary field of cognitive science, where scientists from the most different disciplines cast their opinions upon one and the same thing, namely biological cognition. In doing so, they talk about systems. What from one perspective is a part of the system belongs to the environment from another perspective. When, for example, cognition is studied as the product of a social system, the social environment of an individual belongs to the system, whereas the study of cognition as the product of a bunch of brain cells locates other brains in the system's environment. For the social approach to cognition, parts of the nervous system that are not concerned with, for example, language and memory are located in the system's environment. As long as only hierarchical system relations are allowed, at least one of the two views must be wrong.

For spacetime holism, both descriptions can be useful at the same time. Even the bi-directional causation between system levels can be accepted, as long as the fundamental fault inherent in any division is seen, i.e. as long as no ontological claims are raised for the descriptions involved.

Figure 3 shows one part in the context of a variety of overlapping systems. Only a holistic worldview can explain how this part can usefully be described as a co-producer and product of all these systems.

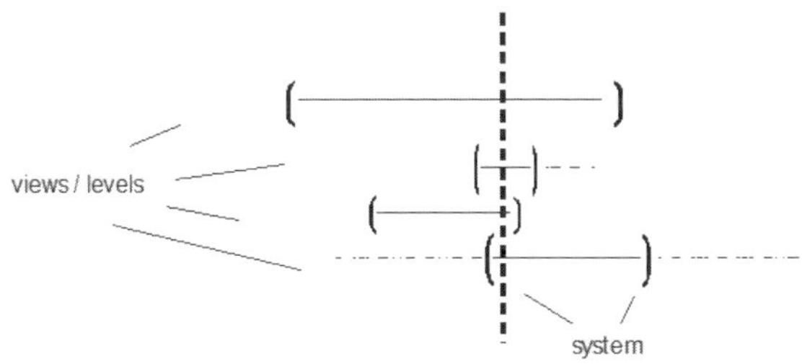

Fig. 3. The spectrum of system contexts for one part.

5. The Cognitive Scenario

Formulating spacetime holism's account of the representation problem and consequently of the observer problem is not much more than summarizing what has already been said. Beside the central features of spacetime holism, the main building block is the shift to systems theory, including the assumption of overlapping systems which allows us to formulate the cognitive scenario.

Figure 4 shows the two intersecting systems involved in the suggested cognitive scenario: *(i)* The cognitive system and *(ii)* the interaction game between (parts of) the cognitive system and (parts of) its environment. In the chosen scenario, the representation problem has to be formulated as follows:

> *In which way can one part, namely the cognitive system, both contain and produce phenomena of the whole system consisting of the cognitive system and its environment in interaction?*

The general answer to this question is already given by spacetime holism's assumption of part-whole containment. It has, though, a counterpart in dynamical systems theory:

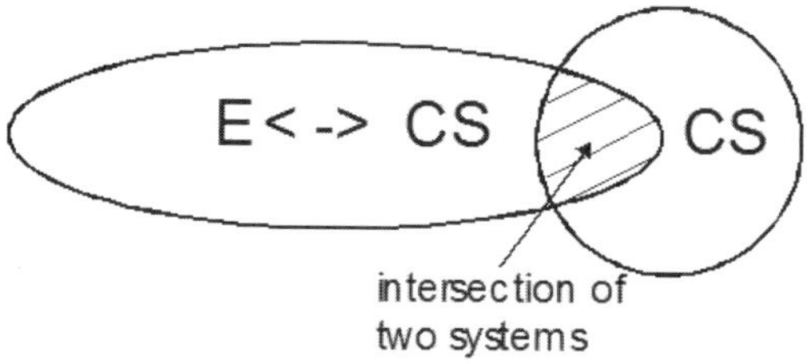

Fig. 4. The cognitive scenario consisting of two intersecting systems: the cognitive system CS and the system formed by interacting parts of CS and its environment E.

According to Taken's theorem of embedding (Takens[7]), it is possible to reconstruct the dynamics of a whole system by analyzing the behavior of single parts[g]. In this sense, the part can be said to represent the dynamics of the whole system.

In our cognitive scenario, the represented phenomenon, of course, is the spacetime pattern of the interaction game between the cognitive system and its environment and not the object as such. As an example, think of the concept of an *obstacle*, which is an entity in space for an outside observer. For the inside observer (i.e. a cognitive system interacting with the obstacle), the obstacle can in the first place only be a set of temporal links between states of the cognitive system, e.g. "when I try to walk through, it will hurt". This temporal relation within the cognitive system is the representation of the external, spatial relation between obstacle and cognitive system, which has been described as the wave aspect. This wave aspect makes up the *meaning* of the representation. On the other hand, we can only speak of a computable representation of the obstacle, when the cognitive system is able to transform the temporal rule (*walk* ➔ *hurt*) into an entity in some internal representation space. As an internal spatial entity, the representation "obstacle" shows stability in time (it is memorized) and can be used as the basis for calculations that guide the behavior of the cognitive system (e.g. "avoid walking through obstacles").

[g] This statement holds also for chaotic systems which produce dynamical behavior of infinite complexity.

6. Conclusions

By its formulation, the Heisenberg cut already points to a holistic world view. If the world and the observer are regarded as the products of a cut, there must be a deeper wholeness connecting them. When a cut is made, there is always something lost or destroyed. The observer problem addresses the loss of the system consisting of the interaction between observer and world which constitutes the meaning of representations and, more generally, the meaning of models.

In this article, we tried to show that the world view of spacetime holism allows us to overcome the contradictions and shortcomings of traditional reductionist approaches to the observer problem. The argument, however, could be read in the opposite direction, as well: A careful analysis of all the aspects of the observer problem leads directly to a holistic world view.

References

1. F.-G. Winkler. *Spacetime Holism - A Fundamental Approach to the Representation Problem in Cognitive Science.* Dissertation, Vienna University of Technology, (2004).
2. D. Bohm, *Wholeness and the Implicate Order*. Boston, MA: ARK Paperbacks, (1980).
3. F.-G. Winkler, "Spacetime Holism and the Hard Problem of Consciousness", in *Computing Anticipatory Systems,* Daniel Dubois (ed.), American Institute of Physics, Woodburg, New York, Conference Proceedings **839**, 465-473 (2005).
4. F.-G. Winkler, "Spacetime Holism and the Passage of Time", in *The Nature of Time: Geometry, Physics and Perception*, R. Buccheri, M. Saniga, W. M. Stuckey (eds), NATO Science Series II, **95**, 393-402 (2003).
5. F.-G. Winkler, "The Normalization Problem of Special Relativity", Physics Essays, 15 (2) 199-210 (2002).
6. F.-G. Winkler. *Die Spezielle Relativitätstheorie- anschaulich erklärt, neu interpretiert.* Peter Lang, (2005).
7. F. Takens, "Detecting strange attractors in turbulence", in *Dynamical Systems and Turbulence*, Warwick 1980, Lecture Notes in Math. **898** (Springer, Berlin, 1981), 366-381 (1981).

CHAPTER 5

A SYSTEMS-THEORETICAL GENERALIZATION OF NON-LOCAL CORRELATIONS

Nikolaus von Stillfried

University of Freiburg, IUK
79106 Freiburg, Germany
E-mail: n.stillfried@web.de

Non-local correlations between quantum events are not due to a causal interaction in the sense of one being the cause for the other. In principle, the correlated events can thus occur simultaneously. Generalized Quantum Theory (GQT) formalizes the idea that non-local phenomena are not exclusive to quantum mechanics, e.g. due to some specific properties of (sub)atomic particles, but that they instead arise as a consequence of the way such particles are arranged into systems. Non-local phenomena should hence occur in any system which fulfils the necessary systems-theoretical parameters. The two most important parameters with respect to non-local correlations seem to be a conserved global property of the system as a whole and sufficient degrees of freedom of the corresponding property of its subsystems. Both factors place severe limitations on experimental observability of the phenomena, especially in terms of replicability. It has been suggested that reported phenomena of a so-called synchronistic, parapsychological or paranormal kind could be understood as instances of systems-inherent non-local correlations. From a systems-theoretical perspective, their phenomenology (including the favorable conditions for their occurrence and their lack of replicability) displays substantial similarities to non-local correlations in quantum systems and matches well with systems-theoretical parameters, thus providing circumstantial evidence for this hypothesis.

1. Introduction

C.G. Jung's and W. Pauli's "Theory of Synchronicity"[1,2], W. v. Lucadou's "Model of Pragmatic Information"(MPI)[3,4] and H. Atmanspacher's, H. Römers's and H. Walach's "Weak Quantum Theory" (WQT)[5,6] (among others) have proposed that non-local (and in that sense acausal) correlations could underlie phenomena in psychology, parapsychology and alternative medicine which have thus far resisted a satisfying explanation within a purely causal theoretical framework. Due to the similarity and relatedness of these theories (in particular MPI and WQT), I will refer to them collectively as Generalized Quantum Theory (GQT). One of the central ideas of GQT is that in addition to the laws of causality there is a non-causal principle at work in shaping our reality. This principle was first scientifically formalized in quantum theory (hence the reference to quantum theory) but, according to GQT, its applicability is by no means limited to the quantum realm in the traditional sense (subatomic particles, etc.). Instead, non-locality is viewed as a much more universal principle which arises from the way in which systems function, given that they are (self)organized in the appropriate way. Of this universal principle, the non-local phenomena observed in quantum systems in the traditional sense are just one particular manifestation. Just like causality does not only work on apples, non-locality is thought to underlie phenomena in all kinds of systems, given that they fulfil certain systems-theoretical parameters.

In the following, I will introduce the concept of a systems-theoretical generalization of quantum theory with a particular focus on non-local correlations. Assuming that the readership of this book will be a rather interdisciplinary one, I will attempt to use as little technical language as possible. To begin with, let us first clarify what we mean by causal and acausal, local and non-local.

I will use the term causality in the limited sense in which it is used in the natural sciences today[a], the sense we are used to describing as "cause

[a] A more comprehensive, yet equally rigorous, concept of causality has, for example, been defined by Aristotle. The modern scientific definition of causality equates to only one of four Aristotelian types of causality, namely the *causa efficiens* ("efficient causality").

and effect"- relationship between events whereby the cause-event precedes the effect-event and where there has to be a space-time connection between the events.

If we chose to accept as a fundamental axiom Einstein's postulate that the speed of light in vacuum (c) is the upper speed limit of propagation in space (and currently there is no evidence to the contrary), this leads to the inevitable conclusion that two events which are separated in space will also always be separated in time if one of them is the cause of the other. The time by which they are separated (Δt) is at least their distance in space (d) divided by the speed of light (c).

This means that cause and effect can never be simultaneous, unless they happen in exactly the same space, in which case, we would not recognize them as distinct events. Locality thus means that simultaneously occurring events can therefore not be causally (and therefore predictably) correlated unless they are connected via a common causal 'root'-event, which again has to precede its effects.

So far so good, but not so in the case of non-local correlations as observed in quantum physics: Here we find situations where two (or more) events occur in such a way that the outcomes of the events are reliably correlated, even if they are separated in space and occur simultaneously. This is basically saying that two events, even though there is no possibility for them to influence each other, nevertheless always occur in a correlated fashion. The only way in which we can classically imagine something like this to work is by a common causal root for both of these events. But quantum physics says that there cannot possibly be, not even in principle, a way in which any third event in the common past of these events could determine them to happen in this particular way, unless we are willing to assume that the *entire* universe and *all* events therein are precisely predetermined in exactly this way by a common root cause.

2. Non-local Correlations in Quantum Mechanics

To understand these weird claims a bit better, let me first introduce quantum physics to those readers who are not yet familiar with it. Despite the outlandish appearance of the phenomena such as the one described above, quantum physics is not concerned with some peculiar fringe area but actually with the very fundamental units out of which our entire physical reality is composed. Quanta are sometimes defined as the naturally occurring units of physical reality; they can be visualized as waves of energy or particles of matter. Usually quanta are considered to be of extremely small scale. For example, an atom can be thought of as consisting of electrons, protons and photons, which are all considered quanta. There is, however, no definite boundary in terms of size. Depending on the conditions, atoms and even large molecules are also considered quanta[7]. More important than actual size is the possibility to talk about the object in question as a truly distinct unit, which means it must be possible to completely isolate it from the rest of the universe, at least regarding certain properties. The object can then be considered a quantum with respect to these properties.

The phenomenon of non-local correlations is also known as the Einstein-Podolsky-Rosen (EPR-) paradox as it was originally deduced in 1935 on a purely theoretical basis from the quantum-theoretical formalism by these authors. It led them to claim that a formalism which allows for such clearly inconceivable correlations must be incorrect or at least incomplete[8]. Proponents of quantum theory, however, maintained that physical reality should behave in precisely this way. It took almost two decades until this question became at least in principle decidable through the ingenious work of J. Bell[9] who thought up an experimental setup for which the predictions of quantum theory were clearly different from the predictions based on classical physical theory (resulting in the so-called Bell's inequalities). Another 18 years later, physics was finally advanced enough to allow A. Aspect to carry out the experiments with sufficient precision[10]: And lo and behold: nature violated common sense and quantum theory proved to be correct.

The experiments which have up to now been replicated in many different forms with fundamentally analogous results usually consist of the following setup[11]: two quanta (most often photons or electrons) are generated in such a way that there is a fixed overall quality which describes the pair, such as, for example, the total energy or the total spin. (This can, for example, be achieved by splitting one photon with a known energy or spin into two photons) After that, the two quanta are kept causally isolated from the rest of the universe with respect to this particular quality. When, for example, considering spin, the photons must not collide or interfere with other quanta in such a way that spin could be exchanged in the process. The quanta are now separated in space. Next, the spin will be measured on both of them simultaneously (or at least within a time span $\Delta t < d/c$). This means that there can be no 'cause and effect'-type relationship between the two measurement events. What is more, the outcome of this measurement cannot, in principle, be predicted based on the history of the quantum, which means it is usually considered undetermined or random. This means that there is a high degree of freedom regarding the quantity of the particular quality (e.g. 'spin') in each of the quanta, while the total quantity of that quality is known and cannot change. (This is expressed in the so-called conservation laws, for example, the laws of conservation of energy and momentum.) When the spin measurements are analyzed, it becomes apparent that the individual spins on each of the quanta always add up to give the correct total. While this would not be surprising if we could assume that the distribution is already determined at the moment when the two quanta are generated, careful analysis of the experiments reveals that this cannot be the case, unless one assumes that the state of the quanta before the measurement is not independent of the state of the measuring device. Since the state of the measuring device is, however, dependent on a large number of factors including the experimenters' will and, more likely than not, the entire rest of the universe, this would mean that the entire universe would have to be determined in this particular way.

Thus, according to J. Bell[12,13] (and largely undisputed since), there are four possible interpretations of this experimental data. If we ignore for the moment the (rather unlikely) option that the observations are an artifact due to some error in the experimental equipment and if we resist the conclusion that it is futile to say anything at all about the nature of reality, we are left with two possible world views, both of which are somewhat hard for the common sense to digest: We either have to assume that quantum events (and therefore all of reality) are indeed fundamentally undetermined, which means they occur in a certain way for absolutely no reason whatsoever while, at the same time, they are entirely reliably correlated with other such undetermined events, which means they must be connected in some non-causal way. Or we have to assume that there are no such things as undetermined events and that even events which we would like to consider as independent (for example, an experimenter's decision as to when, where and how to conduct an experiment and the spin of a particular photon, let alone the precise relative position of all the rest of the universe) are, in actual fact, absolutely determined and, more than that, they are orchestrated in such a way that they produce exactly such a pattern of spin measurements that does not allow us, even in principle, to differentiate it from the first possible interpretation.

Those of you who have not yet delved into quantum physics may doubt my sanity or at least my competence, but luckily you don't have to rely on either, since non-local correlations are one of the most central features of quantum physics and have been discussed in an overwhelming wealth of literature with which you can follow the matter up further. I am only able to attempt a very rough sketch of an outline of a few relevant issues here.

Even though in recent literature, the second interpretation (truly undetermined events with non-local correlations) is usually preferred over the first (universal causal determinism), there is no rational justification for doing so and, in contexts where philosophical accuracy is demanded, also the most respected mainstream physicists (e.g. A. Zeiliger[14]) admit that both interpretations are possible. The relationship of such mutually exclusive and, at the same time, collectively required interpretations is often termed 'complementarity'. Complementarity is another very intriguing and central feature of quantum physics (for example, with regard to the wave-particle-duality) and also of great interest for systems-theoretical interpretations and generalizations of quantum theory[15,16] but we will have to leave it at that for the purpose of this chapter and refocus on non-local correlations by summarizing that, no matter how one chooses to interpret this phenomenon, it is indeed possible to observe correlations that can not be explained by direct causal interaction between the two or more events involved, nor by a common causal root event connecting them but not the rest of the universe.

Since it may be of interest in the context of temporal structures, I should mention that, depending on the experimental setup, non-local correlations of energy can also appear as non-local correlations of detection time points[17]. It may also be noteworthy that it was the phenomenon of temporal coincidence of meaningfully related but not causally connected events that led C.G. Jung to formulate the Theory of Synchronicity in collaboration with W. Pauli. MPI and WQT have also explored the temporal aspects of generalized non-locality, for example in H. Römer's article on the emergence of time[18] and W. v. Lucadou's on self-organizing temporal structures[4].

Another very important feature of non-local correlations is also best explained from a temporal perspective, namely the impossibility of using non-local correlations for signal transfer, also sometimes referred to as Eberhard's Principle[6,19]. Although space does not translate directly into time, it can be said that instantaneous signaling across space is equivalent to signaling into the past since it means sending a signal outside of the light cone[20]. This would thus imply the possibility to change things in the past, which could lead to so-called 'time-travellers paradoxes', like, for example, preventing one's own birth and hence not existing, hence not being able to prevent one's birth, hence existing and thus preventing oneself being born etc. Fortunately, however, this is prevented by the nature of non-local correlations, which, as we have stated above, are closely connected with the unpredictability (degree of freedom) of the correlated subsystems. This means that while, by observing one of two non-locally correlated events, one can instantaneously know with certainty the outcome of the other event even if it happens far away, one *cannot* influence either of the events. We will return to this principle in the discussion of generalized non-locality.

3. Non-local Correlations as a General Systems-inherent Principle

Although the potential implications may be much more far-reaching[16,21,22], GQT's conception of non-local correlations as a general systems-inherent principle is motivated among other factors by an effort to find a new approach to a longstanding dilemma regarding the controversy around anecdotal and experimental data about so called psi-, parapsychological, paranormal or synchronistic phenomena: Largely unnoticed by mainstream science, phenomena such as telepathy, psychokinesis and precognition have been thoroughly investigated for more than a century. Not least due to the high level of skepticism towards this area within the scientific community, the research was, on the whole, conducted according to quite high scientific standards, especially in more recent decades[23]. The overall result of this research effort is at first sight perplexing: there is up to now still not the slightest sign of a consensus regarding even the very existence of the phenomena in question, let alone an accepted explanation of the underlying mechanisms. This becomes understandable when analyzing the nature of the accumulated evidence: On the one hand, there is a wealth of high quality field reports, experimental observations and meta-analyses which, taken by themselves, make plausible beyond necessary doubt the reality of the phenomena such as psychokinesis[24,25], telepathy[26], precognition[27,28] or extrasensory perception[28-30]. On the other hand, there is a large body of failed replication attempts and, to date, not a single experimental setup or field-case exists which allows reliable replication of the observations.

Unfortunately, it is beyond the scope of this chapter to review in more detail the field of parapsychology. I will thus pick just one example which in my view illustrates the general situation very well.

In this study[31], which was conducted by R. Targ and H. Puthoff and their research group at the Stanford Research Institute, a subject who claimed to have psychic abilities was isolated in a visually, acoustically and electromagnetically shielded room. After the room had been locked, a 'target' picture was produced in an office nearby. The content of this picture was (a) determined by opening a dictionary arbitrarily and drawing the first word that could be drawn (for Experiments 1-4); (b) prepared independently by scientists outside of the experimental group (following the subject's isolation) and provided to the experimenters during the course of the experiment (Experiments 5-7, 11-13); and (c) arbitrarily selected from a target pool decided upon in advance of daily experimentation (Experiments 8-10).

Under supervision through a one-way monitor, the subject then attempted to draw a 'response-picture' which was supposed to resemble as closely as possible the target picture. In 3 of the 13 experiments (experiments 5-7), the subject did not produce a picture. The target and response pictures of all other experiments are presented in Figure 1. (No data was omitted.)

I think it is quite obvious why, on the basis of experimental observations like this, some scientists have become strongly convinced that 'there is something to' the so-called paranormal and have even risked their reputation and careers in an effort to convince others of this matter. And truly, if a person like the one studied here could at liberty reproduce this kind of experiment with the same rate of success for all those of us who need to see with their own eyes, soon only those who refuse to look would be left with doubts.

Such success, however, seems to have remained an unfulfilled hope for this as well as other similarly remarkable experiments, of which there are quite a few. A frequently observed pattern is that as soon as replication studies are undertaken, the effects seem, sooner or later, to disappear.

Several possible explanations have been advanced for this situation. Those who feel that the phenomena are real argue that possibly the psychic subject loses motivation and gets bored or tired by the replication attempts. Others doubt the existence of the phenomena in question, arguing that the successful experiments are due to chance, since trying often enough will here and there produce astonishing results by mere statistical coincidence. In the worst cases, the authors of successful experiments which fail to replicate are accused of fraud.

All of these explanations have their 'pros' and 'cons' and none are entirely satisfying, leaving important questions open. Wouldn't it keep a subject motivated to know that so much depends on a successful replication? How many trials would be required to come across a level of coincidence as high as depicted in Figure 1 by mere luck? And why would scientists risk their credibility and career by publishing fraud which they know will get exposed because others will not be able to replicate the findings?

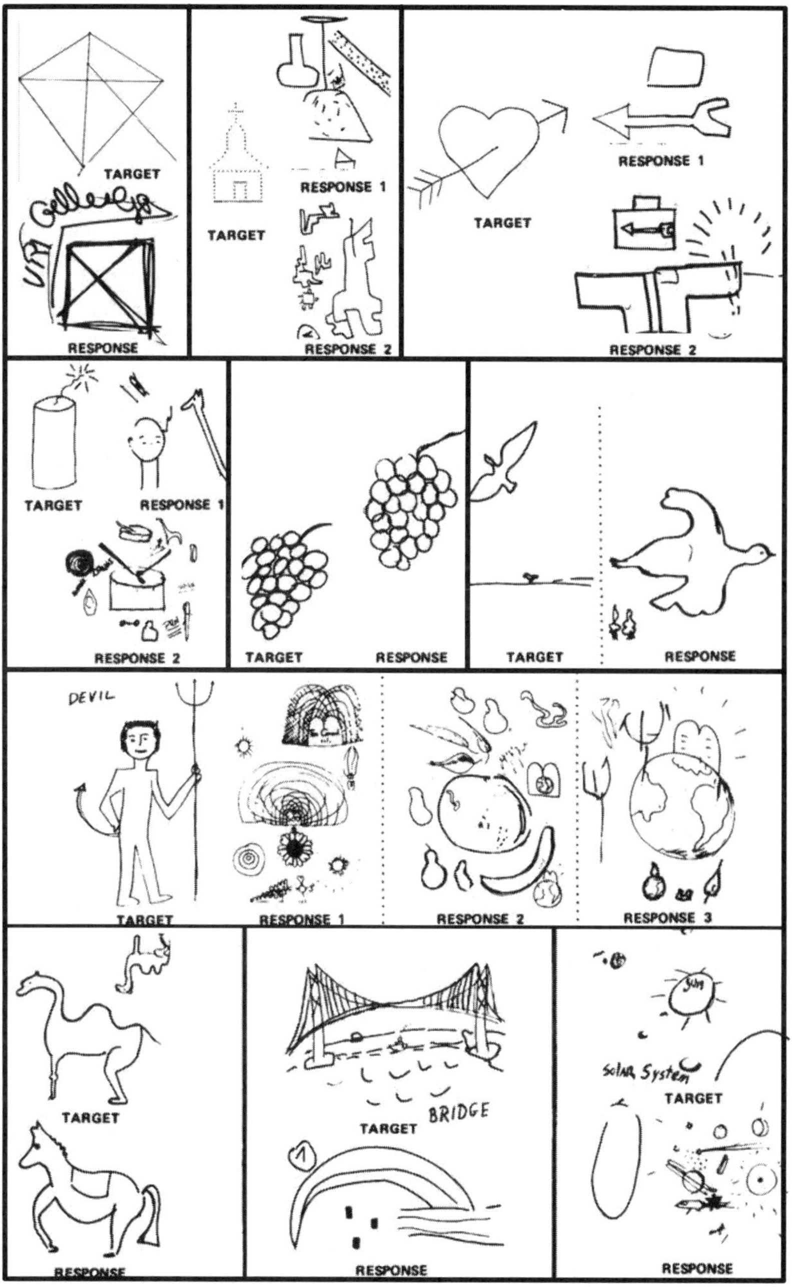

Fig. 1 (Adapted by permission from Macmillan Publishers Ltd: NATURE, © 1974).

Again, there are no unambiguous answers to questions like these and the 'reproducibility-problem' remains at the core of an ongoing controversy. Expectedly, 'believers' and 'skeptics' have focused on the respective part of the evidence which is in keeping with their own convictions. As a consequence, both sides accuse the other of a bias. This can cause personal conflicts which lead to the formation of even more stubbornly divided camps.

Interestingly, the area of research into complementary and alternative medicine (CAM), which has received growing interest in the last decades, seems to be encountering a similar situation[32]: Individuals report substantial benefits from alternative methods such as homeopathy and the CAM market is booming (even though people are often paying out of their own pocket, since CAM is not covered by most health insurances). At the same time, scientific proof and therefore recognition is largely lacking. Individual studies may detect very promising effects of various treatments, but a reliably reproducible paradigm has not yet been discovered.

More and more, the reproducibility-problem is attracting scientific research as a phenomenon in its own right[33-35].

GQT does not provide new evidence to tip the scale to either side but it offers a different interpretation of the phenomena which allows logically consistent sense to be made of *both* sides of the existing evidence. How?

Instead of starting with the question about the existence of the purported phenomena, let us first ask: "If the phenomenon were real, what could the underlying mechanism be?" A thorough assessment of the phenomenology leads to the conclusion that any explanation based on causal mechanisms faces serious challenges: Not taking into account large scale fraud, what causal explanation could there be, for example, for the telepathic-picture-guessing results? Even more difficult to imagine, even in principle, is a potential causal mechanism linking a precognitive dream to the event it foresees. And how should pure intention be able to causally influence physical processes? Or how can a homeopathic remedy possibly have a specific causal effect if it does not even contain a single atom of the substance it was originally prepared from? While arguments of this sort can by no means rule out the possibility of some hitherto undiscovered or overlooked causal explanation, they do make plausible why it may be worthwhile to also consider thinking along totally different lines. In this spirit, let us assume as a working hypothesis that instead of a causal mechanism there is an acausal one underlying this class of (causally) 'inexplicable' phenomena.

If it weren't for quantum physics, probably no one would have thought about that possibility (which seems only fair, since it was physics which demanded the more limited concept of causality in the first place). Since, however, non-local (and, in that sense, acausal) correlations have been thoroughly investigated in quantum physics, we can now even make predictions based upon our working hypothesis: If there are non-local correlations at work also in systems other than those traditionally considered in quantum physics, would it not be plausible to assume that all of those non-local correlations should behave according to the same common principles? We can thus predict that (possibly among other features) non-local correlations should occur in systems which are defined by a conserved global variable and a high degree of freedom of the corresponding variables in the correlated subsystems. And, indeed, there are some indications which support this prediction:

In terms of conserved global variables, we are reminded of 'eigenvalues' or 'eigenstates' of self-organizing systems as they have been described, for example, in the theory of autopoiesis by H. Maturana and F. Varela[36]. From there, it is also known that the more organizationally closed a complex system is, the more stable its eigenvalues will be. W. v. Lucadou[3,4,6] has demonstrated the applicability of the concept of organizational closure to reports of paranormal and synchronistic phenomena and shown that their intensity correlates with the level of organizational closure[37]. For example, in the case of telepathic or precognitive perceptions, the relationship between the people and/or events involved is often characterized by intensity and importance (close relatives, couples, death and accidents). For experimental investigations, those with a high degree of positive personal motivation on the part of experimenters and subjects are known to be more successful[38].

With regard to the necessary degrees of freedom in the subsystems, it is of interest to note that many of the paranormal divinatory techniques such as the I-Ging, Tarot cards, pendula etc. consist of chance processes of very low predictability. What is more, psychological studies have shown that more 'volatile' states of consciousness (such as trance, deep meditation and dreaming) and less predictable personalities marked by high levels of dissociativity, associativity, fantasy proneness, intuition and creativity are associated with a higher prevalence of psychic experiences.

We can further predict that, if psychic or similar phenomena are really based on non-local correlations, it should not be possible to use them for transmitting signals. This is a crucial point in relation to the reproducibility problem, because what a usual experimental setup does, from a systems-theoretical point of view, is to optimize the system under investigation for signal transfer.

By precisely defining dependent and independent variables and eliminating or controlling for confounding variables, ultimately the state of the independent variable can be predicted from observations of the dependent variable. This is equivalent with a signal having been transmitted. With each replication of an experiment, the uncertainty about the precise relationship between independent and dependent variables decreases and thus its suitability for signal-transfer increases[b]. While this is exactly what one is looking for in scientific experiments dealing with causal mechanisms, it may be a fatal hindrance for all attempts at observing acausal non-local correlations, because it would lead to a violation of Eberhard's Principle[c]. In the context of the

---

[b] For illustration: Consider measuring the attention span of a person (dependent variable) after she has consumed, on one occasion, coffee and, on another, beer (independent variable). If, a month later, you measure her attention span again, you may from that not be able to predict with certainty if she just had a coffee or a beer, because there are many possible confounding factors (tiredness, weather, medication etc...). If, however, you repeat these measurements many times and, in addition, start controlling for confounding variables, you will soon be able to tell exactly. This means, in theory, someone else could send you a (one bit) message by deciding what drink to buy that person before she comes to have the measurements taken.

[c] The problem could also be formulated in a different way: in the process of investigating a non-local phenomenon in a classical experimental procedure, one destroys the very prerequisites of the phenomenon to occur, namely the organizational closure of the original system and the degrees of freedom of its subsystems. W. v. Lucadou has used the concept of pragmatic information (first introduced by E.v.Weizsäcker[39]) to formalize the information-theoretical connectedness of sub-systems within a system and between the system and its environment. He has shown that the decline in effect sizes with increasing

telepathy study mentioned above, it is, for example, interesting that the relationship between response and target is not always of the same kind: sometimes, it is a more or less identical copy (e.g. the grapes) and at other times it reflects some meaningful connotations (e.g. dynamite and drum). Thus one cannot reliably transmit a signal from the place where the target is produced to the place where the response is produced. The more replications of the experiment are conducted, however, the more possible this would become and, conversely, the less reliable the phenomenon will have to become, which is exactly what is found to happen[d].

For quantum systems in the traditional sense, the same principle applies, but does not lead to quite the same problem because one is dealing with *absolutely* unpredictable variables and *completely* isolated systems. A rigorous experimental proof of the existence of quantum physical non-local correlations was therefore possible. Nevertheless, the limitations imposed by the nature of non-local correlations are also noticeable there, for example, in the fact that each of the correlated quanta can only be used for one measurement: After it interacts with the measurement apparatus isolation and unpredictability break down and the correlations are no longer discernible.

In macroscopic systems, unpredictability and organizational closure are only ever possible to a *relative* degree. Since their facilitation is, however, a crucial prerequisite for non-local correlations, the most promising observer perspective for observing non-local correlations in such systems is, therefore, the one that applies the least external constraints on them. This may possibly be a more subjective rather than objective one: becoming part of a system and, as part of the system, behaving in accordance with the required parameters, i.e. having a very flexible state of mind while being strongly and authentically motivated and, at the same time, in keeping with higher-order system dynamics.

---

replications corresponds to the increase of pragmatic information extracted from the system[40].

[d] As an interesting aside, it is precisely this pattern that is reported in the CIA's final report on their decades-long psi-research program, which concludes that "psychic power is real, but no good for spying"[41]

Summing up, we can say that the view of non-local correlations as universal system-inherent processes offers an alternative interpretation of otherwise hard-to-explain and highly controversial observations, which is grounded in well established scientific concepts and does not require additional metaphysical assumptions. The lack of replicability of the phenomena in question not only becomes more understandable, but actually provides additional circumstantial evidence for this hypothesis. The similarity between non-local quantum correlations and some so-called paranormal phenomena does not provide further evidence for the existence of the latter but it can explain the specific restrictions on their observability as well as indicating ways in which these phenomena can be experienced and dealt with constructively in real life situations[42, 43].

References

1. C. G. Jung, *Synchronicity: An Acausal Connecting Principle,* ed. (Princeton University Press, 1973).
2. C. G. Jung and W. Pauli, *The Interpretation of Nature and the Psyche,* ed. (Published for Bollingen Foundation by Pantheon Books, 1955).
3. W. von Lucadou, *European Journal of Parapsychology*, **11**, 58–75 (1995).
4. W. von Lucadou, *AIP Conference Proceedings*, **863**, 293 (2006).
5. H. Atmanspacher, H. Römer, and H. Walach, *Foundations of Physics*, **32**, 379-406 (2002).
6. W. von Lucadou, H. Römer, and H. Walach, *Journal of Consciousness Studies*, **14**, 50-74 (2007).
7. L. Hackermüller, et al., *Physical Review Letters*, **91**, 090408 (2003).
8. A. Einstein, B. Podolsky, and N. Rosen, *Physical Review*, **47**, 777-780 (1935).
9. J. S. Bell, *Physics*, **1**, 195-200 (1964).
10. A. Aspect, J. Dalibard, and G. Roger, *Physical Review Letters*, **49**, 1804-1807 (1982).
11. A. Aspect, in *Quantum [Un]speakables – From Bell to Quantum information*, R.A. Bertlmann and A. Zeilinger, Editors. 2002, Springer. (Springer, 2002). p.
12. J. S. Bell, *Dialectica*, **39**, 103–106 (1985).
13. J. S. Bell, Speakable and unspeakable in quantum mechanics. Collected papers on quantum philosophy, ed. (University Press, Cambridge, 1987).
14. A. Zajonc and Z. Houshmand, The New Physics and Cosmology-Dialogues with the Dalai Lama, Edited and narrated by Arthur Zajonc with the assistance of Zara Houshmand, (2004).
15. H. Walach and H. Römer, *Neuroendocrinology Letters*, **21**, 221-232 (2000).

16. N. von Stillfried and H. Walach, *International Journal of Computing Anticipatory Systems*, **17**, 137-146 (2006).
17. S. Viciani, A. Zavatta, and M. Bellini, *Physical Review A*, **69**, 53801 (2004).
18. H. Römer, *Mind and Matter*, **2**, 105-125 (2004).
19. P. H. Eberhard, *Nuovo Cimento*, **46**, 392-419 (1978).
20. J. S. Bell, in *Between Science and Technology*, A. Sarlemin and P. Kross, Editors. 1990, Elsevier Science Publishers, North-Holland. (Elsevier Science Publishers, North-Holland, 1990). p.
21. H. Walach, N. von Stillfried, and H. Römer, in Proceedings of *VIII. International Leibnitz Congress*. 2006. Hannover: Gottfried-Willhelm-Leibniz-Gesellschaft e.V.
22. N. von Stillfried and H. Walach, in Proceedings of *VIII. International Leibniz Congress*. 2006. Hannover: Gottfried-Wilhelm-Leibnitz-Gesellschaft e.V. 1026-1038.
23. M. C. Mousseau, *Journal of Scientific Exploration*, **17**, 271-282 (2003).
24. D. I. Radin and R. D. Nelson, *Foundations of Physics (Historical Archive)*, **19**, 1499-1514 (1989).
25. H. Bösch, F. Steinkamp, and E. Boller, *Psychol Bull*, **132**, 497-523 (2006).
26. D. J. Bem and C. Honorton, *Psychological Bulletin*, **115**, 4-18 (1994).
27. C. Honorton and D. C. Ferrari, *Journal of Parapsychology*, **53**, 281-308 (1989).
28. F. Steinkamp, J. Milton, and R. L. Morris, *The Journal of Parapsychology*, **62**, 193-195 (1998).
29. C. Honorton, D. C. Ferrari, and D. J. Bem, *The Journal of Parapsychology*, **62**, 255-257 (1998).
30. R. G. Standford and A. G. Stein, *The Journal of Parapsychology*, **58**, (1994).
31. R. Targ and H. Puthoff, *Nature*, **251**, 602-607 (1974).
32. H. Walach, *J Altern Complement Med*, **11**, 549-59 (2005).
33. H. Atmanspacher and R. G. Jahn, *Journal of Scientific Exploration*, **17**, 243-270 (2003).
34. J. E. Kennedy, *The Journal of Parapsychology*, **67**, 53-75 (2003).
35. W. von Lucadou, Information Dynamics. NATO ASI Series, Series B: Physics, **256**, 143-151 (1991).
36. F. J. Varela, in *Self-organizing systems*, G. Roth and H. Schwengler, Editors. 1981, Campus: Frankfurt. (Campus, Frankfurt, 1981). p. 14-23.
37. W. von Lucadou and F. Zahradnik, in Proceedings of *Annual Conference of the PA*. 2004. Wien: Parapsychological Association
38. M. D. Smith, *Journal of Consciousness Studies*, **10**, 69-84 (2003).
39. E. v. Weizsäcker, in *Offene Systeme I*, E.v. Weizsäcker, Editor. 1974, Klett: Stuttgart. (Klett, Stuttgart, 1974). p. 83-113.
40. W. V. von Lucadou, *Journal of Parapsychology*, **65**, 3–16 (2001).
41. J. Wolf, ISNCI-Flash, 29. November 1995.
42. W. von Lucadou and M. Poser, *Geister sind auch nur Menschen,* ed. (Herder, Freiburg, 1997).

43. W. von Lucadou and F. Zahradnik, Aktuelle Tipps, Hintergrundmaterialien zum Themenbereich sog. Sekten und Psychogruppen, September, **3**, 01 (2001).

CHAPTER 6

BRAIN TIME AND PHYSICAL TIME

Uri Fidelman

Department of Humanities and Arts and
Department of Education in Technology and Science
Technion, Israel Institute of Technology
Haifa 32000, Israel
E-mail: uf@techunix.technion.ac.il

The hemispheric paradigm verifies Kant's suggestion that time and space are our subjective modes of perceiving experience. Time and space are two modes of organizing the sensory input by the left- and right-hemispheric neural mechanisms, respectively. The neural structures of the left- and right-hemispheric mechanisms force our consciousness to perceive time as one-dimensional and propagating from the past towards the future, and space as a simultaneously perceived multidimensional structure. The introduction of temporal propagation from the future towards the past by Feynman and other physicists caused the transfer of the concept time from the left hemisphere (which cannot perceive this change of the temporal direction) to the right one. This transfer requires and allows for the introduction of additional temporal axes in order to solve paradoxes in physics.

1. Introduction: The Hemispheric Paradigm

There are several dichotomies which distinguish between the cognitive roles of the left- and right-cerebral hemispheres: verbal versus spatial, analytic versus synthetic, discrete versus continuous, etc. Ben Dov and Carmon[1] suggested that all these dichotomies can be obtained from a single information-theoretical dichotomy, which describes the modes by which the two hemispheric mechanisms process data: the extraction of a single datum by the left hemisphere and the integration of several data by the right hemisphere into a new whole.

This dichotomy is applied to construct a model of the cognitive functioning of the brain. According to this model, the cognitive structures are constructed in stages. At the first stage, the left hemisphere extracts one datum after another analytically and transfers each of them to the right hemisphere. Then the right integrates a new whole out of several data. After this, the left hemisphere considers this new whole as a new single datum, extracts it, and transfers it to the right hemisphere to integrate it together with further such new wholes into a more comprehensive new item. This process continues, and more and more complicated cognitive structures are constructed. For example, the process of reading begins by extraction of one letter after another by the left hemisphere. Then these letters are transferred to the right one where several letters are integrated into a word. After this, each word is transferred to the left hemisphere which reads them one after another in temporal succession. These words are then transferred to the right hemisphere, where they are integrated into a sentence, and so on.

According to this model, the left hemisphere organizes the extracted data one after another in time, while the right one organizes the data one next to another in space. Moreover, the same sensory data are available to both hemispheres, and time and space are no more than the modes by which the left- and right-hemisphere, respectively, present the sensory data to our consciousness. This model is in line with the suggestion of the philosopher Kant that time and space are our subjective modes of perception, and not things in themselves.

This model explains the existence of the verbal versus the spatial dichotomy. All the lingual functions are performed one step after another in temporal succession. Speaking and hearing spoken words are performed one syllable after another, reading and writing are performed first one letter after another and then one word after another. All these operations are performed by the left hemisphere, while the perception of spatially presented objects is performed by the right one.

The analytical nature of the left hemispheric perception of data implies that data are perceived like beads on a string and not as a continuous line. This implies the quantization of time. Therefore, the now is a discrete, finite and larger-than-zero interval of time.

On the other hand, the simultaneity of the operation of the right hemisphere causes it to perceive phenomena within continuous space. That is, the perception of two phenomena as discrete and distinct from each other requires *the operation of the left hemisphere which extracts each of these two phenomena and then sends them back to the right one, to be perceived as beads on a string or as mosaic in space*. When the left hemisphere does not participate in the processing, the right hemisphere alone does not perceive distinct parts of space. Thus, between any two points there is another point. This is the mathematical definition of continuity.

2. The Direction of Time

Our daily experience is that time moves always from the direction of the past towards the future. We cannot perceive time differently. However, we can imagine time as going backwards, namely from the direction of the future to the direction of the past. There are physical theories suggesting that at least the time of microscopic particles may really move backwards, from the direction of the future towards the past. The most important one is Feynman's[2] suggestion that an antiparticle is a particle that moves backwards in time. There are some theoretical and empirical studies which support Feynman's suggestion. It seems that there is a contradiction between our intuitive psychological time, which we may call brain time, and physical time. In the sequel of this study, we shall clarify the reason for this contradiction.

3. The Direction of Brain Time

Elsewhere,[3] I have suggested a model describing how the temporal order is registered in the short-term memory (STM) of the brain. According to Hebb, each cognitive detail is represented in the brain by at least one neural circuit, known also as *engram*. The temporal order of two stimuli is also such a cognitive detail, which is represented by engrams.

According to neurological studies which I have reviewed3, every sensory stimulus (at least a visual stimulus) is registered in the brain by three engrams. The first and the third engrams are strong, comprise large neurons (magnocellular engrams) and are transient. They represent the beginning and the termination of the sensory stimulus. The second engram is weak, it comprises smaller neurons (parvocellular engram) and it is sustained. It endures as long as the sensory stimulus is presented.

It is well known that temporal processing is performed more efficiently by the left hemisphere than by the right one. Verbal processing is part of the cerebral temporal processing, and it is known to be performed in Broca's and Wernicke's areas, where the temporal order of lingual stimuli is processed. I have reviewed[3] the neuroanatomical literature and found the following facts:

The magnocellular neurons in the left-hemispheric Broca's and Wernicke's areas are larger than the neurons in the symmetric right hemispheric areas. In addition, the difference between the size of the left hemispheric large and small neurons is larger than the difference between right hemispheric large and small neurons. Moreover, the number of neurons in the right hemisphere is larger than in the left one.

These differences between the left- and right-hemispheric neurons may explain the differences between the cognitive roles of the two hemispheres. The left hemispheric larger neurons produce more neurotransmitters, and therefore they can activate, or inhibit, the postsynaptic neurons faster than the right hemispheric neurons. Therefore the left hemisphere has an advantage in the processing of data presented at a fast rate to our senses. On the other hand, the right hemisphere has larger number of small neurons. Therefore, it can construct a larger number of STM engrams, and can keep a larger number of details simultaneously in STM. This difference explains why the right hemisphere is specialized in perceiving space that comprises many simultaneously presented details, while the left one is specialized in perceiving data that are presented rapidly one after another.

The larger difference between the size of the large and small neurons in the left hemisphere may explain why, and how, the temporal order of two stimuli is registered in the left hemisphere, rather than in the right one. I have suggested[3] that small neurons are more easily inhibited than large ones, therefore the larger difference in size between the large and small neurons implies that average inhibition is faster in the left hemisphere than in the right. Therefore, the determination and registration of the temporal order of two visual stimuli may be performed in the left hemisphere.

Suppose that two visual stimuli are presented one after another, with some small temporal gap between them. Let us designate the three STM engrams representing the first stimulus by I1, S1 and T1, that designate initiation, sustaining, and termination, respectively. The similar STM engrams representing the second stimulus are designated by I2, S2 and T2. The first and the last engram related to each stimulus are strong and transient magnocellular engrams, while the second engram related to each of the two stimuli is a weak and sustained parvocellular engram. Since engrams decay, each engram of the second stimuli is weaker than the corresponding engram of the first engram.

The temporal order of the two stimuli is registered in the brain by some STM engram, designated by TO. TO can be activated only after the termination of the first stimulus and the initiation of the second. We have assumed that TO is a left-hemispheric engram, the activation of which requires some fast inhibition of a STM engram in the left hemisphere. The fastest possible related inhibition is the inhibition of the parvocellular engram S1 by the magnocellular engram T2. Indeed, S1 endures as long as the first stimulus endures, and T2 is the first engram related to the second stimulus that is activated after the cessation of the first stimulus. Therefore, the inhibition of S1 by I2 is the most likely inhibition to the triggering of the activation of TO. Indeed, as long as the first stimulus is present, S1 is active, and no temporal order exists before the activation of I2, designating the initiation STM engram of the second stimulus. We may conclude that the operation of S1 prevents the activation of TO, that can be activated only after the inhibition of S1.

Now we may ask: Why can no perceptual inversion of the temporal order be registered in the brain, although physicists like Feynman believe that it exists (at least in the microscopic world)? Our model of the temporal order determining engram TO in the left hemisphere may solve this problem. If the temporal order is inverted, then the order of the activation of the engrams representing the two stimuli is inverted, and the new order is (from left to right): T2, S2, I2, T1, S1, I1. The inverted order should be represented by a STM designated by TO2. In order that TO2 will be activated, the parvocellular engram S2 should be inhibited by the magnocellular engram T1. However, there are no existing inhibitory neural paths between T1 and S2, therefore, once the order of the events is determined by the left hemisphere, and it is represented by an engram TO, it cannot be inverted by it.

For example, a certain event and its temporal direction can be identified by the left hemisphere as an electron that goes from the past towards the future, but not as a positron that goes from the future towards the past even if, according to Feynman, these two events are identical.

4. The Multidimensionality of Time

As stated above, Feynman[2] suggested that a positron is an electron that travels backwards in time. The meaning of the inversion of the temporal direction of an electron is that the direction of the electrostatic force between it and other charged particles is inverted and the direction of motion due to this force is inverted. That is, an electron that moves backwards in time behaves like a positron; this is the origin of Feynman's suggestion. More generally, inverting the sign of the electrical charge of a particle is equivalent to the inverting of its temporal direction.

However, in addition to the electrostatic force, gravitational force too operates between the electron and other particles. This gravitational force does not depend on the particle's electrical charge but only on its mass. Therefore, the inversion of the charge does not influence the gravitational force. Thus Feynman's suggestion implies the following paradox, which we call Feynman's paradox: How is it possible that the inversion of the temporal direction inverts the electrostatic force, but not the gravitational force?

I have discussed this issue[4,5,6] and suggested there that a solution to this problem is that time is not one-dimensional, but at least two-dimensional. The observation that the inversion of the time related to the electrostatic force does not have any influence on the gravitational force means that both forces may change along two independent temporal dimensions, which can be represented geometrically by two perpendicular axes.

This observation may explain why Einstein and other theoretical physicists failed in their efforts to construct a unified field theory which unites the electromagnetic and the gravitational forces. Maxwell unified the electrical and the magnetic force. He could do this because they share the same temporal axis, and the same temporal coordinate is applied in all of Maxwell's equations. A similar unification cannot be done regarding the electromagnetic and gravitational forces, since these forces operate along two independent temporal axes.

The success of string theory in unifying the gravitational and the electromagnetic forces may be explained by the addition of a temporal axis, disguised as an additional spatial axis. Carroll[7] and Dubois and Nibart[8] found that TCP transformations can transform the three spatial axes into three temporal axes. Similar transformations regarding some of the many additional spatial axes of string theory may explain its success in unifying gravitation and electromagnetism. String theory may have introduced to physics at least one additional temporal axis, disguised as a spatial axis.

5. Physical Time and the Brain

We saw above that the left hemisphere processes time, and it can perceive only one temporal direction: from the past towards the future. Therefore, the left hemisphere cannot handle the inversion of the temporal direction that has been introduced into physics by Feynman. I have suggested[6] that this inversion forces the brain to transfer the conception of time from the left hemisphere to the right one. The left hemisphere has fewer neurons than the right one, and therefore fewer STM engrams. Therefore, it can present to consciousness only a limited number of items, and it presents them one after another serially, along one temporal dimension.

On the other hand, the right hemisphere processes the sensory input in a different mode than the left one. It has many more neurons, and therefore more STM engrams, and it can store more details than the left hemisphere. This enables it to perceive many details simultaneously, and organize them in complex shapes and in several spatial directions and dimensions. Therefore the right-hemisphere spatial mode of presenting the sensory input to consciousness can, and does, present Feynman's temporal directions (which cannot be processed by the left hemisphere) to consciousness in a two-dimensional temporal plane, instead of a one-dimensional temporal line. This means that Einstein's unified field conjectures can be proved only by applying right-hemispheric time, instead of linear left-hemispheric time.

6. Further Temporal Dimensions

I have suggested[4,5,6] that if there are additional physical forces which cannot be unified neither with the gravitational force nor with the electromagnetic force, then more temporal dimensions may be introduced into physics. Moreover, the contradiction between the discrete microscopic quantum world and the continuous macroscopic world of both special relativity theory and general relativity theory, due to Heisenberg's principle of uncertainty, can be removed by assuming that each of the two temporal axes: the gravitational and the electromagnetic one, split into two axes, one microscopic temporal axis along which the quantum events occur, and a second macroscopic temporal axis, along which the electromagnetic events of special relativity and the gravitational events of general relativity occur. The apparently contradictory events occur in different subspaces of the general space-time. Each such subspace has different temporal axes, therefore the apparently contradictory events are entirely independent of each other. Thus, contradiction is avoided. In conclusion, in order to achieve unity of physics and to avoid contradictions in it, we have to assume at least four temporal dimensions, in addition to three spatial dimensions.

7. Is There an Independent Physical Time?

According to Kant, time, as well as space, are no more than our subjective modes of perceiving experience. This view is in line with the hemispheric model of Ben Dov and Carmon[1] according to which time is the mode by which the left hemisphere organizes the sensory input.

However, in this study, we have seen that the left-hemispheric one-dimensional time, which can change only from the direction of the past towards the direction of the future, is not sufficient for describing and unifying all physical phenomena. Therefore, we have replaced the one-dimensional left-hemispheric time with a multidimensional right-hemispheric temporal "space", which comprises at least four independent and perpendicular axes.

It should be noted that the right-hemispheric time is as subjective as the left-hemispheric time. Both are modes by which our brain presents sensory phenomena to our consciousness. Is there, somewhere outside our consciousness, some objective physical time which exists independently of our consciousness? In spite of the feeling of most physicists that there is such a time, we really do not know. This leads us to Kant's view that we cannot know anything about things as they are in themselves. In this study, we have seen that the concept of time changes according to the cerebral mechanism that deals with it. This proves that time is really a subjective notion.

References

1. G. Ben Dov and A. Carmon, A. *International Journal of Neuroscience,* 7, 29-33 (1976).
2. R. Feynman, *QED: The strange world of light and matter.* Princeton University Press, Princeton, NJ (1986).
3. U. Fidelman, *Kybernates*, 31, (3/4), 432-495 (2002).
4. U. Fidelman, *International Journal of Computing Anticipatory Systems*, 19, 236-406. (2006).
5. U. Fidelman, *Physics Essays* 19 (4), (2006).
6. U. Fidelman, *Journal of Mind and Behavior* (2006).
7. J. E. Carroll, The *AIUP Journal of Science and Engineering* (AJSE), 4 (1), (2005).
8. D. M. Dubois and G. Nibart, *International Journal of Computing Anticipatory Systems*, 19, 25-46 (2006).

Part II

Identifying Temporal Observer Perspectives

CHAPTER 7

SIMULTANEITY IN EMOTIONAL MOMENTS

Gerald L. Clore

Department of Psychology, University of Virginia
Charlottesville, VA 22903 USA
*E-mail:*gclore@virginia.edu

Emotions are described as emergent states, which exist only to the extent that multiple affective reactions to the same object occur at the same time. Emotions are thus the confluence of thoughts, feelings, expressions, desires, and so on. They emerge as meta-cognitive representations of embodied affective reactions. Emotions may be initiated by low-level, automatic, unconscious affective reactions, which are then iteratively re-processed with ever greater cognitive involvement until they become elaborated into emotional states. Affective and emotional reactions act as information about the value of objects of judgment and of accessible cognitions and inclinations during tasks. They influence judgment and thought when they are experienced simultaneously with sensory data about the world. Affective influences thus depend on our inability to disentangle affective from descriptive perceptions. To the extent that affective reactions reflect different, incommensurate sources of value (e.g., utilitarian, moral, aesthetic), perceived persons or objects may be experienced as being transcendently good or evil. Experiments varying people's attributions for their affective experiences allow the separate roles of affective and descriptive information to be examined. However, it is the inability to parse everyday experience into its separate sources of evaluative and descriptive information that gives rise to a colourful and transcendent reality.

1. Introduction

In the Alfred Hitchcock film "Rear Window", the actor Jimmie Stewart plays the part of a man confined to his Greenwich Village apartment with a broken leg. He passes the time observing snippets of his neighbors' lives through a telephoto lens as they move past the windows of their apartments. As his neighbors' actions appear in a succession of images framed by their windows, the audience, along with Stewart, begin to construct narratives of their lives by filling in what they cannot see. Students of film (for example, Seeley[1]) suggest that the movie was intended to dramatize not only the mystery that unfolds in the plot but also the process of movie watching itself, a process in which audiences construct seamless narratives from the images they see on the screen. In that regard, the film also dramatizes the process of everyday perception.

Cognitive models of perception (for example, Thompson & Kosslyn[2]) maintain that, like Jimmie Stewart's telephoto lens, the sensory data from our eyes supplies an underdetermined image, so that we must supply possible meanings. The possible conceptual matches generate a pattern of activation in the primary visual cortex, which acts like a hypothesis, tuning the visual system to expect a match to that pattern. Through a "perceptual cycle"[3], the hypothesis guides attention so that new sensory data are collected, which refines the hypothesis, and so on until the process provides a sufficient match between sensation and conception. From overlaying sensory data with expectations in progressive iterations, we come to see. Once fantasy and fact have met in the middle in this way, no one can disentangle the mental from the sensory.

Of course, we think of fantasies more often as escapes from reality rather than as assisting in its realization. When that is the case, it is reasonable to ask how anyone distinguishes self-generated mental events from those that are externally-generated. Part of the answer is that self-generated images are typically less vivid than sensory images, and also that the sensory context is likely to conflict with fantasy images[4]. Thus, one would not confuse a fantasy of being on a ski slope with actually being there, because during such a fantasy one's sensory surround would probably look more like a lecture hall or an office than like a snowy mountain. Dreams, on the other hand, do seem real because we are not aware of any context outside the dream while we are asleep.

This volume is concerned with the role of simultaneity in the structure of phenomenal moments or what has been referred to as successive "nows." There are many psychological phenomena that depend on a temporal overlap of different but confusable information in the construction of the moment. Hypnotism provides a relatively dramatic example. After inducing people to relax, hypnotists often suggest that their eyes will get heavy and their lids will close, that their outstretched hand will feel heavy and begin to lower of its own accord, that they will feel resistance when trying to unclasp their tightly clasped hands, and so on. Most people comply with such strong suggestions. The hypnotist speaks in the passive voice and directs attention to the feelings in the arm, or hands, or eyelids as though they are acting on their own. Gradually the suggestions may become more complex – not noticing pain, having impulses to act, being unable to remember, and so on. Most people have a difficult time becoming fully hypnotized, because they are able to keep the suggestions of the hypnotist clearly located in time and space outside of themselves. A technique for dealing with that obstacle is to have two hypnotists giving the same or similar hypnotic suggestions simultaneously. Once the person is unable to keep the source of each word and thought distinct, it is easier to begin to confuse the impulses that come from hypnotic suggestions with those that are self-produced. When that happens, people often report feeling hypnotized.

Another example comes from psychological studies of a phenomenon known as "semantic priming." In an early study of priming, participants heard a story describing a person in ambiguous terms. He could be seen either positively as a person who was self-confident, adventurous, and independent or negatively as a person who was conceited, reckless, and aloof[5]. People's impressions of the ambiguous character in the story were influenced by whether they had previously seen those same positive or negative words earlier in an apparently irrelevant task. In this and similar experiments, an initial subtle priming of one of the two sets of word meanings then influenced the impressions that judges formed of the ambiguous character, because the primed meanings are mentally accessible as they try to form an impression. However, in such studies, nothing happens if the priming manipulation is more obvious so that participants are able to focus on the source of the primed concepts and keep them distinct. The secret, therefore, lies in the subtlety of the initial priming process, which allows the person to experience the primed concepts as coming into mind unbidden, as though it represents one's true opinion. Again, as in the hypnotism case, such effects can be disrupted simply by changes in attention that make more or less confusable the simultaneous appearance in mind of the person to be judged and the primed concept.

In summary, we have suggested that a number of psychological phenomena depend on the simultaneous presence of two or more experiences. We have also emphasized that often the similarity of the experiences, in addition to their simultaneity, makes them difficult or impossible to distinguish. Under those conditions, a new phenomenon may emerge, such as visual perception, hypnosis, or semantic priming. We turn now to the idea that emotions are emergent phenomena, arising again from a concert of simultaneous affective elements.

2. Emotions are Emergents

The phenomenal moments on which our own work focuses involve emotions. Two different kinds of simultaneity are required for emotional experiences. First, as we examine below, emotions are emergent states, which means that they exist only as the intersection of multiple, partially redundant affective representations[6,7]. Second, human emotional states involve meta-experiences, in which one experiences oneself having affective reactions[8,9]. Thus, emotional experiences might be thought of as requiring both horizontal and vertical kinds of simultaneity. That is, they may require the simultaneous presence of some component parts and also a meta-experience that unifies the experiences of these component parts. Let us examine each briefly.

The commonsense view of emotions is that they are biological entities evident in distinctive physiology, feelings, thoughts, expressions, and so forth. The view that emotions are internal entities that cause these various emotional manifestations has guided most scientific research on emotion. However, that view is currently under challenge (for example, see Barrett[10]). More than a hundred years ago, the American psychologist, William James[11], also challenged this view when he famously proposed, regarding the emotion of fear, that we are afraid because we tremble and run, rather than that we tremble and run because we are afraid. In a traditional model of emotion, such claims make no sense. But if we look through the telescope from the other end, things get clearer. Rather than thinking of emotions as latent entities, it is useful to see them as emergent constructions. Then, expressions, behaviors, feelings, and cognitions become parts of emotions, rather than being things caused by emotions. They are constituents of emotion as opposed to being products of emotions.

In terms of how they are experienced, emotions have generally been treated as having two important dimensions – valence and arousal[12,13]. Valence refers to the fact that emotions vary along a pleasantness-unpleasantness dimension, and arousal refers to the fact that they also vary along an excitement-calm dimension.[a] The positive-negative aspects of affect carry information about value (goodness vs. badness), whereas the arousal component carries information about urgency or importance[14]. Thus, affective reactions reflect the detection or perception that something good or bad has been encountered, to which it is more or less urgent to attend. There are several different kinds of affective reactions, so that people end up expressing their affective perceptions in multiple ways.

Emotions are emergent in that they exist when several such affective reactions co-occur. That is, emotions emerge when the same goodness or badness is represented simultaneously in multiple ways – as thoughts, feelings, expressions, postures, tones of voice, physiology, behavioral inclinations, and so on. It is not that some pre-existing entity of fear, for example, causes distinctive fearful reactions, so much as that the co-occurrence of these fearful reactions constitutes a state of fear. From this view, fear has no independent existence. It is the emergent reality from

---

[a] For some purposes it is more useful to separate valence into separate positive and negative dimensions, each of which varies in arousal. It is not necessary to take a position on that issue to make our present point, which is that affective experiences carry information.

components such as thoughts and feelings about threat, a gaping mouth, wide eyes, tense muscles, the secretion of cortisol and acetylcholine, and so on. All we mean by "fear" is a set of threat reactions or the co-occurrence of multiple representation of threat. There is no other fear that stands apart from and causes these reactions. Rather than being caused by fear, they are fear[6,7].

Expressing a related view, the psychologist James Russell[13] has asked whether emotions really exist at all. He concludes that they do, but only in the sense that the big dipper exists. The stars that make up the big dipper do exist, of course, but their particular constellation, the pattern that gives it unity is illusory, existing only when viewed from the incredibly distant Earth. Similarly, the components of emotion are real enough, but the emotion itself is simply the special properties that emerge from the confluence of various affective parts.

The resulting emotion can nevertheless be a powerful, intractable force. However, it is an emergent force. Just as a crowd may be more persuasive than a single person, emotions and other states are potent because they have multiple, partially redundant sources. Emotions are simultaneous representations of goodness or badness from multiple systems, which may include facial "expressions," thoughts, feelings, postures, vocalizations, physiology, behavior, and so forth. Neither making a face, adopting a posture, or having a particular thought by itself constitutes an emotion. But together, these related and simultaneously activated forms of evaluation can take on an emergent reality, becoming palpable and compelling. When, for example, we believe, feel, and express the presence of threat, the experience of that combination of fear elements can be powerful, because it is more than mere knowledge or feeling or expression; it is an emergent, psychological state. At that point, the big dipper of fear becomes real.

One might have imagined that an emotional feeling would be sufficient. But raw feeling, without a supporting cognitive context may not be experienced as an emotion. Feelings characteristic of fear, rather than clearly indicating that one is threatened, might be attributed to illness, dizziness, or self-generated anxiety without basis. But the triangulation of perception, cognition, feeling, and bodily tenseness makes one know fear.

3. Emotions as Meta-Experiences

A second kind of simultaneity important in understanding the experience of emotions is the idea that they involve "meta-experiences".[8] The true nature of emotion is always in dispute, but recently several writers have emphasized the importance of distinguishing two things that are sometimes conflated – affective reactions and emotions (for example, Baumeister, DeWall, Vohs, & Zhang[15]). Low level affective reactions may be widely shared across species, but full blown emotions of the kind seen in humans are probably rare in other species.

An example of a low level affective reaction would be piloerection, in which the hair stands up on the back of one's neck when threatened. I experienced it recently when I was startled by the loud snort of an unseen deer that I happened to surprise one evening while walking in the woods. I was startled by the noise, but almost instantly realized that it was just a deer. Nevertheless, a moment later, I felt an odd feeling on the back of my neck as the minimal hair that grows there stood on end. The fact that the hair standing on end occurred after the cognitive all-clear signal that it was only a deer, illustrates the partial independence of low level affective and high level cognitive factors in creating emotion.

Consistent with this view is a recent model of emotional appraisal that emphasizes the iterative nature of emotional processing[9]. The idea is that exposure to something emotionally significant initiates a cycle of evaluative processes. The cycle may begin as something quite low level, such as my startle at the loud snort of the deer I encountered. But that stimulus can then get continually reinterpreted as more levels of the context in which it appeared are processed. After initial affective reactions are triggered, the sensory information gets processed, sent to the amygdala, and then is continually reprocessed. Each time the amygdala is reacting to a progressively more differentiated, contextualized stimulus. As the iterations continue, one's own visceral reactions may also be reprocessed so that one's autonomic state becomes cortically represented along with the stimulus itself. The first few iterations are unconscious, but later reprocessing may become relatively reflective, as additional regions of the cortex get involved. The full blown emotion is then a product of a series of recursive feedback loops in which stimulus representations and one's affective reactions get progressively refined, finally resulting in conscious emotional experiences[9].

The proposal of an iterative model of emotional appraisal has the happy property of accommodating the discrepant positions of many otherwise warring scientists. Controversy has always been intense in the study of emotion because, as this model indicates, both low level and high level processes may be involved. A strategy on the part of some has been to point to the continuity in emotional processes across species, and to conclude on that basis that emotion should be explained in very primitive terms. Others, pointing to the richness of human emotion, often speak about emotion in cognitive terms. It may therefore be helpful to distinguish unconscious affective reactions, many of which are subcortical in nature, from emotional states, in which such reactions have been continually reprocessed and cognitively elaborated.

For our present purposes, such iterative models help flesh out the idea of simultaneity as a critical building block of people's experiences of a moment. Human emotional experiences are rich and powerful because they are layered, involving an instantaneous experience or meta-experience of oneself engaged in perceptual and affective reactions ranging from the most primitive to the most sophisticated. Emotions emerge from the co-occurrence of multiple affective reactions. They are made rich by multiple layers of processing, which yields a perception of ourselves as purposeful individuals in a meaningful world.

In summary, in this section we illustrated two ways in which the nature of emotional states illustrates the structure of momentary experience. But beyond the multi-layered nature of emotion, we need also to consider the effects of emotion on experience. Most of or own work concerns questions about how such simultaneity accounts for the impact of emotions and moods on everyday judgment and thought, a topic to which we turn now.

4. Mood Influences on Judgment

A typical experiment in our lab might examine how emotion colors judgment. We might begin by inducing a momentary mood (by playing happy or sad music, showing happy or sad film segments, or asking participants to write descriptions of happy or sad events in their recent past). Subsequently, we might ask them to make evaluative judgments of some kind. These could include anything from estimates of risk, ratings of consumer products, or judgments of life satisfaction. Results often show that happy moods make risks seem less likely, consumer products more desirable, and life as a whole more satisfactory than do sad moods[16]. That is, when individuals ask themselves, "How do I feel about my life as a whole?" the positive affective feelings from being in a happy mood, for example, are difficult to discriminate from any actual feelings of satisfaction as they consider their life as a whole. People have only one window on their momentary affective experience, so that in positive moods, the overlapping pleasant affective cues can lead to more positive ratings than the person would otherwise give.

An example of the experimental paradigm just described involved telephoning people on warm and sunny or cold and rainy Spring days[17]. People generally report being happier on sunny than on rainy Spring days, and we were interested in how affect influences judgment. We found that people's ratings of their lives as a whole showed an influence of their momentary feelings of mood, so that people said their lives were more satisfactory when they made the judgment on sunny than on rainy days. This result suggested that people sometimes use their feelings as information for making evaluative judgments. To confirm that the effect was due to the information conveyed by the positive feelings, rather than simply to the presence of the feelings, we asked some respondents first about the weather before we asked about their life satisfaction. When that was done, feelings of mood occasioned by the weather no longer had any effect on life satisfaction ratings. The weather question had not changed the feelings themselves, but it did change the apparent object of the feelings. They were then experienced as information about the weather, which they were, rather than about life satisfaction. Answering the weather question before the topic of life satisfaction arose enabled them to parse their experience in a way that was otherwise not possible.

This kind of experiment has been done many times with different sources of mood and different objects of judgment.[18] In general, when people focus attention directly on their feelings as an object in their perception before rendering a judgment, they can prevent affect from marrying with perception. Indeed, as we have seen, such effects can readily be produced in the lab. But when a judgment is made on the basis of a combined affective-cognitive percept, then such experiential surgery is not possible.

The example of the weather study involving a misattribution of affect from the weather to life satisfaction makes such affective influences appear undesirable and irrational. However, outside the laboratory, people's feelings are generally occasioned by the objects to be judged, so that they are judgment-relevant. Moreover, research demonstrates that using affective information is essential in decision-making. Individuals who are unable to do so as a result of certain kinds of brain damage are utterly debilitated[19].

The research on mood and judgment illustrates a primary mode of emotional influence. In similar ways, our experiences of people, places, ideas, and activities are appropriately colored by our feelings. Thus, as we walk along we may see, not simply people, trash, buildings, and trees, but perhaps rude people, disgusting trash, graceful buildings, and beautiful trees. It is the impossibility of segmenting perceptual and emotional experiences when they are simultaneous that gives interest and vitality to everyday events.

5. Mood Influences on Thought

Affect influences cognitive performance and thought in very much the same way as it influences judgment. When a person focuses on a task rather than on an object of judgment, positive and negative affective feelings may be experienced as information about the adequacy of one's ideas for performance. Positive feelings may seem to validate accessible cognitions, expectations, and response inclinations, whereas negative feelings inhibit reliance on such self-generated interpretions. For example, people normally adopt a global rather than a local focus in perception, a phenomenon known as the "global superiority effect." In happy moods, people also tend to adopt such a global perspective. However, at other times sad affect can cast doubt on the value of their inclinations so that they no longer focus on the big picture, but instead focus on local, stimulus-specific details.

The textbook experiments in cognitive psychology were designed to showcase the fact that people's sensations are normally heavily interpreted. Rather than simply registering sensory details, we interpret them and relate them to what we already know. Cognitive psychologists have devised various experimental paradigms to show this constructive aspect of everyday thinking. Typically, they find that various kinds of cognitive intrusions arise in perception and memory. However, when people's feeling states are also assessed, we find that negative affective states inhibit these effects.[20] Such results are interesting in the present context because they provide an interesting opportunity to separate somewhat the conceptual from the perceptual components of experience. By conveying a negative value on interpretive activity, sad promotes perceptual activity, resulting in adoption of a local rather than a global focus.[21]

These performance effects can be reversed by varying the implicit attributions that people make for their source of their affect. As in the weather study discussed earlier, if attention is drawn to the true source of people's feelings (e.g., happy or sad music, films, or memories), then people generally discount their feelings as information about the value of their inclinations. When people in happy moods no longer see their positive affect as performance-relevant, they tend to discount the information about efficacy that the feelings otherwise convey. As a result, they act like individuals in sad moods. Conversely, when those in sad moods no longer see their negative affect as performance-relevant, they discount the information about difficulty that it conveyed, and engage in the kinds of cognitive processing usually seen in happy moods (for example, Isbell[22]). Again, the importance of such observations in the present context is they allow us to deconstruct the usually unified affective moment in which simultaneity of feelings and sense data create the emotionally tinged world to which we are accustomed.

In summary, in this section of the chapter, we emphasized that the simultaneity of internal affective cues and externally-based sensations gives give color to everyday perception and regulates thought. But the full power of this process is best seen in cases in which the same object of judgment is represented in terms of different kinds of affect, a process that we discuss briefly now.

6. Multiple Kinds of Good

Whatever else they are, emotions are evaluative states. Indeed, evaluation is a defining feature of emotion; there are no neutral emotions.[23] But if emotions are concerned with the goodness or badness of things, what is the basis of such evaluations? That is, what makes something good or bad?

We proposed that there are three different sources of psychological evaluation, including goals, standards, and tastes.[24] The outcomes of events may be judged as desirable or undesirable on the basis of their implications for one's goals. Thus, one may be happy or sad to the extent that an event furthers or thwarts some goal or plan, or one may feel fear, hope, disappointment, or relief at the prospect that not yet known outcomes will be desirable or undesirable. These common emotions all concern the desirability of outcomes for our goals.

But sometimes things are good or bad, not because they affect our goals, but because they exceed or fall short of important standards of behavior. These might involve standards of excellence, exceeding or falling short of normal performance expectations or social standards, or meeting or failing to meet standards of excellence or moral standards. Thus, one could be filled with admiration either at some unusual athletic feat or at some virtuous action. Such emotions as shame, pride, reproach, and admiration are about the praiseworthy or blameworthy nature of actions, rather than the desirability or undesirability of outcomes. One might be proud for having done something well, regardless of the outcome, or one might be outraged or ashamed as a result of an instance of cheating even if success resulted. The specific goodness involved in the approval of praiseworthy actions is thus different than the goodness of being pleased at desirable outcomes.

Finally, something can also be good or bad because it appeals to our tastes or attitudes. The pleasure of eating ice cream has less to do with goals and standards than with taste. One likes ice cream just because it tastes good. People may find certain things disgusting for no obvious reason; they just are. The reaction is not about reasons, but about something being appealing or unappealing. Liking, disliking, and disgust are reactions that can be occasioned by beauty or ugliness, which are ultimately matters of taste.

One can attend either to the outcome of events, to the agency of actions, or to the attributes of objects, each of which implicates a different source of evaluation, including goals, standards, and tastes or attitudes. These three kinds of good are incommensurate, meaning that one cannot be translated into the other. For example, trying to exchange units of moral good for units of utilitarian good is problematic. The Faust legend is a tragic story about the bad consequences of attempting to make such a bargain. But if different kinds of value are not easily exchanged, what happens when they co-exist? When one person exhibits both moral goodness and goal-based success, how do such incommensurate goods combine?

Suppose, for example, that one met a person who was appealing by virtue of being beautiful, handsome, or eloquent (taste-based goods), and who also acted in a praiseworthy manner by engaging in some inspiring behavior or doing something exceedingly well (standard-based goods). Suppose also that this person created circumstances in which one was better able to accomplish one's own goals, to do things that one wanted to do, or was in some other way good for oneself (goal-based goods). It would not be surprising for someone to fall in love with such a person. The simultaneous experience of multiple kinds of good seems likely to result in an experience of the person as unaccountably and hence transcendently good.

One might imagine a similarly potent reaction to a political candidate displaying utilitarian good by having good policies, moral good by some praiseworthy action, and aesthetic good by being dashing and articulate. Such a candidate might seem unaccountably good to the extent that the simultaneous demonstration of multiple goods made precise allocation of one's positive reactions to its various sources difficult. It is also easy to imagine unaccountable hate from a similarly simultaneous activation of multiple kinds of badness. In each example, I have focused on unaccountability as an important ingredient in such emergence. A phenomenon involving similar processes that may make the unaccountability aspect clearer is three-dimensional vision.

7. The Emergence of Depth in Vision

The conditions for three-dimensional vision include two visual streams providing similar, but not identical information. The high degree of similarity in the image from the left and right eyes makes it computationally impossible to separate them. Lizards, with their widely-spaced and independently moving eyes, presumably have separate registers for information from each eye. But in humans, the eyes are close to each other and move together. As a result, the views they provide are too redundant to separate, so that we end up with one, three-dimensional image, rather than two views with two dimensions each. Because the two inputs are slightly different, we end up with a visual model in three dimensions, a byproduct of overlaying the partially redundant right and left eye images. The partial nature of the redundancy is important, as nothing would emerge from total redundancy.[25] The alignment of the images is also crucial. Sometimes, tiredness, intoxication, or muscle weakness can create double vision, but when the separate images are properly aligned, a single image with a magical third dimension makes things seem to project out into space. The third dimension is thus an emergent property. Simultaneity, partial redundancy, and alignment would seem to be the critical elements in such emergence. New structures emerge when two or more partially redundant representations of the same thing are present in the same place at the same time.

In the case of vision, what emerges is a multi-dimensional model that retains the unique information from the two partially redundant sources. The same idea is applicable to various psychological processes, including, for example, certain interpersonal interactions. As two individuals interact and one person speaks, gestures, and adopts a particular posture, the other person may respond with head nods, postural changes, and words of their own that reflect and build on what has gone before. Each develops a model of the interaction so that they embody their own affect as well as the perceived affect of the other.[26] As in binocular vision, the substantial, but not total, congruence in these representations can create an out-thereness or depth. Again, such depth would seem to depend on the partial overlap of one's own experience and one's model of the other's experience, perhaps supplemented by additional levels of meta-representation.

Other experiences may stand out for similar reasons. For example, nostalgia may include simultaneously experiencing oneself in the present remembering something in the past, feeling at once the immediacy of a sweet memory and the sadness of its distance in time. More generally, the process is evident in an operatic duet or a string quartet. Rather than hearing music from separate sources, a new entity emerges that transcends their individual contributions. Such convergence may be experienced as a vital entity. In the interpersonal situation, couples may point to such experiences as a significant moment in their relationship, because of its emergent, entity-like nature. Are such experiences illusions? Perhaps they are no more illusory than three-dimensional visual images are illusions.

In summary, in this section, we suggested that experiencing emotions, falling in love, political loyalty, conversations, and duets may all be powerful in part because a strict accounting of their sources in one's experience is impossible. Failures of emotional accounting, visual accounting, or auditory accounting allow one to experience multiple facets of the same person, place, or event at the same time. In the emotional case, the different sources of good are incommensurate. When they emanate from the same entity and are apprehended at the same time, no accounting is possible, so that a multifaceted, new entity arises with all of the incommensurate but partially redundant features experienced as one emergent reality, the whole of which seems greater than the sum of its parts, which, of course, it is.

8. Conclusion

This chapter was organized in three parts. The first illustrated the role of simultaneity in the construction of various kinds of momentary experience. These included hypnosis, semantic priming, and emotions. We proposed a view of emotions as emergent states that exist only to the extent that multiple affective reactions to the same object occur at the same time. We also proposed that true emotions emerge from an iterative process in which lower level affective reactions are continually reinterpreted with more cognitive context being included each time. The result is a meta-cognitive representation of one's embodied affective reactions to events.

Part two focused on how affect and emotion influence experience by being experienced at the same time as sensory data about the world. Affect thus colors perception, providing information about the value of whatever is in attention at the time. We reviewed research on mood as a source of affective cues and their effects on judgment and cognitive performance. The fact that affective influences depend on our inability to disentangle them from our perceptions was examined in research that altered people's attributions for their feelings before judgments were made or task information could be processed.

Part three concerned an extension of these processes to multiple kinds of evaluative information. We reviewed proposals that there are three distinct kinds of goodness and badness that eventuate in emotions. We indicated that they are incommensurate and analyzed situations in which a single person was seen to display such multiple goods (e.g., utilitarian, moral, and aesthetic goodness). We maintained that some such convergence of experiences of goodness may be involved in falling in love and the development of political loyalty. We emphasized that the power of such experiences, like the power of three-dimensional vision, may depend on being unable to engage in any mental accounting for the sources of good when multiple, incommensurate goods are experienced simultaneously. Through such convergence, the sense of presence of everyday experience in the world may be enhanced, as persons or events seem to take on transcendent qualities.

Acknowledgement

Support is acknowledged from National Institute of Mental Health grant MH 50074 and National Science Foundation grant BCS 0518835. Correspondence can be directed to Gerald Clore, 102 Gilmer Hall, P.O. Box 400400, Charlottesville, VA 22904-4400. gclore@virginia.edu.

References

1. W.P. Seeley, Imagining Film: Seeing with the Mind's Eye. *Philosophical Inquiry,* 17, 3-14 (2005).
2. W. L. Thompson and S.M. Kosslyn, Neural systems activated during visual mental imagery, in Arthur W. Toga and John C. Mazziotta (Eds.) *Brain Mapping: The Systems* (pp. 536-560). New York: Academic Press (2000).
3. U. Neisser, *Cognition and reality.* San Francisco: Freeman (1976).
4. M. K. Johnson and C. L. Raye, Reality monitoring. Psychological Review, 88, 67-85 (1981).
5. E. T. Higgins, W. S. Rholes and C. R. Jones, Category accessibility and impression formation. Journal of Experimental Social Psychology, 13, 141-154 (1977).
6. G. L. Clore and A. Ortony, Cognition in emotion: Never, sometimes, or always? In R. D. Lane & L. Nadel (Eds.). *The cognitive neuroscience of emotion* (pp. 24-61). New York: Oxford University Press (2000).
7. G. L. Clore and A. Ortony, *Appraisal theories: How cognition shapes affect into emotion.* In M. Lewis, J.M. Haviland-Jones, & L. F. Barrett (Eds.). Handbook of Emotions, 3rd Ed. New York: Guilford Press (in press).
8. R. Reisenzein, Appraisal processes conceptualized from a schema-theoretic perspective: Contributions to a process analysis of emotions. In K. R. Scherer, A. Schorr, & T. Johnstone (Eds.), Appraisal processes in emotion: Theory, methods, research. Oxford: Oxford University Press (2001).
9. W.A. Cunninghamand P.D. Zelazo, Attitudes and evaluations: a social cognitive neuroscience perspective. *Trends in Cognitive Science, 11*, 97-104 (2007).
10. L. F. Barrett, Emotions as natural kinds? Perspectives on Psychological Science, 1, 28-58 (2006).
11. W. James, *Principles of psychology*. Mew York: Holt (1890).
12. Wundt, W. (1874). Grundzuge der physiologishen psychologie. Leipzig: Engelmann.
13. J. A. Russell, Core affect and the psychological construction of emotion. *Psychological Review, 110,* 145-172 (2003).
14. G. L. Clore and S. Schnall, The Influences of Affect on Attitude. In D. Albarracín, B. T. Johnson, & M. P. Zanna (Eds.) *Handbook of attitudes and attitude change: Basic principles (pp. 437-490).* Mahwah, NJ: Erlbaum (2005).
15. R.F. Baumeister, C.N. DeWall, K.D. Vohs and L. Zhang, How Emotion Shapes Behavior: Feedback, Anticipation, and Reflection, Rather than Direct Causation. *Personality and Social Psychology Review*. (in press).
16. G. L. Clore and J.R. Huntsinger, How emotions inform judgment and regulate thought. *Trends in Cognitive Science, 11,* 393-399 (2007).

17. N. Schwarz and G. L. Clore, Mood, misattribution, and judgments of well-being: Informative and directive functions of affective states. *Journal of Personality and Social Psychology*, *45*, 513-523 (1983).
18. G.L. Clore and J.E. Palmer, Affective guidance of intelligent agents: How emotion controls cognition. *Cognitive Science.* (in press).
19. A.R. Damasio, *Descartes' error: Emotion, reason, and the human brain*. New York: Avon (1994).
20. G.L. Clore and J. Storbeck, Affect as information about liking, efficacy, and importance. In J. Forgas (ed.). *Affect in Social Thinking and Behavior (pp. 123-142).* New York: Psychology Press (2006).
21. K. Gasper and G. L. Clore, Attending to the big picture: Mood and global vs. local processing of visual information. *Psychological Science, 13,* 34-40 (2002).
22. L. M. Isbell, Not all happy people are lazy or stupid: evidence of systematic processing in happy moods. *Journal of Experimental Social Psychology, 40,* 341-349 (2004).
23. G.L. Clore, A. Ortony and M. Foss, The psychological foundations of the affective lexicon. *Journal of Personality and Social Psychology*, *53*, 751-766 (1987).
24. A. Ortony, G.L. Clore and A. Collins, *The cognitive structure of emotions.* New York: Cambridge University Press (1988).
25. S. Vrobel, Simultaneity and Contextualization: The Now's Fractal Event Horizon. Paper presented at a Conference on dynamical systems in the cognitive sciences. Ascona / Monte Verità, Switzerland (2005).
26. S. Boker, Embodiment of Self and Other: Mutual Regulation of Affective Symmetry in Conversation. Paper presented at a Conference on dynamical systems in the cognitive sciences. Ascona / Monte Verità, Switzerland (2003).

CHAPTER 8

ON TIME EXPERIENCE IN DEPRESSION

Hinderk M. Emrich

Department of Clinical Psychiatry & Psychotherapy, Medical School Hannover
Carl-Neuberg-Str. 1, D-30625 Hannover, Germany
E-mail: emrich.hinderk@mh-hannover.de

Catharina Bonnemann

Department of Clinical Psychiatry & Psychotherapy, Medical School Hannover
Carl-Neuberg-Str. 1, D-30625 Hannover, Germany
E-mail: bonnemann.catharina@mh-hannover.de

Detlef E. Dietrich

Department of Clinical Psychiatry & Psychotherapy, Medical School Hannover
Carl-Neuberg-Str. 1, D-30625 Hannover, Germany
E-mail: dietrich.detlef@mh-hannover.de

Emotionality of subjective time experience in depressed patients is related to Michael Theunissen's concept concerning the dominance of the past in relation to the spontaneity of the subject within a present moment of time. Electrophysiological data using the event-related potential method point to the view that negative emotions cannot use a value-related cognitive/emotional blockade of valuation processes. In addition, the subjective time experience is substantiated by reports of patients with an affective disorder. Furthermore, the role of forgetting to overcome such memory-related impairments is discussed within the context of possible functions of consciousness.

1. Introduction: The Brain as a Time Machine

"Time is the way how we perceive the world around us."[1]. This fundamental sentence by Ehlers gives the opportunity to raise the question how the brain may act as a "time machine" to enable this. Indeed one may envisage the brain as containing aspects of a "time machine" in three different ways of understanding: Firstly, the brain is a neurobiological physical system like other systems – existing in time – being an object of understanding by physics, by biophysics, especially of neurochemical oscillators and thus is subject to the concepts of chronobiology. Secondly, in a completely different way of understanding, the human brain is a "time machine", since it can memorize "stored regularities" and can recall them. The brain thus contains aspects of "time transcendence" – to use a term introduced by the Berlin philosopher Michael Theunissen [2]. Human beings are – in the sense of the phenomenology of Edmund Husserl – able to perform protension, i.e. the conceptualization of the future as well as retention, a reference to the past. Furthermore, we are able to generate the time experience of the "now", of the "presence within time". One special problem herein is how to relate, how to synchronize the "outer", the so-called Aristotelian or "physical" time to the "time of inner experience", i.e. the time of Augustinus, and the Bergsonian time. Psychiatric disorders as well as extraordinary experiences of normal life tend to induce problems within this type of coordination, and depression represents a mood-disorder, within which, also experimentally, a disturbance of the inner organization of time in relation to objective time can be demonstrated. Thirdly, there is the aspect of a "time machine" in a Kantian understanding, namely that the human mind – apparently on the basis of the brain – can construe the universals of perception, especially the category of "time". How are the functions of such a "time machine" realized within ourselves?

2. Time and Affect: Something about Anxiety

One of the main ideas within neuropsychological research is that affective changes are great determinants in the construction (and deconstruction) processes of cognition and thus of the constitution of consciousness[3]. This has been demonstrated in illusion research, in psychosis research, in the understanding of the action of psychedelic drugs as well as in "synaesthesia", a fabulous way of healthy subjects to construe hyperintegrative inner perceptual worlds[4]. Interestingly, this emotional determination ("emotional tone") behind cognition and self- and world-experience, contains fundamental aspects of the neurobiology of time as well as the philosophy and psychopathology of time. Pöppel[5] as well as Singer[6] in physiology and Theunissen[2] in philosophy have made great contributions to this.

We would like to demonstrate the main principles of the basis of a neurobiological theory of anxiety and depression in relation to time, established by the neuropsychologists Gray and Rawlins[7]. An understanding of this concept, however, requires a representation of the basic principles according to which our brains function.

As the neurophysiologist von der Maalsburg[8] stated, the brain – as a machine – represents a "significance detector"; that means that it is able to detect – within an enormous amount of noise – a particular event which contains the quality of being "meaningful". To exert this function, the brain is composed – as a multi-systems parallel processor – from many more or less independent "modules", which interact in an intricate way. There are systems of conceptualization within the frontal lobe, systems of processing of sensory data within the occipital brain and systems of emotionality, of valuation, within the inner temporal lobe and limbic system. These functions can be demonstrated by functional brain imaging techniques, e.g. in positron emission tomography (PET) or functional nuclear magnetic imaging scans (fMRI). The neuropsychologist Desimone[9] has proposed a general organization scheme of the flow of processing such that so-called top-down processes are realized from prefrontal to occipital and that "bottom up processes" function from occipital to prefrontal (compare Figure 1). What does this mean from the point of view of experience of time? Interestingly, in this understanding, the theory of time starts from the future (in analogy to Heidegger), from the conceptualization of "what comes next?" It is the world of predictions and desires, from which we start in the sense of the neurobiologist Martin Heisenberg[10], who construes the world of activities of a biological system not from reflex, from reaction to outer changes, but from novelty-seeking, a mode of action he calls “initial activity”. The presence, the “now” is, accordingly, construed from actual sensory data in relation to these conceptualisations, i.e. from bottom-up information in relation to expectations. This is realized within the intermediate part of the brain, namely the temporal lobe. However, very interestingly, here the entrance to limbic structures and to memory structures is realized. How do we come to the past? How do we enter Marcel Proust's[11] inner memory world within his “recherche” for the “lost past”?

We have to go into the physiology of memory, which is a physiology of the hippocampus, as a temporal organizer, in relation to cortical feedback loops.

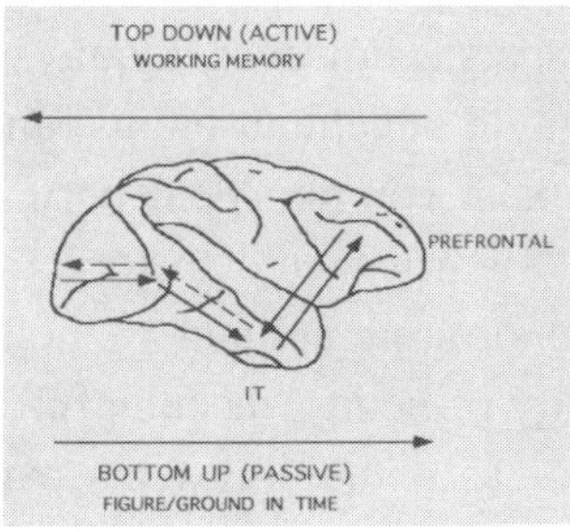

Fig. 1: From Desimone et al. (1995).

From these basic principles, we can understand how anxiety is organized within our brains. It is an extreme form of “significance detection” of something which turns out to be aversive. The model is called the “hippocampal comparator system”. We demonstrate this concept since the components of the generation of inner time are represented there (compare Figure 2).

We start again – in the sense of Heisenberg[10] – from the future, from the “what comes next?”. *The predictions dominate the perception of time*. It depends on the predictions – and we can add: on the emotionality, the emotional climate, the “emotional tone” of the predictions – what the “measurement of significance”, in the sense of von der Maalsburg[8], will find out within the present situation. These

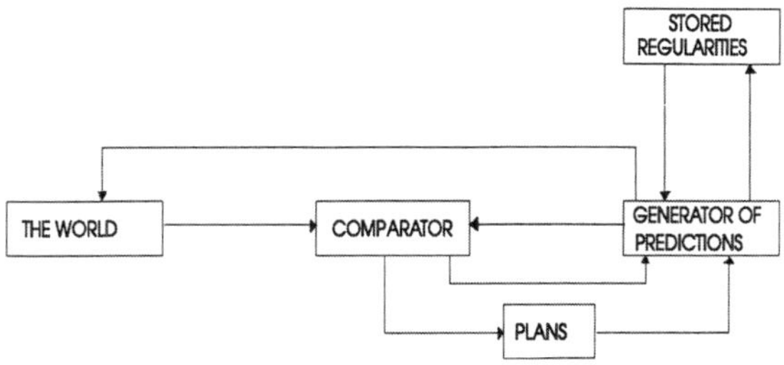

Fig. 2: Hippocampal comparator system (Gray & Rawlins 1986).

predictions are possible from the special type of actualisation of the "stored regularities" of the past only. Here the cortico-cortical loops containing memory traces come into play. And now, from the sensory data of the occipital lobe the presence, the data of "now", of the world, ask for the "comparator" function: if there occurs an unexpected possibly dangerous situation, the system generates an "alarm signal", which activates the limbic system. This input creates "arousal", "motor inhibition" and increased attention via a behavioural inhibiting system. Interestingly, this is the place where the famous anxiolytic drugs, the benzodiazepines, act, since they can down-regulate the intensity of this internal alarm signal and thus reduce the anxiety reaction. If there is a malfunction of these binding sites, anxiety and epilepsy can occur.

This concept of constitution of time within our internal "timemachine" is also plausible from another point of view: the neuropsychology of synaesthesia[4]. Here an additional intermodal integration within one moment of presence is realized. This integration apparently is manifestated by a so-called "limbic bridge", in which a synthetic "fusing" integration of e.g. a digit and a colour is realized. Whether this limbic integration is performed by chronobiological oscillators[5,6] or by a special neurobiological "wiring", is a problem to be solved[4,6].

3. Internal Balance of Values

Disturbances of the inner equilibrium of the components of the self which shape identity, brought about by processes of re-presenting by remembering, and the consequences these disturbances have, clarify the potential for explanation inherent in a theory of the processes of identity formation, based on the psychology of the self.

It is important to consider the structural principles of forgetting in relation to overvaluation with respect to the stability of "promise contexts". The sociologist Hans-Dieter Gondek[12] has emphasised the specific value of promise contexts in evaluating personal identity, something which is essential in connection with the problem of forgetting: a forgotten promise, after all, is not really a promise. To put it differently: how does identity change when forgetting takes place? Identity would seem to imply the promise that one is – and will remain – the person one is at the moment, quite the opposite of Nietzsche's "You must become the person you are." Of course, people do not generally behave in this way; they do not accept that they and others are, or can be, only what they manifestly are at the present moment. They do in fact intuitively behave in the way Nietzsche demands; they expect developments in other persons which they assume would also take place within themselves. We arrange to meet the person whom the other person will have become by the time the appointment takes place, at least we do if we are sensible. We form an opinion of others according to the available possibilities, according to tendencies rather than according to "faculties". Falling back or surpassing oneself can both occur; both must be possible and are indeed necessary. Within the framework of psychoanalytic therapies, but also spontaneously in everyday life, people can in fact take themselves completely by surprise. During the therapeutic process, it may happen – indeed it is the intention – that alien-image and self-image approach one another or at least interact with one another, something which can lead to the most radical of changes.

From what we have seen so far, it is clear that the formation of identity is always accompanied by complex processes of adjusting the inner balance of values, so that it is to be expected that any disturbance of the processes aimed at restoring such equilibrium will lead to psychopathological phenomena. Disturbances in the internal balance of values normally lead to "coping" processes, i.e. processes of management and increased performance, which ensure that one's self-demands and one's self-image fit together better once more. In cases where this is not successful, or when it cannot succeed, a whole range of changes in behaviour are possible, ranging form grieving to processes during which the balance of values is distorted, to processes of denial, of encapsulation or repression. Nietzsche's famous statement[13] is appropriate here: "'I did this', says my memory. 'I cannot have done it', says my pride, and remains relentless. Ultimately, it is my memory which gives in."

4. Inner Time Consciousness in Depression

Quite a few authors have been examining the phenomenology of altered time perception in patients suffering from mood disorders, among them psychiatrists such as L. Binswanger, E. Minkowski and C. V. E. von Gebsattel. Presupposing that consciousness of the world is an integral part of self-consciousness, one runs across a common experience of daily life, namely, that self-consciousness is inextricably connected to our particular time consciousness. While experiencing the current world, we experience ourselves in the flow of world time. It is a trivial but remarkable fact that the only possibility to experience reality is to experience the "now"; the ongoing situation. Memory, on the other hand, is generally considered to be a sheer construction of the brain. Put simply, time consciousness is conditioned by our present state, which is determined by factors like attention, motivation and mood. For this reason, time experienced by persons suffering from mood disorder is of special interest for a phenomenological approach to the topic of time consciousness.

During periods of severe melancholia or in depression, an uncoupling of the self from the world is often described by the persons concerned. The cause for this divergence between self-consciousness and consciousness of the flowing "world" time is mosttimes hard to comprehend. The result is an experienced discrepancy between self and environment, between "inside-" and "outside-world", and therefore between individual- and world time.

Under normal conditions, we experience our world in a way that occurrent moments of conscious "nowness" are kept in an intentional currency. On the basis of Edmund Husserl's analysis of "inner time consciousness", experience of any sort has a certain temporal structure such that any moment of experience contains a retentional reference to past moments of experience, a current openness to what is present, and a protentional anticipation of what is just about to happen. Everything is embedded in the flow of time which pre-reflectively determines our individual perception of the current world, whereas reflection is only possible if a temporal horizon has been established.

Referring to this concept of inner time consciousness, one can state that, if this temporal horizon becomes affected, the abilities to immediately experience and consciously reflect the world both get disturbed. An altered "inner time consciousness" brings about an altered self-consciousness, this alteration being pronounced in patients suffering from mood disorders.

The following description by one of our patients suffering from bipolar disorder illustrates the altered time perception during a depressive period.

"I can't always tell why depression starts. Last time it was severe. Over the worst period of time, I am totally apathic. Then I feel like a totally flat auto battery. You know exactly that time advances and that I am putting off things that have to be done. I do still live within real time, which does not only overstress me – it smothers me. I am overaware of what I have to do and what I should do, it is like standing in front of a gigantic, steep wall. I spend my time in bed, lying there day by day – and day by day – there is no other way, and even movements have to be avoided. I am not able to move or to go on. It generally is a progress for me, when I reach the point when I am able to read again. So I just continuously keep on reading, sometimes for weeks. I do perceive time, but I avoid thinking about the real time. At this point, I preferably read books from another period that represents a different reality. History books, for example, or books about other times, e.g. the post-war-time. I have been told a lot about those times. And I am reading old newspapers, but serious ones, which are 1-2 years old. It is because, like this, by reading historical stuff, I am far away from everything here and can escape from the "now". Like this, I want to find a connection to the past – as to establish a relationship to live again. My aim is then to gain a realistic insight into past times. Reading old newspapers is, in this context, a control check, how far my own and the authors' assessments concerning past times agree with my personal estimation of the past, for example, I am asking myself if predictions came true. This process is like very carefully being put into operation again, it is like newly gained activation during melancholic episodes after apathic phases. I am seeking contact to the real world by dealing with the past. By being engaged in literature about the past, I abscond from reality, so I do not have to confront the present situation. It is indeed kind of an escape. But also a rapprochement to life, like a bridge or an auxiliary contact."

This exemplification points out several aspects of altered time- and self-perception in depression, as well as depicting the resulting incapability to deal with daily life. A general feeling of deceleration or even stagnation can lead to altered perceptional capabilities. Some patients describe this kind of state as feeling a "heaviness" or "weight", whereas the passing time is perceived as kind of "viscous". The depressive person gets stuck in this state and is exposed to his circling thoughts which attach him to the past. Meaning and content of the recurrent thoughts become portentous and lead to a feeling of stagnation as well as to an increased "disconnection" from the present world. Quite often, the reduced exchange with the social environment produces a complete isolation, as preoccupation with the person's own problematic privacy severely disturbs communication with others. In clinical practice, symptomatic isolation often turns out to be one of the most painful experiences for depressive patients. Concerning the altered self-consciousness, it is important to emphazize that intentions are turned away from the environment and are instead directed inwardly.

Reduced motivation is another typical symptom that results from altered subjective temporality during depression. Due to the fact that intentions are directed to a lived life, chances which have passed and missed possibilities, the future becomes unattainable for the patient. An orientation towards things that will be forthcoming in the future is impossible. This often produces feelings of guilt and of a general finality. It seems impossible to come to terms with the past as well as to create a new self. Properties like spontaneity, flexibility and decisiveness are disturbed in depressive persons, who often describe an "inner paralysis".

It can be recorded that besides the experienced deceleration and the feeling of isolation during depression, the person concerned becomes a "prisoner of the past" within a stagnating now. The fleeting time of the "outside world", as well as the world itself, is perceived as being uncoupled from the subjective time of the perceiving self which is committed to the past. The experience of a disconnection between self and environment perpetuates itself at it comes to consciousness, leading to a more or less solitary paralysis.

5. System Theory and Forgetting

How should – from the perspective of a system theory model – such a process of coping with alteration of the past and thus also the process of "forgetting" be seen? How does such a system behave in a concrete situation when what is involved is affecting plans and relating them to the past? Suggestions as to such decision-making processes can be found in the theory of nonlinear dynamic processes. As Friedrich Cramer[14] demonstrated in his book on the theory of time, complex systems which are evolving constantly display successive branching, with "the mutual influence of three aspects of dynamic systems, namely structure, function and fluctuations, leading to unexpected phenomena". If one applies this description of a complex system in the field of physical chemistry as a metaphor for the biographical development of a biological system, one can compose – with Peitgen[15] – a tree diagram in which it is not possible to predict how the final state of the system will actually be filled in. After the evolution of a complex system, a large number of unstable final situations arise whose variety can be greatly reduced by impaired development ("trauma"). Instead of sixteen final situations (as shown in Figure 3), only three or four such states can actually be filled in.

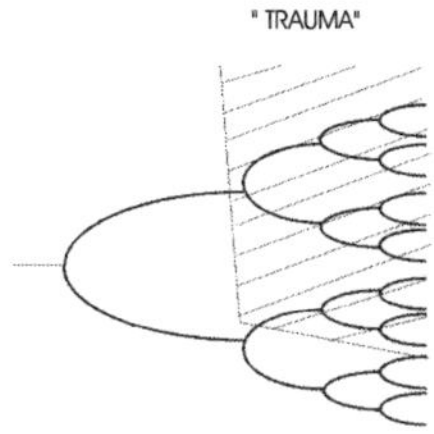

Fig. 3: Scheme of reduced evolution of a complex system due to impairment by "trauma", Emrich (1999).

Examining this disturbance chronologically and using the metaphor of the "path travelled", one finds that influences on the system have caused the path to be a specific and extremely restricted one. The question which arises is how, looking back, one should deal with this enforced path in a "coping" manner. If one decides to use this method of description as a metaphor for the reconstruction of the "path" along which an individual develops from a plurivalent initial situation to a final situation moulded by the various factors relevant to development, one can also describe the influences – both favourable and harmful – which have served to determine this path. Early traumas may have had such a powerful negative influence on the available choices in a vulnerable phase of development that these events – which are later generally "forgotten" later on – entirely determine, modify and restrict the rest of the subject's life. A whole range of options, lives and choices as potential system states have thus been eliminated. Thus there are three ways for the subject to deal with the consequences of this development. The first deals purely with certain points: it is only a result as it actually is which is considered: "one must base oneself on the facts." The second possibility is – in Hegel's sense – to consider the truth in the context of both the path and the results. The third and final way of looking at things not only involves reconstructing the path but also reconstructing the other options which would have been possible at each given turn-off. This way of looking at things might be understood in terms of Proust's[11] "recherche du temps perdu" as an associative recalling of subliminal truths from the past; this is dealt with in Michael Theunissen's book on the "negative theology of time"[2]. Theunissen says that "Proust's 'spontaneous remembering' uncovers, to put it traditionally, the eternity element in the past. What he finally calls 'time regained' is time only as the self's own other."

6. Neurobiological Experience

The fact that the dimension of time is an extremely relevant aspect within the experience of depression is indicated e.g. by the time-dependent typical course of the intensity of the symptoms in endogenous kinds of depression. Therefore, the dimension of time in melancholia was recognized to be a dominant factor, to perform a special kind of "dominance of time". Another important aspect, however, regarding the pathogenesis of depressive diseases, was also recognized: this is the impact of the neurophysiology of memory systems. Is depression a type of memory deficit, a lack of forgetting in the sense of Nietzsche? Regarding this, depression may be suggested as an extreme and rigid keeping to the "facts of the past". As a consequence, this may lead to an intense pressure without having creative possibilities or options of future intentional acting.

We have tried to measure phenomenological aspects of the concept of Theunissen's time philosophy, using objective neurophysiological parameters. It should be possible to gain more information about a main problem in psychiatry: the coupling between emotional and cognitive processing. How do thoughts and emotions interact? What part does the memory system play with this regard?

For this purpose, we made use of a continuous word recognition experiment[16], which we modified specifically regarding the emotional connotation of the presented words. We presented words with negative, positive and neutral emotional content on a video monitor, and in parallel, we measured the electrophysiological activity, the EEG. One method, however, of measuring the stimulus-related electrical activity to measure the activity which is related to the presentation of the words, is to record so-called event-related brain potentials (ERPs). These are minute voltage fluctuations which can be recorded non-invasively from the intact human scalp and occur in response to stimulus events in parallel with their cognitive processing. These ERPs have been shown to be sensitive to memory processes and to emotional factors on cognitive processes, e.g. face recognition. In addition, a frontal positive activity starting around 250 ms after stimulus presentation was described for emotional stimuli, which were more pronounced when subjects were instructed to focus on the emotional connotation of the stimulus.

We identified ERPs as a tool for investigating memory processing in depressive patients with a special emphasis on electrophysiological correlates of the emotion/cognition-coupling.

For our experiment, one ERP component is of specific importance and should be explained in more detail, namely, the N400. Kutas & Hillyard[17] investigated ERPs evoked by the last word of a sentence which was presented to build a certain (semantic) context. They reported that semantically inappropriate final words of these sentences (e.g. "he spread the warm bread with *socks*") elicited a large-amplitude negative ERP component with a peak latency of 400 ms (the N400 component, compare Figure 4) relative to the ERPs elicited by semantically appropriate words (e.g. "it was his first day at *work*"). Contrasting, semantically appropriate but physically abberant words (e.g. word printed in a larger type) elicited a positive-leaning potential in the same time window as the N400 ("she put on her high-heeled ***SHOES***"). Kutas & Hillyard speculated the N400 may be an "electrophysiological sign of the 'reprocessing' of semantically anomalous information", but other explanations have been suggested as well. Regarding our experimental approach, it is relevant to realize that the better a word fitted into the preceding context and the more it was expected, the smaller was the N400 and the more pronounced was a positive ERP component in the time range between 300 and about 700 ms.

In our experiment, we presented words on a video monitor, with some of them being repeated after some intervening items in a continuous word-recognition paradigm. For every presentation of the words, the depressive patients or control subjects had to decide whether this word was presented for the first time (first presentation = "new" word) or a second time (repetition = "old" word).

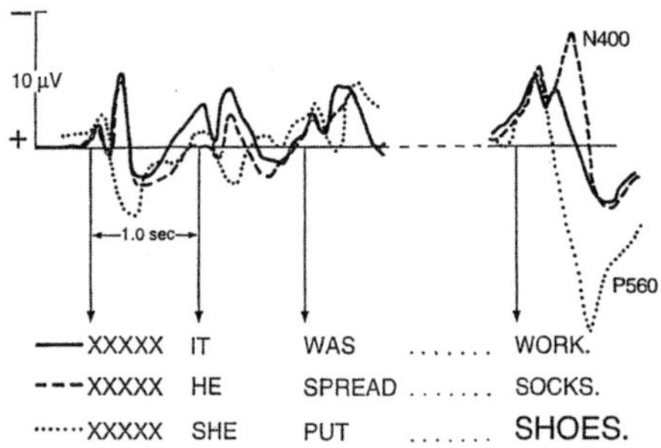

Fig. 4: ERPs to sentences ending with a non-anomalous, semantically anomalous, or physically anomalous word. (Modified from Kutas and Hillyard, 1980).

For the repeated words, the event-related potentials had more positive potentials (deflection) between about 200 – 800 ms post-stimulus, which was the result of a priming or repetition effect (compare Figure 5). This ERP difference is called old/new effect and is mainly a result of a reduced N400 and an enhanced late positive ERP component.

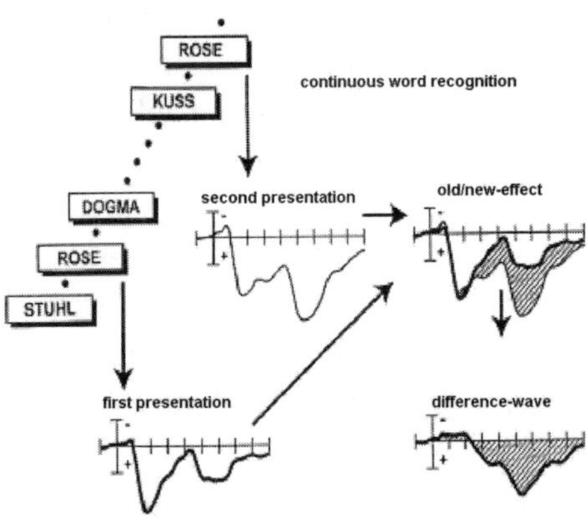

Fig. 5: Continuous word recognition experiment in which a typical old/new effect is evoked.

The presented words in our experiment were classified into three groups of different emotional content: negative words like "to die", positive words like "to kiss" or neutral words like "to go". This differentiation was not known by the subjects.

The question arises, how do depressive patients respond to the presentation of the words with different emotional content compared to healthy control subjects?

Figure 6 shows the average ERPs for the 3 different categories of words for 11 healthy control subjects for the midline electrode positions (from frontal to parietal). Especially frontally, the negative and positive toned words (verbs) induced a bigger difference between the first and second presentation compared to the neutral words. This difference is called old/new effect ("new" for the first presentation and "old" for a second presentation of the words). In the depressive patients (Figure 7) this difference has nearly disappeared.

These findings may be understood by suggesting that the cognitive hippocampal memory system, including the frontal cortical structures, was influenced by the negative emotional memory system: it appears to

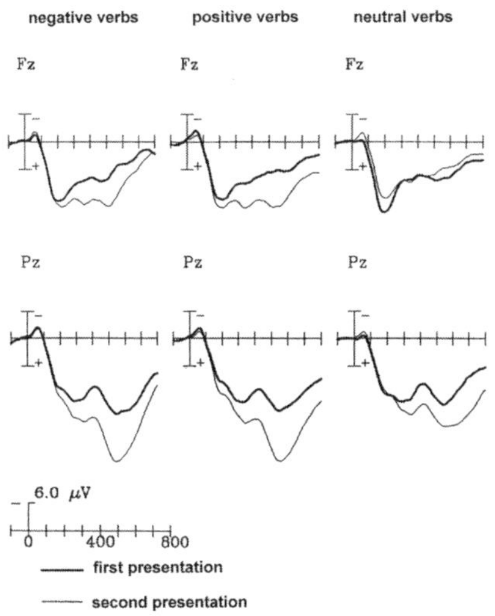

Fig. 6: Grand average (n=11) ERP waveforms to correctly detect first (thick line) and second (thin line) presentations of the words (verbs) for the control subjects. The second presentations of words elicited a more positive waveform starting at about 200 ms after the onset of the stimulus. This old/new effect demonstrated here for the central electrodes (Fz, Pz) is more pronounced for the presentation of the "emotionally toned" negative and positive words.

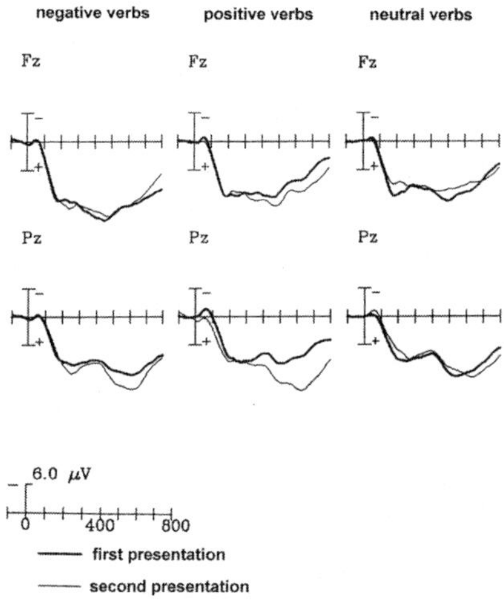

Fig. 7: Grand average (n = 9) ERP waveforms to correctly detect first (thick line) and second (thin line) presentations of the words (verbs) for the depressive patients. The old/new effect is significantly reduced compared to the control subjects.

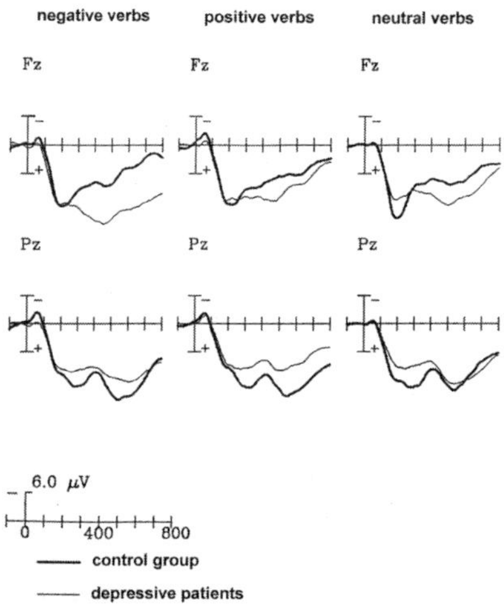

Fig. 8: Grand average ERP's to correctly detect first presentations of the words only for the control group (n = 11; thick line) and for the depressive patients (n = 9; thin line).

be primed and influenced especially by negative cognitions and memories, which leads to this massive reduction of the old/new-difference. The system appears to be precharged with negative cognitions. This may be clarified by demonstrating the ERPs for the first presentations only as shown in Figure 8).

The depressive patients' event-related potentials for the first presentation were so pronounced as if they had been presented before. The N400 component relevant to context-integration processing appeared to be absent. No context integration seemed to happen for the negative items especially. In other words, negative words were much more expected than positive and neutral ones. This is congruent with Theunissen's concept, that depressive patients appear to be under massive pressure by their negatively toned cognitions, by the negatively charged memory system, and by the dominance of time.

7. Creativity, Consciousness and Forgetting

The problems of voluntariness in neurobiological terms have been discussed at the 1994 Dahlem conference on the flexibility and constraints in behavioral systems: "One may define an *axis of voluntariness* along which animals can be classified. Animals should never be perceived to be climbing this axis towards voluntary behavior. Indeed, phylogenetic and adaptive conditions (e.g., change in body size or predictability of the environment) could conceivably cause animals to evolve less voluntary states. Although for the sake of simplicity, one may discuss an unidimensional axis, there is no need to restrict it to one dimension. Finally, one may hope that this axis, and those that evolve from it, will replace the anthropomorphic, human-centered axis that now exists. That is, while primates may occupy one end of the scale, that highly derived animal group should not be the standard against which all animals are judged. Indeed, one may wish to be more objective than are present definitions of voluntariness." (c.f. Smith et al., Dahlem Conference, 1994)[18]

And further: "A concept of voluntariness contains attainment of a goal, which embodies the animal's intentionality. Any useful goal concept would vary considerably with the level of complexity of both the animal and the behavior studied. So a description of the goal state in terms of its richness of representation must be specified in detail. It is trivial to say that a hungry animal has a goal of obtaining food.

A goal of voluntary behavior must contain the purpose behind the movement. It must in a sense reduce the discrepancy between an existing sensory state and a desired state. In a simpler sense, the goal can be represented by external stimuli in the immediate sensory world of an animal. Any behavior to which we can attribute a chain of causes, that is, a series of stimuli each of which serve to release the next few stimuli in the sequence, would clearly not fulfill our predispositions for thinking about voluntariness. Yet that kind of behavior attains a goal. Bacteria, for example, will move within a chemical gradient toward food according to sensory cues that directly regulate the orientation behavior. More complex representations of goal states must be defined. Levels of voluntariness manifest themselves in how goals are formed. One possible concept is the idea of "fitting" between "concepts" and "sensory data", i.e., the consideration that neuronal systems generally perform comparisons between "expected" and actual states. Perception means to find the optimal internal neuronal state ("concept", "model"), which fits to a set of sensory data, whereas goal-directedness means to establish a definite internal pattern and to "quantify", and the difference between these two values is interpreted by the system as "dissatisfaction". (c.f. Smith et al., Dahlem Conference, 1994)[18]

From this point of view, the following relation between the basic mechanisms of perception and goal-directedness would arise. In the case of perception, the aim is to find an optimized internal "model" pattern (as a conceptualization) to fit an external set of uninterpreted sensory data. If there is no optimum fit the system is unsatisfied and has to apply internal "conceptualization-pressure" to generate alternative conceptualizations for a better "understanding" of the external sensory-data-pattern. Satisfaction occurs when the optimum fit between "sensory data" and "model" is realized, which is disturbed in depression.

The principles under which creative new "meanings" are generated may be exemplified in a schematic drawing of the elementary process in relation to a "reactive conceptualization pressure". Reactive conceptualization pressure appears to be a relevant feature in a circuit when no plausible fit is reached between conceptualization and not yet interpretable data (Figure 1). It is assumed that in this situation the system generates "variants" of already established conceptualizations, leading to a special type of "neuronal darwinism" in the sense of Gerald Edelman.

It is assumed that the generation of "consciousness" is the fitting-process (described above) between internal "model" and a set of sensory data in a convergence zone (the concept presented here has partially been elaborated in cooperation with Harald Atmanspacher, Munich). The optimal internal "model" represents the actual state of sensory data ("lottery of models"). In the case of goal-directedness the system pre-establishes an internal model and the sensomotor-system has to play a kind of game to vary the external data until the fitting between the pre-established internal model and the sensory data is optimized, i.e. until the difference between model and sensory data is minimized ("lottery of proposals").

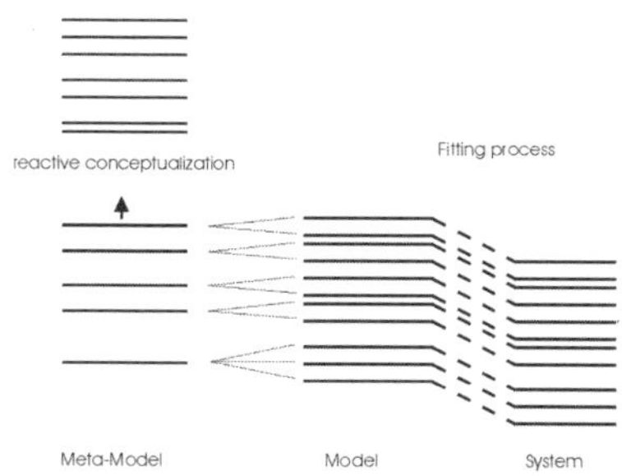

Fig. 9: Schematic representation of the concept of fitting-processes in perception and the role of "reactive conceptualizatio" in formation of new models and meta-models.

Such a concept, however, implies that an internal world of values is constructed by the system in which the differences between different degrees of fit are defined in an order of rank, i.e., the "game rules" imply that the better the fit, the higher the value of the acquired state as well as its goal-directedness. The difference is that in perceptual processes the system ranks and tests the different conceptualizations with the aim of "understanding" what's going on, whereas in goal-directedness the game is to vary senso-motor states to reach an optimal fit between a pre-established internal model and actual sensory data.

Within this context, a goal of human voluntary behavior in depressive patients may be to overcome the disease, to overcome the dominance of the past. In such a situation, creative reactive conceptualisations may help to induce optimized fitting processes, and therefore, should reduce depression.

Following Theunissen's concept of "resistance to the dominance of time"[2], we will deal with a type of "active" dis-acknowledgement as a type of "forgetting process", namely, dis-acknowledgement in the sense of forgetting as a creatively interpretative mode of representing. What is involved is a means of dealing with the past by using the imagination to access the life path that lies behind while including all the options which were not chosen and by using the imagination to bring to life subliminal "experiences" in the context of the "processes of choosing" which took place when travelling along this path. One might term this the "Proustian method". Theunissen describes it as follows: "Proust is able to recall "regained time" precisely by destroying it. Memory, which functions as an organ of reconciliation, is spontaneous in that it leaps out from the solid context of occurrences and opens itself to the unique in which the subject perceives the internal subjective world which he has continually suppressed on the basis of what appeared to be external." With this, it opens the surface of memory. It has something of Hegel's "re-collection"[19]. This type of active forgetting means an imaginative time travel to the past and integrating it into present life. Thereby, it is possible to forget, and thus "solve" depression, which finally appears to be a very creative process.

References

1. J. Ehlers, *Concepts of Time in Physics: Insights Obtained and Open Questions.* Talk held at the International Conference "On Time", Einstein Forum Potsdam. May 2003, Potsdam, Germany (2003).
2. M. Theunissen, *Negative Theologie der Zeit.* Suhrkamp Verlag, Frankfurt a.M. (1992).

3. H.M. Emrich, Konstruktivismus: Imagination, Traum und Emotionen, in S.J. Schmidt (Hrsg.), *Kognition und Gesellschaft. Der Diskurs des Radikalen Konstruktivismus 2*. Suhrkamp, Frankfurt a.M., pp. 76-96 (1992).
4. H.M. Emrich, Synästhesie, Emotion und Illusion. in Kunst- und Ausstellungshalle der Bundesrepublik Deutschland (eds.), *Der Sinn der Sinne*. Schriftenreihe Forum Band 8. Steidl, Göttingen, pp. 126-139 (1998).
5. E. Pöppel, *Die Grenzen des Bewußtseins*. DVA, Stuttgart (1985).
6. W. Singer, Search for coherence: a basic principle of cortical self-organization. *Concepts in Neuroscience* **1**, 1-26 (1990).
7. J.A. Gray and Rawlins, J.N.P., Comparator and buffer memory: an attempt to integrate two models of hippocampal functions, in: R.L. Isaacson, K.H. Pribram (eds.) *The Hippocampus*. Plenum vol. 4, New York, pp 151-201 (1986).
8. Ch. von der Maalsburg, personal communication (1993).
9. R. Desimone, Miller, E.K., Chelazzi, L., Lueschow, A. Multiple Memory Systems in the Visual Cortex, in: M. Gazzaniga (Ed.) *The cognitive Neurosciences*. MIT Press, Cambridge, Massachusetts, pp 475-486 (1995).
10. M. Heisenberg, Über Universalien der Wahrnehmung und ihre genetischen Grundlagen. In: H. v. Ditfurth, E. P. Fischer (eds.) *Mannheimer Forum 1989*. Piper, München, pp 11-69(1989).
11. M. Proust, *Auf der Suche nach der verlorenen Zeit*. Suhrkamp Verlag, Frankfurt a.M. (1993).
12. H.-D. Gondek, Das Versprechen und seine Verbindlichkeit, in K. Holz (Hrsg.) *Soziologie zwischen Moderne und Postmoderne. Untersuchungen zu Subjekt, Erkenntnis und Moral*. Focus, Gießen, pp 39-65 (1990).
13. F. Nietzsche, *Unzeitgemäße Betrachtungen*. Insel Verlag, Frankfurt a.M. (1981).
14. F. Cramer, *Der Zeitbaum. Grundlegung einer allgemeinen Zeittheorie*. Insel Verlag, Frankfurt a.M. (1993).
15. H.-O. Peitgen, *"Fraktale - gezähmtes Chaos"*. Vortrag: XVI. Deutscher Kongress für Philosophie, Berlin 20.-24. Sept. 1993. (1993).
16. M.D. Rugg, ERP studies of memory, in M.D. Rugg, M.G.H. Coles (eds) *Electrophysiology of mind, event-related brain potentials and cognition*, Oxford Psychology Series, No. 25. Oxford University Press, Oxford, pp 130-170 (1995).
17. M. Kutas, Hillyard, S.A. Reading senseless sentences: Brain potentials reflect semantic incongruity. *Science* **207**, 203-205 (1980).
18. B.H. Smith, Altman, J.S.; Bargmann, C.I.; Carew, T.; Emrich, H.M.; Freeman, W.J.; Heisenberg, M.; Homberg, U.; Menzel, R.; Meyrand, P. and Ohl, F. Group Report: How flexible and how constrained is the concept of voluntary behavior? In: R.J. Greenspan & C.P. Kyriacou (Eds.), *Flexibility and constraint in behavioral systems*, John Wiley & Sons, Chichester, pp. 195-208 (1994).
19. G.W.F. Hegel, *Phänomenologie des Geistes*. Suhrkamp Verlag, Frankfurt a.M. (1973).

CHAPTER 9

CONTEXTUALIZATION: MEMORY FORMATION AND RETRIEVAL IN A NESTED ENVIRONMENT

Martina Piefke[1], Hans J. Markowitsch[2]

Physiological Psychology, Department of Psychology and Sports Sciences
Bielefeld University, P.O. Box 10 01 31, D-33501 Bielefeld, Germany
E-mail: [1]martina.piefke@uni-bielefeld.de, [2]hjmarkowitsch@uni-bielefeld.de

Episodic memory functions are highly context-dependent. This is true for both experimental and autobiographical episodic memory. We here review neuropsychological and neuroimaging evidence for effects of differential encoding and retrieval contexts on episodic memory performance as well as the underlying neurofunctional mechanisms. In studies of laboratory episodic memory, the influence of context parameters can be assessed by experimental manipulations. Such experiments suggest that contextual variables mainly affect prefrontal functions supporting executive processes involved in episodic learning and retrieval. Context parameters affecting episodic autobiographical memory are far more complex and cannot easily be controlled. Data support the view that not only prefrontal, but also further medial temporal and posterior parietal regions mediating the re-experience and emotional evaluation of personal memories are highly influenced by changing contextual variables of memory encoding and retrieval. Based on our review of available data, we thus suggest that experimental and autobiographical episodic memories are influenced by both overlapping and differential context parameters.

1. Introduction: The Concept of Episodic Memory

According to the multiple memory systems view[1-3], human long-term memory is composed of at least five memory systems. Two of them, procedural memory and the priming system, act on the implicit (i.e., unconscious) level of information processing. Perceptual and semantic memory system can be involved in both implicit and explicit (i.e., conscious) memory, but are usually regarded as explicit memory systems. Perceptual memory acts primarily on the pre-semantic level of information processing (based on familiarity judgments) while the semantic memory system is engaged in the main semantic processing of materials. The episodic memory system operates on the level of explicit processing of information[1]. Figure 1 characterizes the five major systems of human long-term memory by the specific kind of information processed by each system.

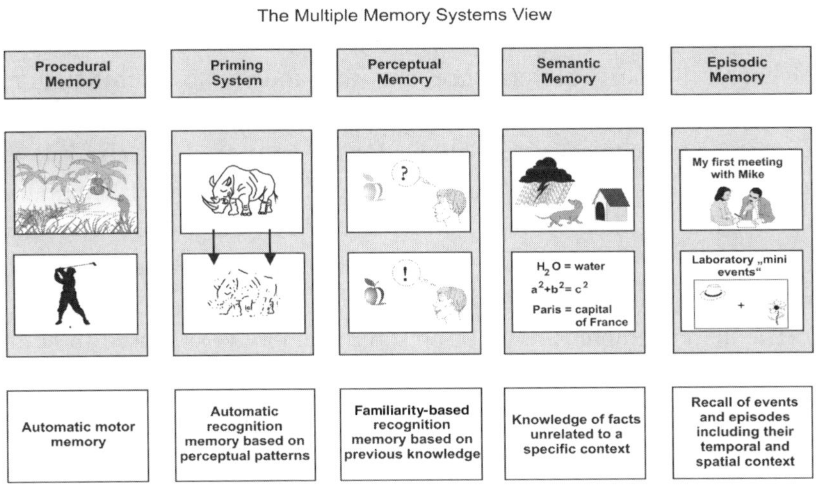

Fig. 1. According to the multiple memory systems view, human long-term memory comprises at least five memory systems: procedural memory, the priming system, perceptual memory, semantic memory, and episodic memory. The figure specifies the distinct memory types by the categories of information processed by each system. Both laboratory episodic memory and autobiographical recollection are included in the episodic memory system. Learning and retrieval of fact knowledge (also autobiographical facts) belong to the domain of semantic memory. Episodic, semantic, and perceptual memory are explicit (i.e., conscious) forms of memory. Implicit (i.e., unconscious) forms of memory are mediated by procedural memory and the priming system.

At least during explicit retrieval of memories, information is newly encoded[4,5]. The brain mechanisms underlying this coincidence of retrieval and re-encoding of stored information probably differ from those operating at initial encoding. However, new information is always integrated into old (i.e., already stored) materials in situations of new learning. This mnemonic mechanism implies that information which is already stored in the brain necessarily provides the context (i.e. the frame of reference) for each newly acquired memory.

This chapter deals with episodic memory, in particular, with its relationship to differential contexts of memory encoding and retrieval. Two subtypes of episodic memory will be discussed: *laboratory episodic memory and episodic autobiographical recollection*. Laboratory episodic memory deals with the recall or recognition of experimental items including their temporal and spatial context at the time of encoding of information[6]. Temporal and spatial context parameters may influence behavioral memory performance and the underlying neurofunctional mechanisms. The contexts of memory formation and retrieval provide the nested environment which allows for the unfolding, but also sets the limitation of our ability of episodic memory performance. In laboratory experiments, episodic memory can be assessed by the use of "remember-know"[7] and "source memory" paradigms[8] which both allow for the experimental control of context variables.

The autobiographical type of episodic memory is supposed to be the most complex form of human memory[1,2,9]. It provides a direct link into the awareness of the time course of one's life history and, typically, includes a synchronous emotional evaluation of the recalled event. It allows for the meaningful reconstruction of one's own past and a sense of self-coherence and self-continuity by creating subjectively significant personal past experiences embedded into one's life history[2,10]. Given these complex associations of information processing, it is conceivable that episodic autobiographical memory processing may depend on more complex contexts of information encoding and retrieval than laboratory episodic memory. For example, the age and emotional state of an individual, uncertainty, complexity of the encoding and/or retrieval environment, and current individual life situations may considerably influence autobiographical memory performance and its neurofunctional correlates. Therefore, context variables which may influence episodic autobiographical memory cannot be reduced to spatial and temporal parameters of the encoding and retrieval environments.

1.1 Self-Reference of Episodic Autobiographical Memory

Autobiographical episodic memory is strongly interconnected with the *ability of taking a self-perspective.* Accordingly, it is likely to play a key role in processes of identity formation and personality development which include both personal continuity and change. Memories of one's personal past experiences constitute the experiential basis for continuous and changing aspects of self-concepts across the time course of an individual's whole life. Vice versa, self-concepts may provide the "mental contexts" which form the processes of re-interpretation of autobiographical events and hence alter the respective autobiographical memories themselves[11,12]. Libby and Eibach[13] reported a tendency in healthy volunteers to use a third person observer perspective when they visualized memories of actions which conflicted with current views of their selves. A similar effect was observed for the imagination of actions with varying self-concept compatibility. Online judgments of an action's self-concept compatibility may thus substantially affect the perspective adopted for visualization of memories.

Pasupathi[12] suggested that social conversation on personal past experiences may provide the context in which humans are able to reconstruct themselves within their physical and social worlds across life-span. Two principles are likely to govern this kind of conversational recounting: *co-construction* (i.e., combined influences of actors and contexts on conversational re-constructions of personal past episodes) and *consistency* (i.e., the effect of conversational recounting on subsequent memory). Pasupathi[12] proposed that the interaction of these two principles in real-life nested environments may account for the social construction of autobiographical memory and the influences of conversational re-construction of a personal past on the development of an individual's personal identity.

Effects of cultural contexts on the elaboration of autobiographical memories and self-descriptions have also frequently been observed. For example, Wang[14] found that earliest childhood events remembered by Americans had occurred approximately 6 months earlier than those recalled by Chinese people. Americans' memories were emotionally elaborated, specific, lengthy, and self-focused. By contrast, the Chinese provided brief accounts of childhood events which were focused on collective activities, general routines, and rather emotionally neutral events. In their self-descriptions, social rules played a predominant role.

1.2 Episodic Memory and Emotional Context

Mood states of an individual may also act as contexts affecting episodic memory. The concept of *mood-congruent retrieval*[15-17] proposes that episodic (and also semantic) memories with an emotional valence corresponding to an individual's mood state at retrieval are more frequently and/or better recalled than memories with a non-congruent emotional valence. The assumption of *mood-dependent retrieval*[18,19] holds that mood at encoding has to correspond to mood at retrieval to yield better recall of the respective episode, even in the case of relatively neutral autobiographical information. The two concepts are used to model specific types of the interaction among memory and contextual emotional states in both laboratory episodic memory and autobiographical recollection

The assumption of mood-congruent retrieval is of particular interest for clinical studies on memory impairments of patients suffering from Major Depression (MD). Yang and Rehm[17] investigated autobiographical memories in depressed and non-depressed elderly subjects. Participants were asked to recall and rate personal past episodes for their emotional valence (happiness versus sadness). Additionally, subjects had to indicate the degree of importance of each episode (1) at the time of its occurrence ("then"-rating) and (2) when looking back to the event from the present perspective ("now"-rating). Non-depressed participants perceived greater positive change in the emotional valence associated with autobiographical events between the "then" and "now" ratings than depressed individuals. MD patients (relative to the control group) also rated more autobiographical memories as sad "now" and perceived both happy and sad memories to become more neutral. Interestingly, retrieval of memories which obtained higher ratings of happiness was associated with less depression, suggesting a self-enhancing function of reminiscing. The data show evidence for close interactions between influences of emotional and temporal context parameters on episodic memory in MD patients.

Philpot and Madonna[20] investigated laboratory mood-dependent retrieval using a serial learning task with subsequent recall for the studied items. Recall was significantly greater when mood was the same during learning and retrieval (in comparison to distinct mood states at encoding and retrieval). Similarly, Kenealy[19] reported that healthy human volunteers who learned and recalled stimulus materials in distinct mood states showed greater decrements in recall than subjects who were in the same mood at study and test. Interestingly, however, when the participants were provided with retrieval cues the effects of mood-dependency disappeared. The data of Kenealy[19] thus support the view that occurrence and erasure of mood-dependent retrieval can be manipulated by experimentally varying the presence or absence of retrieval cues. This finding suggests that effects of mood-related contexts interact with those of situational cueing-contexts during the retrieval of episodic memories. However, some studies contradict the assumption of mood-dependent retrieval. Foa et al.[21] induced anxiety in speech anxious students by giving them the information that they will be required to deliver a speech in the course of the experiment. At encoding, subjects rated the degree of self-descriptiveness of anxiety-related and control adjectives. In the subsequent recall test, no mood-dependency was observed. Rather, data show evidence that subjects who presented with increasing anxiety (assessed by measures of heart rate and by self-report) from encoding to the recall test least often recalled anxiety-related adjectives. These findings may indicate that attention to threat-related information may diminish in mood-dependent retrieval in anxiously aroused volunteers despite increases in anxious mood.

2. Brain Mechanisms and Behavioral Characteristics of Episodic Memory

Lesion studies in animals, neuropsychological assessment of amnesic patients, and functional neuroimaging studies demonstrate that both laboratory episodic memory and autobiographical recollection rely on a distributed network of prefrontal, medial and lateral temporal, as well as posterior cingulate and retrosplenial brain regions (for reviews, see[2,3]). Thus far, no consensus has been achieved concerning the issue of hemispheric lateralization of episodic memory processing (for reviews, see[3,22]).

Laboratory episodic memory differs from episodic autobiographical recollections at least with respect to the following aspects (see also Figure 2):

(i) the *time-window* of memory-related information processing which is restricted to the experimental setting in the laboratory type of episodic memory but potentially covers one's entire life history during autobiographical recollection,
(ii) the (often emotional) *re-experience* of an autobiographical event during its recollection which is usually absent in laboratory episodic memory,
(iii) the (emotionally and cognitively determined) *personal significance* of autobiographical recollections which cannot be attributed to laboratory episodic memories,
(iv) the *self-referential perspective* and *"autonoetic" awareness* (see below) is naturally associated with autobiographical recollection but not with laboratory episodic memory tasks,
(v) the processes of *mental time travel* associated with autobiographical recollection which are absent or marginal during non-autobiographical episodic memory,
(vi) the *reconstructive mechanisms* of autobiographical memory which enable us to reconstruct our personally meaningful past, but are absent (or at least widely reduced) in laboratory episodic memory tasks.

Autonoetic awareness implies the senses "self-experience" and "ownership". It allows us to refer to ourselves as the main contextual parameter of our autobiographical memories. In non-pathological cases of autobiographical remembering we are thus assured to remember events of our "own" personal past[2,3]. The introduction of "remember-know paradigms" aimed at identifying a "surrogate" of autonoetic awareness in laboratory experiments of episodic memory[7,23].

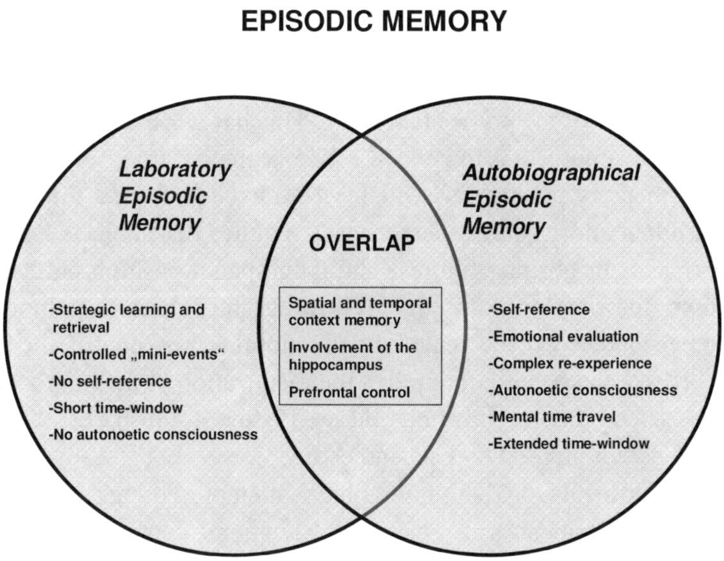

Fig. 2. Functional integration and segregation of laboratory and autobiographical episodic memory. Both forms of episodic memory include spatial and temporal context information, rely on hippocampal functions, and are controlled by (functionally distinct) prefrontal regions. However, they differ from each other with regard to the quality and complexity of intrinsic and extrinsic contexts influencing the encoding and retrieval of information.

Typically, a remember-know paradigm includes a learning phase and a subsequent testing phase during which participants are asked to recognize studied items out of a set of new and learned stimuli. For each recognized item, participants also indicate whether they "remember" having seen the item in the study phase including the respective temporo-spatial context information (e.g., whether it was one of the first or last in a series of items), or rather "know" that it was one of the studied items (in the sense of a "feeling of familiarity") but are unable to provide information about the specific context of the item's presentation. In terms of this approach, only "remember-responses" are indicators of autonoetic conscious awareness and thus for episodic memory of an item. To decide between the presence of a "remember" or "know" memory, task performance in experiments based on this paradigm poses strong context monitoring demands on the participants which are likely to recruit mainly regions of the prefrontal cortex. This has consistently been shown by neuroimaging experiments[24-26]. The same is true for the related "source memory paradigms" which were developed some time later[8]. A standard laboratory source memory task requires that subjects learn items (e.g., objects or abstract figures) and their spatial position on the screen (context information). In the following recognition memory test, volunteers judge whether each shown stimulus was one of the studied items. If yes, the paradigm requires the localization of the item's position on the screen in the learning phase. Figure 4 illustrates the source memory paradigm used by Cansino et al.[8].

In experiments on autobiographical memory, things are far more complicated. Influences of spatial and temporal context parameters go beyond the modulation of monitoring demands and are thus likely to affect not only PFC functions, but also long-term memory- and emotion-related functions of the medial temporal lobe and posterior parietal cortex[27]. Due to mental time travel, the associated reconstructive processes, and continuously emotional re-evaluations across an individual's life time, context parameters of autobiographical recollections cannot easily be controlled. Nonetheless, some neuroimaging experiments successfully manipulated context parameters of autobiographical remembering and yielded new insights into the brain structures and neurofunctional mechanisms mediating the contextualization of memories of one's own past. Piefke et al.[10] performed a functional MRI study on recent and remote autobiographical memories with positive or negative emotional valence. The study thus experimentally controlled two context parameters: the time interval elapsed since the occurrence (and, accordingly, the encoding) of an event and its emotional valence (positive or negative). Overall, bilateral hippocampal activations related to the recollection of personal past episodes were observed in this experiment, albeit with a clear left

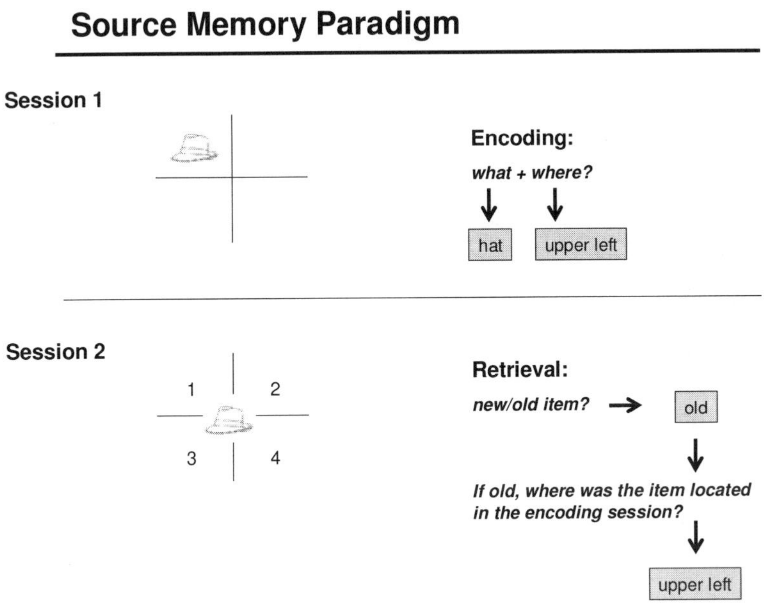

Fig. 3. Source memory paradigm requiring the learning (session 1; upper part of the figure) and retrieval (session 2; lower part of the figure) of objects *and* their positions on the screen (here, four positions are possible). In session 1, participants learn *what* (here: hat) is presented and *where* it is located on the screen (here: upper left). In session 2, subjects see the learned objects and additional new objects located in the center of the screen. Participants (i) decide whether each item is one of the learned objects or a new one. If an item is regarded as one of the learned objects, they (ii) indicate where it was located in the learning session (here: upper left, i.e., quadrant 1). Source memory paradigms allow for the assessment of spatial context-related episodic memory in laboratory experiments.

hemispheric preponderance when all autobiographical memory conditions (recent positive, recent negative, remote positive, and remote negative memories) were compared to a low-level baseline. Interestingly, strong bilateral activations of the hippocampus were evident during the retrieval of recent relative to remote memories (see Figure 4a). These findings suggest that hippocampal involvement in autobiographical memory is modulated by the "age of memories", that is, the temporal context depending on time interval elapsed since information encoding.

Differential Hippocampal Activation During Recent Relative to Remote Emotional Autobiographical Memory Retrieval

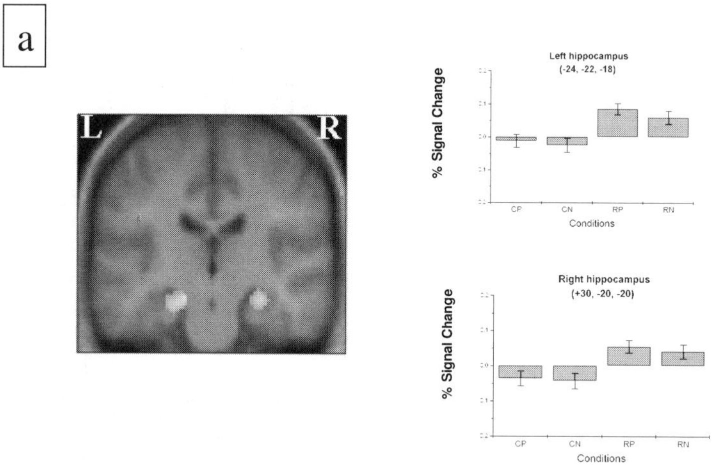

Fig. 4. (a) Bilateral increases of neural activity in the hippocampus during the recollection of recent (relative to remote) emotional autobiographical episodes reported by Piefke et al.[10]. The histograms display the percentage of blood-oxygen-level-dependent (BOLD) signal change for the maxima of significantly increased hippocampal activity as a function of the respective experimental condition. The reverse contrast (i.e., remote in comparison to recent memories) did not yield statistically significant activations in any brain region. R = right; L = left; A = anterior, P = posterior. CP = positive childhood events, CN = negative childhood events, RP = positive recent events, RN = negative recent events. (b) Assumptions of the consolidation theory and the multiple trace theory (MTT) concerning temporal context-related aspects influencing the role of the hippocampus in episodic memory retrieval. The consolidation model (upper part) presumes a time-limited involvement of the hippocampus in the retrieval of recent episodic memories during the consolidation period. Later, upon long-term memory storage in the neocortex, retrieval of information becomes independent of the hippocampus. In contrast, the MTT (lower part) suggests that the hippocampus is always needed for the retrieval of episodic memories, regardless of whether the time-period elapsed since the encoding of information is short or long. It predicts a flat gradient of hippocampal involvement in retrieval processes as time passes, with remote information requiring less engagement of the hippocampus. By the MTT, this is ascribed to the stabilization of information related-to associative processes. Currently, it is still a matter of debate whether the consolidation theory or the MTT may better account for the currently available clinical and neuroimaging data on hippocampal memory functions.

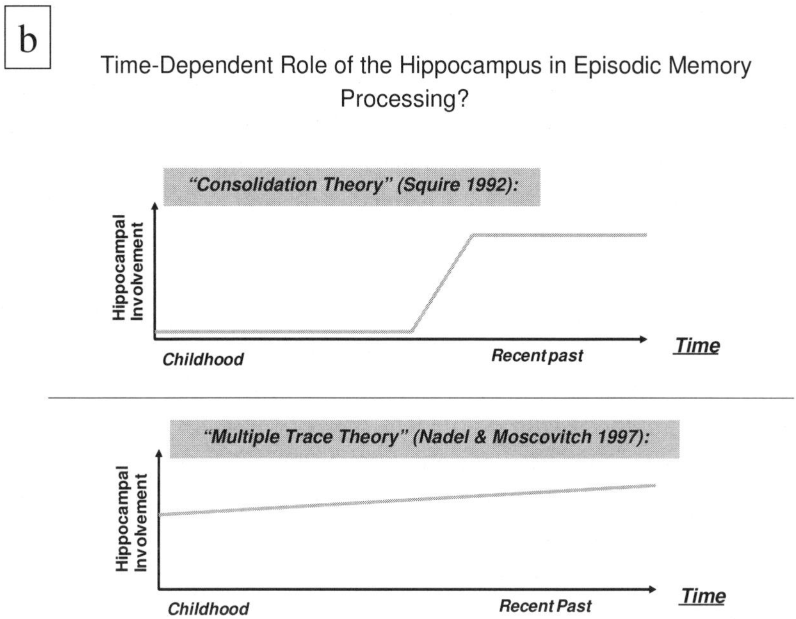

Fig. 4. (*Continued*)

2.1 Contexts of Memory Encoding and Retrieval

2.1.1 The Hippocampus and Temporal Context Memory

The hippocampus plays a core role in declarative long-term memory processing in animals (for a review, see[28]). However, investigators do not agree with regard to its exact role in "conscious" memory functions. The consolidation theory suggests that the involvement of the hippocampus in learning and memory is limited to the learning and consolidation phases of memory processing, during which memories are characterized by instability[28,29]. According to this view, memories are transferred to neocortical areas within a certain time-interval and upon storage become independent of the hippocampal complex. The consolidation theory is thus in good accordance with the functional neuroinaging study of Piefke et al.[10] in that it assumes a dependency of hippocampal functions on temporal context parameters of memory processing. Various lesion studies in experimental animals and neuropsychological assessment of patients with hippocampal lesions also corroborate this model[30-32]. Rempel-Clower et al.[32] reported that the length of temporal gradients of enduring retrograde amnesia may correlate with the increasing extent of hippocampal damage. In two patients with lesions limited to the CA1 field of the hippocampal region, amnesia covered the last 2 years prior to brain damage only. In two other patients with more extended damage to the hippocampal complex, amnesia extended back 15 years in one patient and 25 years in the other.

Importantly, however, some lesion studies investigating memory functions of the hippocampus in both animals and patients failed to replicate the characteristic pattern of temporally graded memory loss[33-35]. Rather, flat gradients with equivalent memory loss for all time periods prior to hippocampal damage were reported. To account for those inconsistent data, the "multiple trace theory" (MTT) has been proposed as an alternative model of hippocampal function in long-term memory processing[34]. The MTT assumes a persistent involvement of the hippocampus in episodic memory retrieval irrespective of the time-interval which elapsed between the acquisition and retrieval of information. According to the MTT, each reactivation of a memory leads to the creation of a new memory trace which is assumed to involve an ensemble of hippocampal and neocortical neurons. The proliferation of the memory traces occurring with repeated reactivation of information is supposed to render older memories less susceptible to hippocampal dysfunction than recent ones. Depending on the location and the extent of medial temporal damage, distinct types of retrograde amnesia may then result. The assumptions of the consolidation theory and the MTT are illustrated in Figure 4b.

2.1.2 Context Memory and Executive Functions of the Prefrontal Cortex

Concerning experimental episodic memory retrieval, evidence has been shown for a high degree of functional segregation within the PFC which most likely depends on the diversity of executive functions involved in laboratory memory tasks[36,37]. However, no consensus has been achieved with regard to the exact roles of distinct PFC areas in episodic memory, yet. Right PFC regions have been implicated in "retrieval mode"[38,39], "retrieval effort"[40], and "post-retrieval monitoring"[41,42]. The demands on "retrieval effort" and "post-retrieval monitoring" can be assumed to depend strongly on the temporal and spatial contexts of information encoding and retrieval. Fletcher et al.[43] assumed that right dorsolateral regions are recruited by increasing monitoring demands while the ventrolateral structures mediate retrieval mode. An alternative account relates differential patterns of PFC activation during laboratory episodic memory performance to a neurofunctional differentiation between processes of "retrieval attempt" and "retrieval success"[44,45]. In addition to the ventrolateral and dorsolateral PFC, orbitofrontal and frontal polar regions are sometimes recruited. Involvement of these areas in laboratory episodic memory has been related to high demands on context memory[24], including emotional components[46] of a given task.

Although ventrolateral and dorsolateral PFC areas are also implicated in the executive control of autobiographical episodic recollection[10,47], orbitofrontal and medial PFC regions play a predominating role in remembering the personal past[27]. In particular, they are involved in self-referential[48] and social cognition[49], as well as emotion and its control[46]. These cognitive and emotional abilities build up major components involved in the processing of contextual parameters of autobiographical recollections.

In a review of functional neuroimaging data on overlapping and differential PFC activations related to laboratory and autobiographical episodic memory, Gilboa[50] proposed that the right mid-dorsolateral PFC is primarily involved in experimental episodic memory while the left ventromedial PFC predominates in autobiographical recollection. According to Gilboa[50] this prefrontal functional segregation might reflect *distinct modes of post-retrieval verification and monitoring of laboratory and autobiographical episodic memories*. Autobiographical memories are likely to be monitored by a subjective "feeling of rightness" which is related to a "self-schema" and recruits the left ventromedial PFC. In contrast, experimental episodic memory requires elaborate strategic monitoring and control processes to avoid errors (e.g., omissions, repetitions), and thus engages the right mid-dorsolateral PFC. It is reasonable to assume that the monitoring of experimental and autobiographical episodic information is built upon distinct cognitive strategies. Nonetheless, PFC involvement in either type of episodic memory can probably not be reduced to differences in monitoring demands. Differences in the quality of context memory, in particular, autonoetic consciousness, autobiographical reconstruction, the awareness of the elapse time, mental time travel, and emotional evaluation (which may all be directly mediated by functionally segregated PFC areas) also need to be taken into consideration.

2.1.3 The Role of the Retrosplenial Cortex in Context Memory

The retrosplenial cortex lies adjacent to the retrocommissural hippocampus, a small tissue stripe extending around the posterior end of the corpus callosum. The retrosplenial area is interconnected with the parahippocampal cortex which possesses a core role in spatial cognition[51,52]. Epstein and Kanwisher[52] suggested that the "parahippocampal place area" mediates the processing of scenes and landmarks. The retrosplenial cortex is anatomically and functionally closely related to the hippocampus and other structures of the medial temporal lobe subserving context memory, spatial cognition, and emotion processing. Accordingly, it may have a crucial function for (i) spatial and temporal episodic context memory and (ii) the typical interplay of emotion and cognition in autobiographical recollections. It is consistent with this hypothesis that severe and lasting amnesia for both laboratory and autobiographical materials following selective retrosplenial damage was reported from clinical studies[53,54]. Moreover, functional neuroimaging studies relate the retrosplenial cortex to *emotional* memory in the non-autobiographical[55,56] and the autobiographical episodic memory domain[10,57], as well as to emotion processing in a more general sense[58].

There is some evidence in support of a specific retrosplenial engagement in autobiographical recollection. Piefke et al.[10] reported *increased neural activity in the retrosplenial cortex in combination with bilateral hippocampal activation during the retrieval of recent (relative to remote) emotional autobiographical memories*. This finding corroborates the view that the retrosplenial region closely interacts with the hippocampus, and that this interaction may be crucial for the recall of emotionally salient and familiar autobiographical episodes of one's recent past. It is thus conceivable that the retrosplenial cortex is in particular sensitive to *memory-related contextual aspects of emotional and temporal salience and familiarity*.

FMRI studies have further been implicated retrosplenial areas in the recognition of personally familiar faces and voices[59], as well as objects and places[60]. Although these experiments did not directly assess autobiographical recollection, the autobiographically familiar stimulus materials employed are likely to have triggered the recall of personal past episodes including their temporal and emotional contexts. Based on a meta-analysis of functional neuroimaging data, Sugiura et al.[60] argued that the retrosplenial and posterior cingulate cortices are composed of a number of functionally segregated subdivisions. They proposed a functional differentiation between the caudal posterior cingulate cortex (cPPC), middorsal PCC (mPCC), posterodorsal PCC (pPCC), and the retrosplenial cortex.

According to the meta-analysis of Sugiura et al.[60], *the pPCC and the retrosplenial cortex are selectively activated during tasks requiring the retrieval of "real life information"*, including autobiographical episodes[10,61], familiar people, and other self-related real world entities[62], and the mental imagery of moving around in a personally familiar environment[63]. *These forms of memory and cognition are all associated with complex contextual aspects of time, space, familiarity, emotion, and self-reference.* Activations associated with the retrieval of materials learned in a neutral, temporally and spatially controlled laboratory context, by contrast, were predominantly located in the mPCC[64,65] suggesting that this area plays a core role in basic aspects of temporal and spatial context memory. Finally, the cPCC has been implicated in memory-related navigation in large-scale space[66] and effortful episodic retrieval[59]. This region could thus specifically mediate executive aspects of temporo-spatial episodic context memory. Figure 5 illustrates the location of retrosplenial activations during the recall of real-life information.

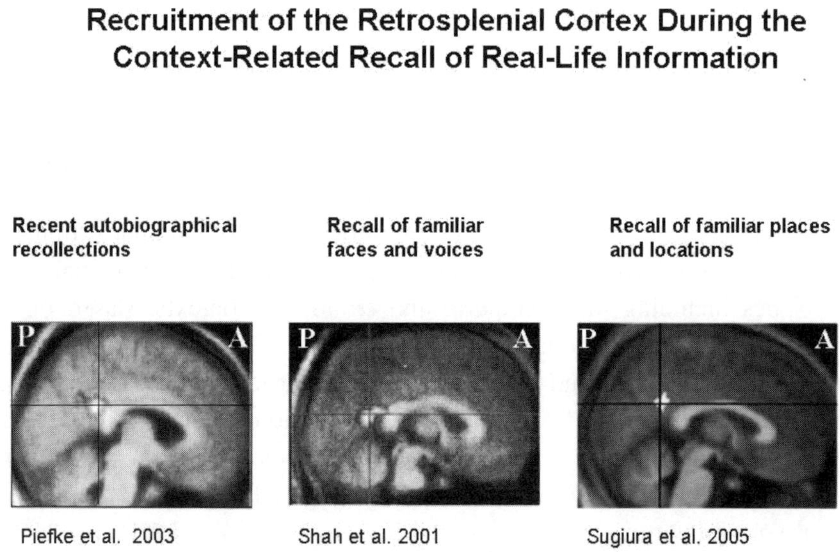

Fig. 5. FMRI evidence for retrosplenial activation during the retrieval of real life information. (a) Piefke et al.[10] demonstrated increased neural activity in the retrosplenial cortex (in combination with bilateral hippocampal activation) during the retrieval of recent (relative to remote) emotional autobiographical memories, suggesting that this brain region is specifically involved in the recall of emotionally salient and familiar autobiographical episodes of one's recent past. (b) Shah et al.[59] implicated a corresponding retrosplenial area in the retrieval of personally familiar faces and voices. (c) Sugiura et al.[60] reported activation of the retrosplenial cortex associated with memories of familiar places and locations.

3. Conclusions and Outlook

The empirical data on contextual aspects of laboratory and autobiographical episodic memory summarized here indicate that neuropsychological performance and neurofunctional mechanisms of both forms of episodic memory strongly depend on parameters of the nested environment of information encoding and retrieval. Context parameters which are relevant in laboratory forms of episodic memory can be experimentally manipulated by the variation of temporal and spatial environments of information encoding and retrieval. Such variation particularly exerts influences on PFC functions. In contrast, the complex and intricate context parameters affecting autobiographical episodic memory cannot easily be controlled experimentally. They include the age and emotional state of an individual, aspects of uncertainty, complexity of the encoding and/or retrieval environment, and one's current life situations. Accordingly, they may affect interactions of prefrontal, medial temporal, and posterior parietal neuronal networks supporting autobiographical recollections. The intricate nested environments of autobiographical memory encoding and retrieval thus remain a challenge for neuroscientific research which calls for elaborate novel cognitive-behavioral and neurofunctional experimental assessment.

References

1. E. Tulving, in *The Cognitive Neurosciences*, Ed. M.S. Gazzaniga (MIT Press, Cambridge, MA, 1995), pp. 839-847.
2. H.J. Markowitsch, in *Handbook of Clinical Neuropsychology*, Ed. P. Halligan, U. Kischka, and G. Beaumont (Oxford University Press, Oxford, UK, 2003), pp. 724-740.
3. M. Piefke, in *Trends in brain mapping research*, Ed. C. Franklin (Nova Science Publishers, New York, 2008), in press.
4. R.L. Buckner, M.E. Wheeler, and M.A. Sheridan, *J. Cogn. Neurosci.*, 406 (2001).
5. H.J. Markowitsch, *Brain Res. Brain Res. Rev.*, 117 (1995).
6. E. Tulving, *Elements of Episodic Memory* (Clarendon Press, Oxford, MA, 1983).
7. G. Dalla Barba, M.C. Mantovan, E. Ferruzza, and G. Denes, *Cortex*, 143 (1997).
8. S. Cansino, P. Maquet, R.J. Dolan, and M.D. Rugg, *Cereb. Cortex*, 1048 (2002).
9. M. Piefke and G.R. Fink, *Anat Embryol*, 497 (2005).
10. M. Piefke, P.H. Weiss, K. Zilles, H.J. Markowitsch, and G.R. Fink, *Brain*, 650 (2003).
11. T. Habermas and S. Bluck, *Psychol Bull*, 748 (2000).
12. M. Pasupathi, *Psychol. Bull.*, 651 (2001).
13. L.K. Libby and R.P. Eibach, *J. Pers. Soc. Psychol.*, 167 (2002).
14. Q. Wang, *J. Pers. Soc. Psychol.*, 220 (2001).
15. C. McFarland and R. Bühler, *J. Pers. Soc. Psychol.*, 1424 (1998).
16. J.D. Teasdale, *J. R. Soc. Med.*, 1038 (1983).
17. J.A. Yang and L.P. Rehm, *Int. J. Aging Hum. Dev.*, 39 (1993).
18. E. Eich, D. Macaulay, and L. Ryan, *J. Exp. Psychol. Gen.*, 201 (1994).

19. P.M. Kenealy, *Q. J. Exp. Psychol. A*, 290 (1997).
20. V.D. Philpot and S. Madonna, *Psychol. Reports*, 203 (1993).
21. E.R. Foa, R. McNally, and T.B. Murdock, *Behav. Res. Therapy*, 141 (1989).
22. A.C.H. Lee, T.W. Robbins, and A.M. Owen, *Crit. Rev. Neurobiol.,* 165 (2003).
23. J.M. Gardiner and R.L. Java in *Theories of memory*, Ed. A.F. Collins, S.E. Gathercole, M.A. Conway, and P.E. Morris (Erlbaum, Hove, UK, 1993). pp. 163-188
24. T. Fujii., M. Suzuki, J. Okuda, H. Ohtake, K. Tanji, K. Yamaguchi, M. Itoh, and A. Yamadori, *Neuroimage,* 1596 (2004).
25. C. Ranganath, M.K. Johnson, and M. D'Esposito, *J Neurosci,* 1 (2000).
26. M. Suzuki, T. Fujii, T. Tsukiura, J. Okuda, A. Umetsu, T. Nagasaka, S. Mugikura, I. Yanagawa, S. Takahashi, and A. Yamadori, *Neuroimage*, 1790 (2002).
27. R. Cabeza and P. St. Jacques, *Trends Cogn. Sci.*, 219 (2007).
28. L.R. Squire, *Psychol. Rev.*, 195 (1992).
29. E. Teng and L.R. Squire, *Nature*, 675 (1999).
30. S. Zola-Morgan and L.R. Squire, *Science*, 288 (1990).
31. H. Bak, N. Antoun, K.K. Balan, and J.R. Hodges, *J. Neurol. Neurosurg.Psychiatry*, 40 (2001).
32. N.L. Rempel-Clower, S.M. Zola, L.R. Squire, and D.G. Amaral, *J. Neurosci.*, 5233 (1996).
33. J.J. Bolhuis, C.A. Stewart, and E.M. Forrest, *Q. J.Exp. Psychol. B*, 129 (1994).
34. L. Nadel and M. Moscovitch, *Curr. Opin. Neurobiol.*, 217 (1997).
35. R. Westmacott, L. Leach, M. Freedman, and M. Moscovitch, *Neurocase*, 37 (2001).
36. A.R. Aron, T.W. Robbins, and R.A. Poldrack, *Trends Cogn. Sci.*,170 (2004).
37. T.D. Wager and E.E. Smith, *Cogn. Affect. Behav. Neurosci.*, 255 (2003).
38. S. Kapur, F.I. Craik, C. Jones, G.M. Brown, S. Houle, and E. Tulving, *Neuroreport*, 1880 (1995).
39. L. Nyberg, E. Tulving, R. Habib, L.G. Nilsson, S. Kapur, S. Houle, R. Cabeza, and A.R. McIntosh, *Neuroreport*, 9 (1995).
40. D.L. Schacter, T. Curran, L. Galluccio, W.P. Milberg, and J.F. Bates, *Neuropsychologia,* 793 (1996).
41. M..D. Rugg, P.C. Fletcher, C.D. Frith, R.S.J. Frackowiak, and R.J. Dolan, *Brain*, 2073 (1996).
42. T. Shallice, P.C. Fletcher, C.D. Frith, P. Grasby, R.S.J Frackowiak, and R.J. Dolan, *Nature*, 633 (1994).
43. P.C. Fletcher, T. Shallice, C.D. Frith, R.S.J Frackowiak, and R.J. Dolan, *Brain*, 1249 (1998).
44. R.L. Buckner, W. Koutstaal, D.L. Schacter, A.D. Wagner, and B.R. Rosen, *Neuroimage*, 151 (1998a).
45. R.L. Buckner, W. Koutstaal, D.L. Schacter, A.M. Dale, M. Rotte, and B.R. Rosen, *Neuroimage*, 163 (1998b).
46. A.P. Smith, R.N. Henson, R.J. Dolan, and M.D. Rugg, *Neuroimage*, 868 (2004).

47. M.M. Vandekerckhove, H.J. Markowitsch, M. Mertens, and F.G. Woermann, *Behav. Neurol.*, 200 (2005).
48. K. Vogeley, P. Bussfeld, A. Newen, S. Herrmann, F. Happe, P. Falkai, W. Maier, N.J. Shah, G.R. Fink, and K. Zilles, *Neuroimage,* 170 (2001).
49. M. Schulte-Rüther, H.J. Markowitsch, G.R. Fink, and M. Piefke, M. (2007). *J. Cogn. Neurosci.*, 1354 (2007).
50. A. Gilboa, *Neuropsychologia*, 1336 (2004).
51. M. Piefke, P.H. Weiss, H.J. Markowitsch, and G.R. Fink, G.R. *Hum. Brain Mapp.*, 313 (2005).
52. R. Epstein and N. Kanwisher, *Nature*, 598 (1998).
53. D. Bowers, M. Verfaellie, E. Valenstein, and K.M. Heilman, *Brain Cogn.*, 47 (1988).
54. G. Gainotti, S. Almonti, A.M. DiBetta, and M.C. Silveri, *Neurocase*, 519 (1998).
55. R.J. Maddock, and M.H. Buonocore, *Psychiatry Res.*, 1 (1997).
56. E.J. Maratos, R.J. Dolan, J.S. Morris, R.N. Henson, and M.D. Rugg, *Neuropsychologia,* 910 (2001).
57. H.J. Markowitsch, M.M. Vandekerkhove, H. Lanfermann, and M. Russ, *Cortex*, 643 (2003).
58. R.J. Maddock, *Trends Neurosci.*, 310 (1999).
59. N.J. Shah, J.C. Marshall, O. Zafiris, A. Schwab, K. Zilles, H.J. Markowitsch, and G.R. Fink, *Brain,* 804 (2001).
60. M. Sugiura, N.J. Shah, K. Zilles, and G.R. Fink, *J. Cogn. Neurosci.*, 183 (2005).
61. E.A. Maguire and C.D. Frith, *Brain*, 1511 (2003).
62. C.L. Leveroni, M. Seidenberg, A.R. Mayer, L.A. Mead, J.R. Binder, and S.M. Rao, *J. Neurosci.*, 878 (2002).
63. E. Mellet, S. Briscogne, N. Tzourio-Mazoyer, O. Ghaem, L. Petit, L. Zago, O. Etard, A. Berthoz, B. Mazoyer, and M. Denis, *Neuroimage*, 588 (2000).
64. R.N. Henson, M.D. Rugg, T. Shallice, O. Josephs, and R.J. Dolan, *J. Neurosci.*, 3962 (1999).
65. S. Konishi, M.E. Wheeler, D.I. Donaldson, and R.L., Buckner, *Neuroimage*, 276 (2000).
66. E.A. Maguire, *Scand. J. Psychol.*, 225 (2001).

CHAPTER 10

COMPLEXITY AND EMERGENT TEMPORAL STRUCTURE

Peter M. Allen

Complex Systems Research Centre
School of Management
Cranfield University, Beds. MK43 0AL
England
E-mail: p.m.allen@cranfield,ac,uk

We describe the mechanisms of complex systems evolution and show how this both generates and is generated by a hierarchical spatio-temporal structure. The mechanisms behind "evolutionary drive" are described, together with the important role that temporal and spatial hierarchy plays – where the mitigation of selective forces by different levels of structure make creative evolution inevitable. Several examples in human systems are presented to illustrate the ideas.

1. Introduction

The study of open systems and the temporal and spatial structures that they could generate led to new paradigms: Nicolis and Prigogine[1,2,3] to dissipative structures, Haken[4] to synergetics and Mandelbrot[5] to fractals. All of these opened the path to complexity and complex systems that were concerned with inherently dynamic, non-linear, evolutionary and non-equilibrium situations. In turn, this implied that time itself was deeply involved in the way that systems evolve and structure, and indeed that our own perception of time played an important role in phenomena that we observe and wish to understand. The older traditions of natural science had, through the use of either mechanical or equilibrium assumptions, divorced themselves from disequilibrium and history. The complex intertwining of processes and events and the consequent interplay of chance and necessity were missing from traditional views of scientific explanation, and so ecology, economics and much of social science were seen in terms of equilibrium solutions representing post-evolutionary optima. Of course, in the real world, evolution is on-going, and the changes that are produced are driven by an emergent selection process. This operates on partially incoherent lower level structures and, as we shall see, a remarkable property emerges by which it is the partial separation of time-scales that allows us to define entities and discuss structures even within an overarching, evolutionary context. In this chapter, we shall present several models that clarify these ideas, and show how they can provide a more subtle and correct representation of the world and of the dilemmas and ambiguities that really characterize our situation.

The very idea of an "observer" and a "perspective" implies that there is someone present who is more persistent than the phenomena that are being observed, and indeed that presence itself is only possible as part of a still longer time-scale biosphere. It also means that although observers can actually see phenomena that are more rapid than their own lifetime, slower, longer term processes and changes must be understood as the result either of some theoretical construction or through historical comparisons. Furthermore, if the purpose of observations is to try to create an understanding of world and its workings, then this can only be achieved by developing a nested view of the structure of phenomena – nested both in spatial and temporal scales.

Understanding itself implies some degree of "stationarity" at the scale concerned, within which there is an observed change or movement that is "explained" by the overarching stationarity. Indeed, the idea of understanding or sense-making is dependent on finding or creating a suitable framework within which we can order apparent changes and "explain" them within some seemingly unchanging concept. In a universe that is really evolving over time, where structural changes will continue occurring for billions of years, we see that complete explanation is really an illusion and that our understanding and explanations rely on the acceptance of a hierarchy of nested temporal and spatial scales. We therefore only have observers and perspectives, providing that they accept or are unaware of the fact that the things they regard as true and fixed are really not completely so. This implies that in an evolving universe there can never be a completely satisfactory explanation of all things, and that our interpretive frameworks are really created because we find them useful. When we find them inadequate, then we try to modify or add to them in order to reflect some perceived evolutionary emergence in the real world. This therefore also tells us that our interpretive framework will always be a post-rationalization with some lag behind the world. We are forced to "boldly go" and therefore must face inevitable uncertainties to some degree. Fortunately, as products of evolution we have evolved naturally to do just this, and to live with some level of uncertainty in our personal lives. We run into trouble when we are dealing with the artificial worlds of financial, political and legal institutions where it is often difficult to introduce methods that allow for the essential uncertainties of reality.

In this paper, we shall attempt to present a view of complexity as naturally leading to "Evolutionary Drive" – a theory put forward some years ago by Allen and McGlade[6]. In this view, evolution is driven by the interplay of processes that create micro-diversity within systems on a short term scale and the selection operated by the differential dynamics and different temporal scales that emerge. This is seen as a continuous, on-going process and not one that has already "run its course" as in the Game Theory approach of Evolutionary Stable Strategies of Maynard-Smith[7]. Instead, we shall see that it as an on-going process that is occurring in real time, and which involves innovations and qualitative changes of various kinds. History is still running. Each behavioral type is in interaction with others, and therefore evolutionary improvements may lead to greater synergy or conflict between behaviors, and in turn to a chain of responses without any obvious end. And if there is no end, then the most that can be said of the behavior of any particular individual or population is that its continued existence proves only that it is, and has been, good enough – but not that it is optimal.

Our aim here is to show how the underlying mechanisms of evolution involve micro-diversity within a system, with divergent temporal rates and different time-scales. This, in turn, drives an evolving, emergent system structure that is characterized by a changing level of structural diversity. It also explains exaptation, an idea raised by Gould and Vrba[8] and Gould[9] since the micro-diversity pre-exists the "uses" and "niches" that they later are "explained" as fitting. This will take into account both the "selective" effects of interactions between species and the simultaneous mechanism that discovers new "strategies" or "niches". In Darwinian thinking, the micro-diversity that occurs is considered to be "random" and independent of the selection processes that follow, while in human innovation we may think that there is intention, calculation and belief that may "channel" diversity into some narrower range.

2. Evolutionary Drive

Diversity is a measure of the number of qualitatively different types of entity present corresponding to individuals with different attributes, including different speeds of response and temporal behaviors. It also implies that they are relatively stable entities that can be "labeled" over some lifetime. It may be that they share some dimensions, but differ on others. This is an important point because it refers to a fundamental issue for evolution – it concerns both the qualitative changes that occur in systems and structures over time as well as the temporal signatures and rhythms. The models that we shall discuss then concern the growth and decline of populations of individuals of different types, even though in reality each individual underneath this description is in fact changing continuously with age and experience. In other words, our models are abstractions that consider the interaction and growth and decline of populations of individuals that are considered fixed, even though in reality they are not. Of course, at a different scale, individuals may be described by populations of cells of different kinds, and each individual may undergo an evolution as the numbers of different types of cells change over time.

This introduces an important issue – that of multiple levels of description which encompass both spatial and temporal dimensions. In evolutionary systems, the internal nature, including behavioral response times, of the interacting individual entities changes over time, as does the configuration of the interactions between these types, leading to a changing overall system performance within its environment. This presents us with a view in which individuals are bundles of their internal components, the local community or organizations they form are bundles of these individual types, and ecosystems and larger structures they form are bundles of these local communities. The essential feature is that of the co-evolution of successive layers of interacting elements both horizontally and between levels. The diversity of the different levels of structure arises through these co-evolutionary processes that are in turn driven by the generation of micro-diversity – diversity at the level below. To illustrate this, let us consider the simplest possible example. Let us consider how a population evolves. It evolves if new behavior both invades a population and also grows to a significant level in the system.

2.1. Micro-Diversity – the Basis of Evolutionary Drive

The dialogue between population dynamics – the simple reduced model of an ecosystem – and mutations or innovations, is particularly interesting in that it gives rise to what is usually referred to as evolutionary ecology.

Consider, as an example, the simplest possible ecosystem, a single species growing according to the logistic equation:

$$\frac{dx}{dt} = bx(1 - \frac{x}{N}) - mx \tag{1}$$

This equation, describing the growth of a species x in a system with limited resources has a stable, stationary state, $x_0 = N(1 - m/b)$. This supposes a hierarchy of time scales concerning the parameters such as b and m related to the nature of "x", and N that reflects the environment. Both the nature of x and the environment are considered fixed.

If evolution is about the changing nature of "x", which might or might not be in the context of a changing environment, then we need to ask what new behavior could invade the stationary state of equation (1). Let us consider the effect of an individual x that explores the pay-off to be had for a new behavior – that of a "mutant", x'. The new behavior competes with x to an extent β for the limiting resource N ($0 < \beta < 1$). The mutant is characterized by some other birth rate b' and death rate m', and the system equations become:

$$\frac{dx}{dt} = bx(1 - \frac{(x + \beta x')}{N}) - mx \tag{2}$$

$$\frac{dx'}{dt} = b'x'(1 - \frac{(x' + \beta x)}{N}) - m'x' \tag{3}$$

Now, we ask whether or not the pay-off for x' is positive. If so, it will invade the system. This question is decided by testing the stability of the pre-existing state, $x' = N(1 - m/b)$; $x' = 0$. If it is stable, then x' cannot invade the system. If it is unstable, invasion can proceed.

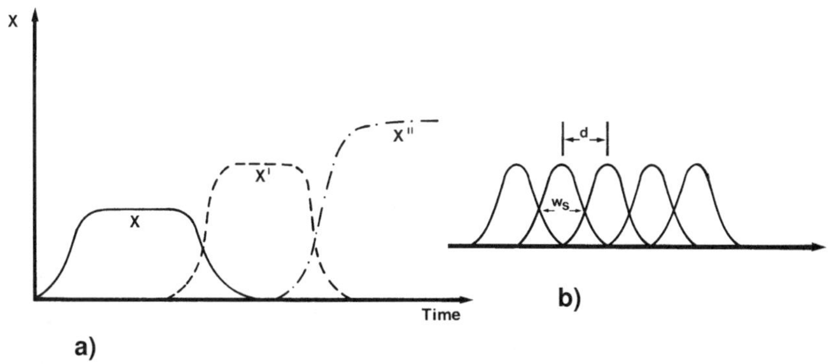

Fig. 1. a) Within the same "niche", we find successive replacement by more "effective" populations. b) A resource spectrum will be filled by different populations.

A simple stability analysis shows that the condition for x' to invade is

$$N'(1 - m'/b') > \beta(N(1 - m/b) \tag{4}$$

When this condition is fulfilled, x' will grow. Thus, if we supposed that different mutants are produced by some microscopic mechanisms, with randomly scattered values of b', m' and N', then only those mutants whose parameters satisfied condition (4) can "invade" and grow in the system. It would reject all the others.

Hence, as a result of random mutations, evolution within a given "niche" ($\beta = 1$) will lead to increased "exploitation", or increasingly efficient use of the resources. An important additional point is that the condition that allows x' to grow in a given niche also ensures that the preceding type, x, must decrease and disappear, as portrayed in Figure 1a). In other words, within a given niche, the condition (1) ensures successive replacement and "improvement" of the population.

When overlap is not total, $\beta < 1$, invasion is easier, since the value of $N'(1 - m'/b')$ need not be as high. What we shall observe, therefore, in a system with limited resources is that over a long period of time an initially empty resource spectrum will gradually be filled by different populations as in Figure 1b), each adapted to a certain range of resources. Also, within any particular range or type of resource, the efficiency of exploitation will increase irreversibly. This result can be extended to situations where genetics are explicitly considered, but these slightly more complicated equations do not lead to a different qualitative result.

This approach can be generalized as shown by Allen[10] and was applied to different ecological systems[11]. The method can allow us to calculate the total population diversity that will evolve on a given resource base. The temporal implications are naturally that if the resource base is stable over long times, then the populations will evolve over time and develop an ecology of behaviors that will exploit it. For example, the population diversity of Darwin's finches that occupy different sized islands in the Galapagos provides an example. These islands are home to some 14 species of finch, which are generally not found elsewhere and different sized groups or "bundles" of finch fill different sized islands (Lack, 1947)[12].

The slowly cumulative effects of fluctuations in the reproduction of the finches will lead to "predictions" about what diversity of behaviors will later be observed. We can link the degree of resource volume, environmental fluctuations (May[13]), and morphological diversity and "predict" the volume of niches available to different populations, assuming the structural stability of the overall ecology. In other words, evolution leads to a given amount of "coherent" diversity. As an example, Table 1 shows the actual total variance in beak measurements (in mms) for the particular species that inhabit that island. But the motor that drives this evolutionary filling of niches is actually that of micro-diversity generation – diversity produced at the level below that of the ecosystem.

Table 1. The relationship found between beak diversity and resource volume for the transitional vegetation found on all the islands (Bowman, 1961)[14].

Island	**Species**	**Resources**	**Total beak Variance (mms)**
Indefatigable	Magnirostris	Large Seeds	21.6
Albemarle	Fortis	Medium Seeds	20.6
Charles	Fuliginosa	Small Seeds	18.9
James	Scandens	Cactus	18
Chatham	Fortis	Large Seeds	17.6
	Fuliginosa	Medium and Small Seeds	
	Scandens	Cactus	
Tower	Magnirostris	Large Seeds	13
	Difficilis	Small Seeds	
	Conirostris	Cactus	
Narborough	Magnirostris	Large Seeds	11.4
	Fortis	Medium Seeds	
	Fuliginosa	Small Seeds	
Hood	Conirostris	Large Seeds and Cactus	10.2
	Fuliginosa	Small Seeds	
Culpepper	Conirostris	Large Seeds	9.6
	Difficilis	Small Seeds and Cactus	
Wenman	Magnirostris	Large Seeds	7.16
	Difficilis	Small Seeds and Cactus	

2.2. *Evolution as the Result of Micro-Diversity*

A study of the temporal hierarchy of fluctuations, reproduction, stable population types and resource patterns allows us to predict the emergent ecological structures.

Let us now consider the workings of this "micro-diversity" and see how it both drives evolution – and is also selected for by evolution.

$$\frac{dx_i}{dt} = \{(b_i * x_i * f + .5 * b_{i+1} x_{i+1} (1-f) + .5 * b_{i-1} x_{i-1} (1-f)\} * (1 - \frac{\sum_j x_j}{N}) - m * x_i \quad (5)$$

This equation expresses the outward diffusion of populations from any particular type, and their amplification or suppression according to their *b*/*m* parameters through selection. Let us consider the logistic equation and a "population character space" of different possible behaviours for the population *x*. If we consider that initially we have a single "pure" population type, then it will sit on one particular behaviour, *i*. However, if there are any "errors" or "explorations" made in the reproduction of this behaviour, then over time this will correspond to diffusion outwards in that space – to the generation of micro-diversity. So:

$(1-f)xi$ flow from behaviour i to $i+1$ and $i-1$
$.5*(1-f)\nu xi+1$ flow from $i+1$ into i
$.5*(1-f)\nu xi-1$ flow from $i-1$ into i.

ν represents the fraction of explorations that are viable. The first term corresponds to "normal" reproduction, the second and third to the production of micro-diversity. The assumption is made, however, that the diversity merely results in a distribution of birth and death rates.

This experiment tells us that the population increases in fitness – climbs the hill – because of random, unintentional processes of "exploration" in character space. Ignorance and error making are very robust sources of such exploration. Clearly random changes in the design of any complicated entity will mean that most experiments are simply non-viable, and only half (randomly) of those that remain are "better". This effectively tells us that there is an "opportunity cost" to behavioral exploration. This is illustrated with the "before" and "after" pictures of Figure 2 with a species climbing the fitness landscape by simply making random "errors" in its reproduction.

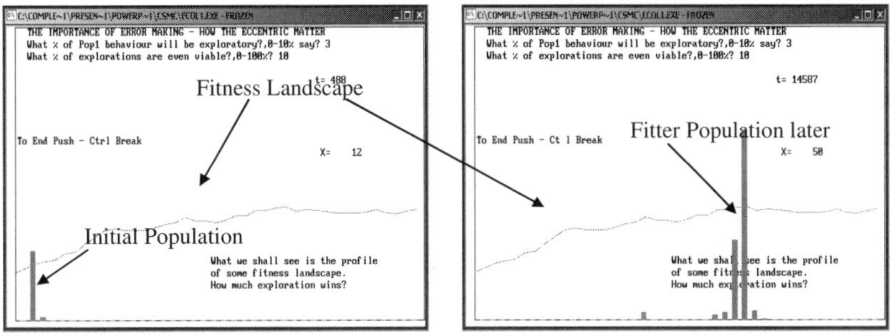

Fig. 2. Micro-diversity production will lead to many non-viable types, but still the population will evolve greater fitness and climb the fitness landscape.

In a second experiment, we can examine "how much" diversity creation (error-making) wins in the discovery of better performance. In this experiment, we launch two populations simultaneously at the foot of the fitness hill, and see how successfully they climb it. Here population 1 is assumed to have a 5% exploration rate in character space, while population 2 has 10%. However, we also make the assumption that of these "random" experiments, only 2% are actually viable, which means that there is a considerable "opportunity" cost in exploring rather than making perfect copies. Initially, population 2 wins, because, despite its cost in non-viable individuals, diffusing faster is rewarded by the fitness slope but later, when the hill is climbed, faster diffusion in no longer rewarded and population 1 dominates. This sequence of events is shown in Figure 3.

Our model shows that when we are in a new domain, and there is much to learn, then high rates of exploration pay off. However, when we are in a mature system that has already been thoroughly explored, there is no point wasting effort on further exploration. This is something like an r to K switch in ecology, but our model is generating it endogenously. Of course, we can only know that there are opportunities or not by actually engaging in exploration, but clearly, unless there is some structural change, the value of exploration falls with sector maturity, and this will lead exploration behavior to switch to exploitation.

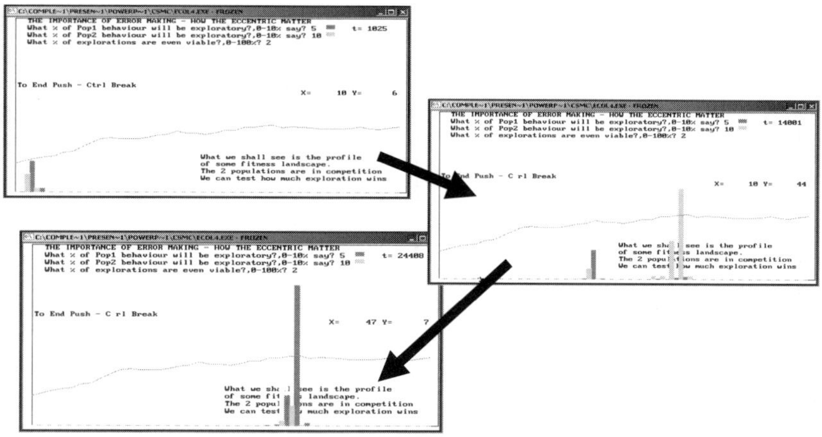

Fig. 3. If we have two populations that have different rates of "exploration" then we find that the relative success changes from early exploration to late exploitation.

2.3. *Probabilistic Evolutionary Drive*

In order to describe the stochastic dynamics taking place within our sub-volume, V, we need to consider the basic processes that must be included in our description that increase or decrease the numbers of any particular population present. Also, we must take into account that "potentially" there are many different types of each population that could, in theory, exist or come into existence. Which ones actually exist in the system will depend precisely on the evolutionary path the system follows.

The processes which are present in the system and which cause the change of the probability distribution over time are of three basic types: "increase", "decrease" and "diffusion" across the boundaries of the small volume V. In our study, we shall also distinguish between "replication" of an existing type and the "creation" to a new type of individual element. We can write down (Allen and Ebeling[15]) a linearized birth and death equation governing the probability distribution of some new individual or design prototype that considers the probabilistic dynamics of each prototype "population" (which may be only a single initial example) in the context of the pre-existing situation. The rate of production of prototypes depends on their "expected performance", and on the available investment and size of the company. This equation captures all the processes, A, that would tend to increase the new type, $x2$, linearized in $x2$, and evaluated in existing (or expected) circumstances. We also have the term of "decline" which contains all the terms, D, tending to reject the new type $x2$. This equation can be solved for all times t, and we can calculate the probability that $x2$ will in fact not take off.

Thus in general, if the growth terms are less than those of decline, then the probability of extinction is 1, while for $A > D$, the probability of extinction is still, D/A. This means that a new entity which "really" has a 10% better "fitness" ($A/D = 1.1$) than the previous type, nevertheless still has only a 1 in 10 chance of surviving.

We can write this as:

$$A / D = 1 + \delta \tag{6}$$

where δ measures the increase or decrease in efficiency of the new design, then the probability of survival for a time t is:

$$t = A^{-1}.n \tag{7}$$

If now we evaluate this expression for a time t, which is the inverse of the average rate of production of n prototypes,

$$P_{\text{sur}}(t) = 1 - P_{\text{ext}}(t) = \frac{\delta}{1 + \delta - e^{-(A-D)t}} \tag{8}$$

The probability of a mutant population, initially of size 1, which is δ% better or worse (δ can be negative) surviving for n prototype generation times is:

$$P_{sur}(n) = \frac{\delta}{1 + \delta - e^{-n\delta/(1+\delta)}} \tag{9}$$

This quite general form is true for any "novelty" that is created with the possibility of taking a place in the macroscopic ecosystem or economy and is shown as a function of δ in Figure 4.

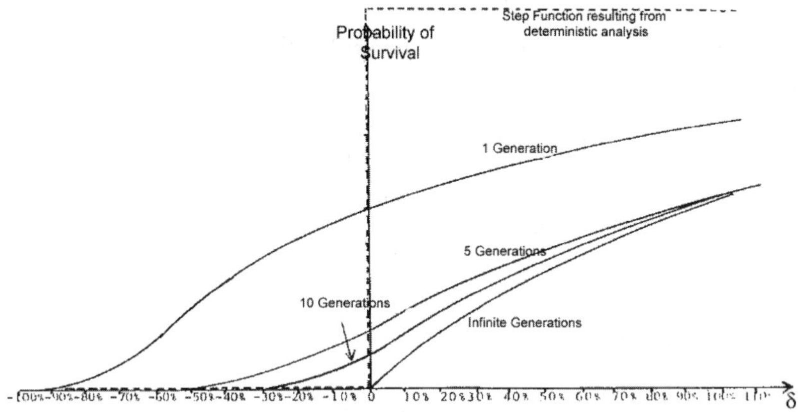

Fig. 4. The probability of survival for a novelty of relative fitness δ.

The curves shown in Figure 4 confirm the conclusion that the very sharp distinction that appears in the deterministic analysis between favorable and unfavorable designs, is smoothed out in this more correct stochastic picture.

Our stochastic analysis thus enables us to explore how evolution really occurs when we take into account the difficulty of passing through the initial sages of any "take-off". A period of stochastic dynamics must necessarily precede the establishment of a new population either in an ecosystem or in an economy. Only if it reaches sufficient numbers can it be described by deterministic equations such as in section 2.2 above If the probability of the appearance of a particular design of given effectiveness, δ compared to the "normal", is given by a Gaussian distribution, then we can see how the "fitness" of the new product affects the probability that it will in fact survive to find a place in the system.

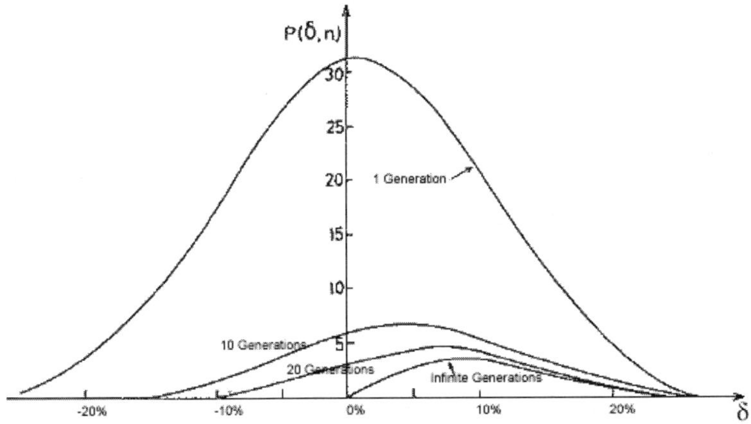

Fig. 5. If novelties are essentially of random relative fitness, then this shows the dynamic probability function for the evolutionary process.

The key point about all this is that it shows us that evolution is not a "knife-edge" process. Our calculations show that even quite inferior performing designs can survive for quite a time, and even very high performance ones still require some luck to make it through to success.

In other words, evolution is messy – not clean cut, and the contrast is not about new forms either being proved to be lethal or successful, but about a muddling through type of reality in which much that is mediocre can survive for quite a time, and much that is good falls by the wayside.

3. Complexity and Temporal Emergence

In the section above, we showed that although the diversity creating mechanisms in the replication of entities leads to the successful discovery of ecological or economic niches, the more detailed study of the probabilistic effects involved in the initial stages of any new entity lead to a much more mitigated view of the "effectiveness" of evolution in selecting new designs in or out. However, this does not throw any doubt on the power of evolution as a creative mechanism, because the messiness of the process enables an exploration of possible designs that may exist only on "the other side" of valleys of low fitness. Not only that, but the fact that non-linear systems for structures with emergent properties means that the selection mechanisms are often operating mainly on these collective, systemic capabilities, and that they cannot detect internal differences in an organization, providing these are not too affected.

In other words, the emergence of collective modes of behavior in complex systems makes exploration and diversity creation inevitable, because the forces of selection cannot act directly on the interior of the system, and therefore errors, exploratory behaviors and deviations will necessarily occur. In order to understand this idea, we can think of the fact that as individuals we can think of all sorts of ideas, and providing that we do not speak them, or put them into operation, then we cannot know how good they really are. Indeed, the whole idea of setting up a studio or a lab to experiment embodies this idea precisely, since it allows explorations and trials to be made, for ideas that may prove to be initially unpromising, but can be developed further to great success.

Again it is the emergence of a hierarchy of structural and temporal levels which is both the result and the motor of evolution. If "selection" could reach right in immediately and judge the fitness of some new microstate, then nothing other than continuous improvement could occur. But because each level of structure provides a partial shield within which exploration can and will occur, then valleys of unfitness can be crossed and entirely new organisms, markets, and concepts created.

The presence of firms with different levels of exploration and exploitation (error-making and accuracy) will automatically lead to evolution selecting whichever is most appropriate. So, evolution will be driven by the amount of diversity generation to which it leads. Evolution selects for an appropriate capacity to evolve, and this will be governed by the balance between the costs of experimental "failures" (the non-viable individuals created) and the improved performance capabilities discovered by the exploration. This is what holds "total diversity generation" in check.

Having examined the simple examples of section 2, we find a clear outcome. Evolutionary Drive tells us that evolution is driven by the noise to which it leads. Rapid fluctuation will drive longer term change. Not only that, the slow process of selection will automatically select for individuals and groups that exhibit rapid fluctuations. This process will select for evolvability as well as for particular types of micro-diversity at a given time. We may remark in passing that this means that there will never be a completely clear understanding of any evolving system at a given time, because it will always contain elements that may or may not turn out to be successful.

We understand situations by making creative, but simplifying assumptions[16]. We define the domain in question (the boundary) and then making some assumptions about the temporal stability of the entities that we observe, establish some rules of classification (a dictionary) that defines the "content" of the system at a given time. But over time, these elements may change, some new ones emerging, some old ones disappearing and others changing internally. This "evolutionary tree" is an abstraction concerning types of thing rather than things themselves. In order to make practical decisions at a given time, however, we then consider the current system and how is it operating now. This is "operational" not strategic and makes temporal assumptions of structural stability, taking us away from open, evolutionary change to the predictions of running a fixed set of processes. If the events considered are discrete, then the running is according to a probabilistic dynamics, and we have what is called stochastic non-linear dynamics, where different regimes of operation are possible, but the underlying elements never change nor learn, nor tire of their behaviors. If we assume that we can use average rates instead of probabilities for the events, then we arrive at deterministic, system dynamics. This is in general, non-linear dynamics, and may be cycles or chaos or at equilibrium, but what happens is certain, simple and easy to understand.

A succession of models, of levels of understanding, arise from making successive, simplifying assumptions, and therefore models on the right are increasingly easy to understand and picture, but increasingly far from reality. Structure and organization are seen as temporary emergent properties of change, as suggested by Tsoukas and Chia[17]. The operation of a mechanical system on the right hand side of Figure 6 may be easy to understand but that simplicity has assumed away the more complex sources of its ability to adapt and change.

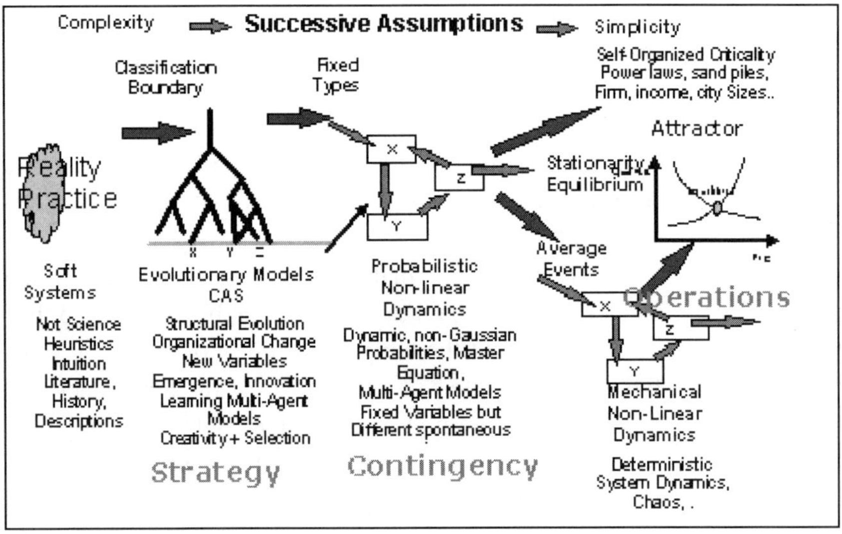

Fig. 6. The "deconstruction" of our understanding of complex, evolving reality.

A mechanical model is more like a "description" of the system at a particular moment, but does not contain the magic ingredient of micro-diversity that constitutes Evolutionary Drive. The capacity to evolve is generated by the behaviors that are averaged (and hence removed) by Assumptions 3 and 4 – that of average types and average events. Organizations or individuals that can adapt and transform themselves do so as a result of the spontaneous generation of micro-diversity within their local contexts, though without knowing which novelties will prove successful or how the larger system will be affected. This captures for us the meaning of Prigogine's phrase[2] concerning complexity – "From Being to Becoming" between a reality that is dead, a fixed mechanical representation, and the unpredictable, living reality of our experience, a reality that is continually "becoming".

4. Complexity in Human Systems

Let us briefly mention some of the applications that have been made of these ideas in understanding human systems and behavior.

4.1. A Fisheries Example

A detailed model was developed of Canadian Atlantic fisheries (Allen and McGlade[18], Allen[19,20]). This consists of a spatial model capturing the movements of fishing fleets and the resulting dynamics of fish. The fishermen's movements are informed by their "mental maps" of potential profit, and this is based on the information they have from other fishermen. Clearly, members of their own fleet will exchange information, but this may not be true between "competing" fleets. The model therefore describes fishing as a result of the acquisition and exploitation of knowledge about where profits are being made by other fishermen. This means that there is a spatial positive feedback mechanism that structures fishing patterns. Of course, in deciding which zone to go to, fishermen take into account the distances involved to go there, and to return to port and the cost of fuel.

In addition to these effects, however, our equations explore the degree of "synchronization" that the boats within a fleet will display. A parameter R is used to represent the degree of "rationality" the skippers within a fleet will display in responding to the information about the apparent location of fish stocks. This can vary from essentially random behavior, small R, to very concerted, rational response to information, with high R. The former we have called "stochasts" and the latter "Cartesians".

Running these fishery simulation models demonstrates the importance of a temporal hierarchy. "Cartesian" fleets seeking profit with short term economic rationality cannot fish successfully because they focus on exploitation at the expense of exploration. Profit is actually generated in the longer term by behavior that does not seek profit maximally but which allows exploration and learning! This is because in order to fish for any length of time, it will be necessary not only to exploit current information about fish stocks, but to generate new, currently unknown fish stocks, and exploit them. So exploitation alone is inadequate and some mix of exploration and exploitation is required. This tells us that the ability to learn is more important over time than the ability to exploit efficiently. This model also shows us that there is no such thing as an Optimal Strategy. As soon as any particular strategy becomes dominant in the system, then it will always be vulnerable to the invasion of some other strategy.

4.2. Emergent Market Structure

Evolutionary models of a market system can explore the realistic growth and decline of firms attempting to produce and sell goods. These competing products will be seen by potential customers according to their particular desires and needs, but in the simple case described here, customers will be differentiated simply by their revenue, and therefore their differing sensitivities to price. Again there is an important assumption about a hierarchy of temporal scales. It is assumed that the preferences and needs of customers are long lasting, and that the products once produced and sold also last for some time characteristic of the sector under study. However, customers' decisions to buy occur rapidly and so the stock and the number of product users change dynamically over time. Firms last for as long as they can keep their bank accounts above a critical level, and customers come back into the market after some time, when the product they bought has worn out.

The structure of the model is shown in Figure 6. Inputs and labor are necessary for production and the cost of these added to the fixed and start-up costs. Each produces goods that are sold by sales staff who must "interact" with potential customers in order to turn them into actual customers. The potential market for a product is related to its qualities and price, and although here we have assumed that customers all like the same qualities, they have a different response to the price charged. The price charged is made up of the cost of production (variable cost), to which is added a mark-up. The mark-up needs to be such that it will turn out to cover the fixed and start-up costs as well as the sales staff wages. Depending on the quality and price, therefore, there are different sized potential markets coming from the different customer segments.

The model calculates the relative attractivity of a particular product (of given quality and price) for a customer of a given type (poor, medium or rich). This calculates the "potential market" for each firm at each moment, and their sales staff must interact with these potential customers if they are to actually make "sales" and bring in revenue.

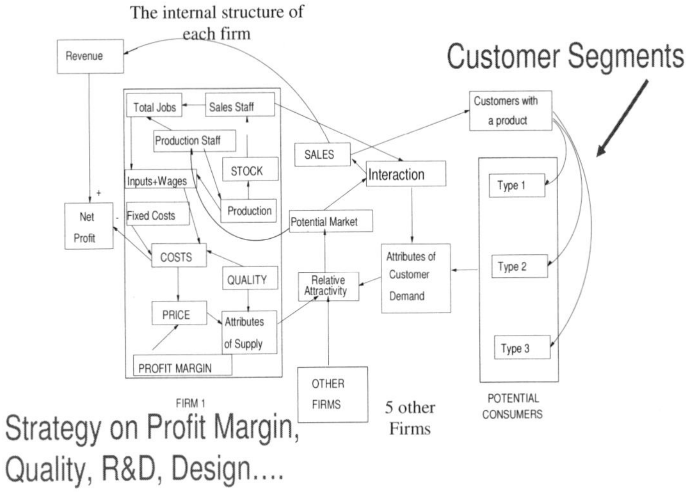

Fig. 7. A model of 6 interacting firms/agents whose choice of mark-up and quality represent their strategies.

When a sale is made, then the potential customer becomes a customer and disappears from the market for a time that depends on the product lifetime. The revenue from the sales of a firm is used to pay the fixed and variable costs of production, and any profit can be used either to increase production or to decrease the bank debt if there is any. In this way, the firm tries to finance its growth and to avoid going near its credit limit. If a firm exceeds its credit limit, then it is declared bankrupt and is closed down.

In traditional economic theories, firms are supposed to act, or to have acted, in such a way as to obtain maximum profit. But, here, our model shows us that the market is the place where firms will learn how much profit their current product and price strategy will bring in – given the current strategies of their competitors. Because every new market starts with an investment in new, unproven products, then initially firms will experience negative profits. Only by exploring the return on a particular behavior can a firm know what it is – and this will be a function of time and cannot be judged instantly. Only by taking part in the market can a firm discover how much market can be captured, and how fast this can happen. A lower price mark-up will lead to less profit per sale, but perhaps to more sales, while a high price mark-up may lead to high profits per sale, but to fewer sales. The market is the arena in which this knowledge can be encountered, and the responses of firms, and the mutual impacts on one another's sales and profits can be played out[21,22,23].

Whatever firms expect, in reality, clearly some of them get it wrong, because a large fraction go bankrupt. In our model, therefore, we simply have assumed that managers believe in the strategy that they have, and that if it doesn't work, they either change it according to their adaptive rules, or they go bankrupt. Our model shows that it is the economies and diseconomies of production and distribution that will determine the number, size and scale of the niches that may be discovered. Our models offer us a similar capacity to predict the different niches that may exist as was described in the initial section on Darwin's finches in the Galapagos. Our model can be used to explore the effect of different learning strategies of firms. The strategy space in view here is that of what % profit to charge and what quality of product to make. We shall consider 4 different learning strategies for firms.

(i) Darwinian Learning: In this case, we launch firms with random strategies, and if a firm goes bankrupt, we replace it with a new firm, with a random strategy. In this way, in theory, the performance of the market will improve as unsuccessful strategies are eliminated, and only successful ones remain after a long time.
(ii) All Imitate: Here, firms are launched initially with random strategies, but firms adopt the strategy that is winning. In this way, in theory, the resulting market should evolve to a collection of firms all using a very successful strategy.
(iii) All Learn: In this case, after being launched with random strategies, firms each explore the effects on profits of changing quality and mark up. They then move up the profit slope – if possible. In this way, they demonstrate the effect of a learning strategy.
(iv) Mixed Strategies: Here, after a random launch, we have two Darwinists, two imitators and two learners. This leads to an

evolution of the market structure gradually exploring and changing as profit and loss select for the winners and losers.

The results of these simulations are shown in Figure 8.

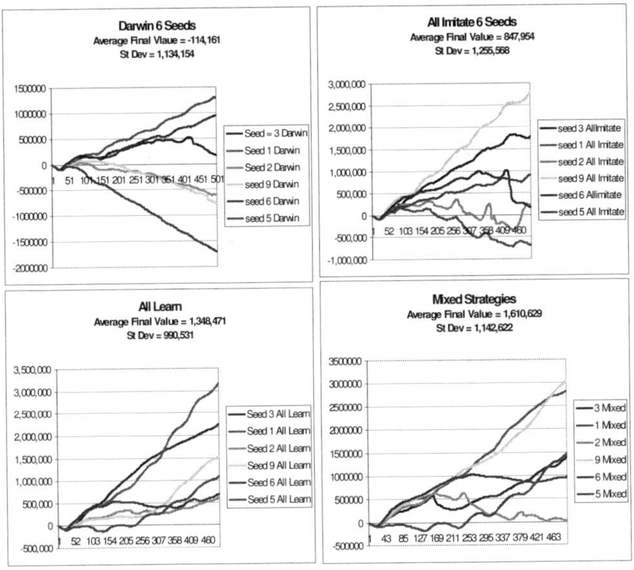

Fig. 8. Simulation results.

i) The model shows us that in learning method (i), average market value achieved by the process, including the cost of bankruptcies, is actually negative. The different cost lines correspond to different random sequences expressing the possible effects that luck can have on the outcome. Darwinism, applied to market evolution, using bankruptcy as the selection process, is so costly on average that the industry as a whole makes a loss.

ii) This does perform better than Darwinian learning on average, with an average final value of over 800,000 compared to -114,000. But the strategy seems to provide the most unstable trajectories, with some examples of market crashes and severe setbacks in the general pattern of improvement.

iii) When firms are prepared to learn from experience by probing their "profit" landscape and moving in the direction of increase, then we have the best results for the market.

iv) When there are 2 Darwinian forms, two imitators and two learners, the results are also good, though the learners tend to beat the others.

Our model tells us how learning within the individual firms can lead to the emergence of a stable market structure. So, processes of change and adaptation underpin the appearance of stability. The model shows once again that the basic process of "micro-variation" and differential amplification of the emergent behaviors is the most successful process in generating a successful market structure, and is good both for the individual players and for the whole market, as well as its customers (Metcalfe[24], Foster and Metcalfe[25]).

4.3. Evolution of Manufacturing Organisations

The study of organizational change and strategy can be looked at by reflecting on the organisations in terms of their constituent practices and techniques. The changing patterns of practices and routines that are observed in the evolution of firms and organisations can be studied using the ideas of Evolutionary Drive. We would see a "cladistic diagram" (a diagram showing evolutionary history) showing the history of successive new practices and innovative ideas in an economic sector. It would generate an evolutionary history of both artifacts and the organizational forms that underlie their production (McKelvey[26,27], McCarthy[28], McCarthy, Leseure, Ridgeway and Fieller[29]). Let us consider manufacturing organisations in the automobile sector.

With these characteristics (see Figure 9) as our "dictionary", we can also identify 16 distinct organizational forms: 1) Ancient craft system, 2) Standardised craft system, 3) Modern craft system, 4) Neocraft system, 5) Flexible manufacturing, 6) Toyota production, 7) Lean producers, 8) Agile producers, 9) Just in time, 10) Intensive mass producers, 11) European mass producers, 12) Modern mass producers, 13) Pseudo lean producers, 14) Fordist mass producers, 15) Large scale producers, 16) Skilled large scale producers.

Standardisation of Parts	1	TQM sourcing	27
Assembly Time Standards	2	100% inspection sampling	28
Assembly line layout	3	U-Shape layout	29
Reduction of Craft Skills	4	Preventive Maintenance	30
Automation (Machine paced shops)	5	Individual error correction	31
Pull Production System	6	Sequential dependency of workers	32
Reduction of Lot size	7	Line balancing	33
Pull procurement planning	8	Team Policy	34
Operator based machine maintenance	9	Toyota verification of assembly line	35
Quality circles	10	Groups vs. teams	36
Emloyee innovation prizes	11	Job enrichment	37
job rotation	12	Manufacturing cells	38
large volume production	13	Concurrent engineering	39
mass sub-contracting by sub-bidding	14	ABC Costing	40
exchange of workers with suppliers	15	Excess capacity	41
Training through socialisation	16	Flexible automation of product versions	42
Proactive training programmes	17	Agile automation for different products	43
Product range reduction	18	In-Sourceing	44
Automation (Machine paced shops)	19	Immigrant workforce	45
Multiple sub-contracting	20	Dedicated automation	46
Quality Systems	21	Division of Labour	47
Quality Philosophy	22	Employees are system tools	48
Open Book Policy with Suppliers	23	employees are system developers	49
Flexible Multifunctional workforce	24	product focus	50
set-up time reduction	25	Parallel processing	51
Kaizen change management	26	Dependence on written rules	52
		Further intensification of labour	53

Fig. 9. 53 Characteristics of manufacturing organizations.

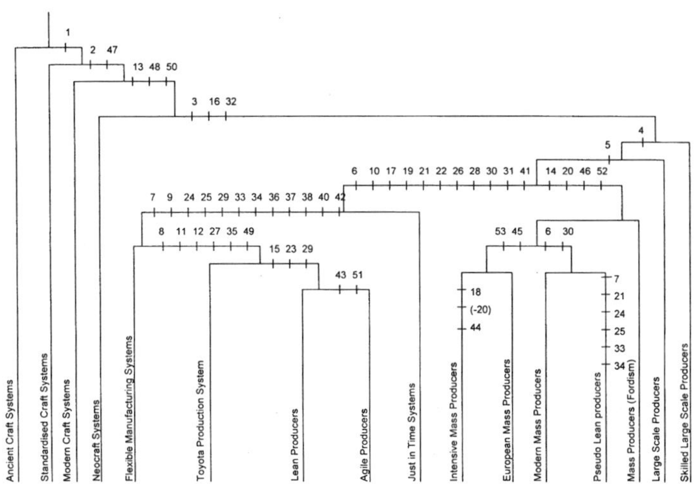

Fig. 10. The cladistic diagram for automobile manufacturing organizational forms.

Cladistic theory calculates backwards the most probable evolutionary sequence of events (Figure 10). Again, in agreement with the ideas of Evolutionary Drive, we shall look at this as being the result of micro-explorations, and then a differential amplification of systems with emergent capabilities. We have studied the evolution of the automobile production industry by conducting a survey of manufacturers, and obtaining their estimates of the pair-wise interactions between each pair of practices. In this approach, the microscopic explorations consist in the attempts to connect in new practices to an existing system, with the object of improving performance and creating positive emergent capabilities. As has been reported before, we can understand and make retrospective sense of the evolution of the automobile industry.

We have then been able to develop an evolutionary simulation model, in which a manufacturing firm attempts to incorporate successive new practices at some characteristic rate. The "receptivity" of the existing complex determines which new practice will in fact be amplified or suppressed if it tries to "invade".

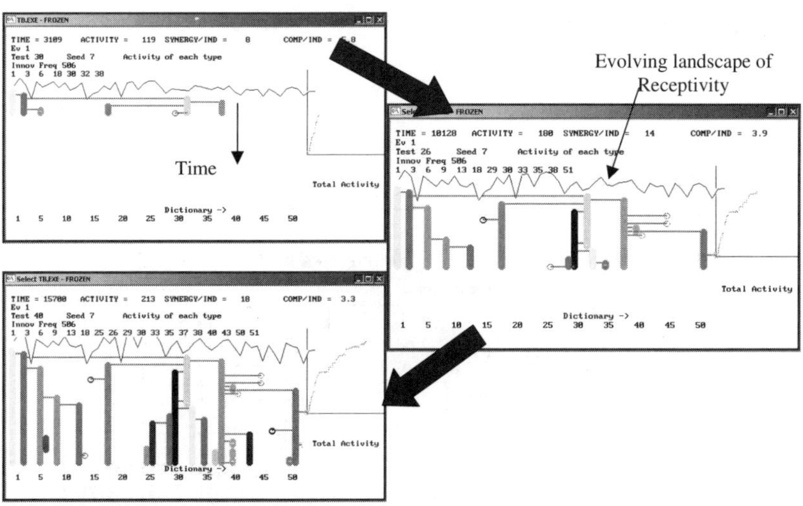

Fig. 11. Successive moments (t = 3000, 10000 and 15000) in the evolution of a particular firm. The evolutionary tree of the organization emerges over time.

Figure 11 shows us one possible history of a firm over the entire period of the development of automobile production. The particular choices of practices introduced and their timing allows us to assess how their performance evolved over time, and also assess whether they would have been eliminated by other firms. As a result of the different firms experimenting over time, there is an incredible range of possible structures that can emerge, depending simply on the order in which practices are tried. But, each time a new practice is adopted within an organization, it changes the "invadability" or "receptivity" of the organization for any new innovations in the future. This illustrates the "path dependent evolution" that characterizes organizational change. Successful evolution is about the "discovery" or "creation" of highly synergetic structures of interacting practices.

The model starts off from a craft structure. New practices are launched with an "experimental" value of 5. Sometimes the behaviour declines and disappears, and sometimes it grows and becomes part of the "formal" structure that then conditions change which innovative behavior can invade next. Overall performance is a function of the synergy of the practices that are tried successfully. The particular emergent attributes and capabilities of the organization are a function of the particular combination of practices that constitute it. Different simulations lead to different structures, and there are a very large number of possible "histories". This demonstrates a key idea in complex systems thinking. The explorations/innovations that are tried out at a given time cannot be logically or rationally deduced because their overall effects cannot be known ahead of time. Therefore, the impossibility of prediction gives the system "choice".

The competition between different firms' exploratory pathways through time means that those who for one reason or another fail to find synergetic combinations of practice will be eliminated. Once again, we find the principle of Evolutionary Drive, where the micro-explorations involving the testing of new practices leads to microscopic diversity among firms, and in turn these are either amplified or suppressed by the economic competition.

4.4. Research Communities

Another example of these ideas is that the structural attractors such as the market configurations and organizational forms of the sections above can be seen in considering the research literature that expresses how communities of researchers create, and evolve successive different structural attractors. In a study of medical research papers concerning the treatment of a particular form of heart disease, the literature of successive years in a particular domain have been analyzed, and the keywords and contents have been analyzed into different groups of research approach (Ramlogan et al.[30]). From this, we can see how ideas lead to "clusters" of researchers and papers with particular set of dimensions or keywords, and how these evolve onwards to new structures over time. In this example, we see how an initial idea develops, and other different attractors emerge over time. In general, in science we would see the activity of some attractors waning, and others emerging and temporally taking over. This illustrates the concepts of structural attractors as resulting from the connection of different concepts from a possible dictionary of concepts that is also itself being extended by microscopic exploratory thoughts. Sometimes a new idea will be tried out, with a single "outlying" paper starting off. However, this may either be picked up and become a new cluster, or may die out as an unsuccessful exploration.

Figure 12 shows us that evolution is about the discovery and emergence of structural attractors that express the natural synergies and conflicts (the non-linearities) of underlying components. Their properties and consequences are difficult to anticipate and therefore require real explorations and experiments to be going on, based in turn on diversity of beliefs, views and experiences of freely acting individuals.

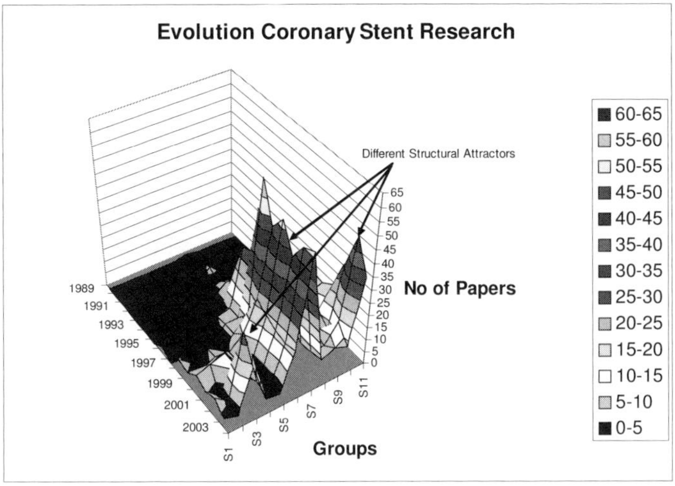

Fig. 12. The emergence of different research clusters over time as multi-dimensional structural attractors.

5. Conclusions

From the discussions and models presented above, we can derive some key points about evolution in biological and human systems and the importance of temporal hierarchies in these creative processes.

Evolutionary Drive is summed up by saying that evolution is driven by the noise to which it leads. And this is also saying that evolution and the creation of structures of seeming stability and large temporal spans is driven by the existence of internal temporal hierarchies of process. Indeed, evolution will select for systems with these internal hierarchies because they will confer adaptability on the system, and therefore provide it with an evolutionary capability which will be selected for.

Of course, this also means that aggregate descriptions corresponding to the seemingly stable structures will be useful shorthand characterizations of systems in the short term, but will fail to provide a real indication of the potential for instability and structural change that is really present. Succinct and understandable descriptions of organizations and structures and their current, mechanical functioning may blind people to the fact that these are temporary systems which may change spontaneously in reality.

Complexity must therefore lead us to permanently distrust our current knowledge and maintain a constant doubt as to its truth. We need to behave as evolution does, and make sure that mechanisms of exploration and experiment are present in the organization. Although they may not be profitable in the short term, they may be the only guarantee of survival into the longer term. In considering the evolution of knowledge in whatever domain over time, we see that knowledge involves articulated words, concepts and variables that inhabit certain dimensions and which together are deemed to provide some emergent capability. Over time, the cluster evolves as new connections are made involving new variables and dimensions. These either run in parallel, catering to different parts of the "market" or actually replace them as being the "improved theoretical view".

These temporary islands of stability of emergent clusters of ideas, components, practices or firms are what we have called Structural Attractors – dynamical systems that emerge and persist for a time. For products, it is bundled technologies; for markets, it is certain bundles of co-evolving firms; for organizations, it is bundles of co-evolving practices and techniques; for knowledge more generally, it is bundles of connected words, concepts and variables that emerge for a time.

In hard science, a new theory must be capable of being falsified, and therefore must produce testable predictions, which if not falsified lead to cumulative knowledge. In complex systems however, clean predictions are no longer possible and the knowledge evolves under less severe selection criteria. In human systems, emergent structural attractors can occur simply because there is demand for their particular emergent capabilities. Fashion, lifestyles, art, artifacts and communities of practice can emerge and survive, providing that there is a clientele for them. They are not about being "true or false" but simply about whether there is a "market" for them.

Living systems create a world of connected, co-evolved, multi-level structures which may be temporally self-consistent, but will evolve and change over time. New ideas (technologies, strategies, practices, and concepts) feed into this world of self-organized structures, and can either be amplified or suppressed. They may either replace a previous attractor, or find their own niche after some adaptation and change, or they may be eliminated.

Acknowledgement

This work was supported by the ESRC RES-000-23-0845 Aerospace project with Sheffield University.

References

1. G. Nicolis and I. Prigogine, *Self-organization in non-equilibrium systems*, WileyInterscience, New York, NY. (1977).
2. I. Prigogine, *From Being to Becoming,* W.H. Freeman, New York (1981).
3. I. Prigogine and I. Stengers, *Order out of Chaos*, Bantam Books, New York (1987).
4. H. Haken, *Synergetics*, Springer Verlag, Berlin (1977).
5. B.B. Mandelbrot, *The Fractal Geometry of Nature*. Freeman, New York (1982).
6. P.M. Allen and J.M. McGlade, *Evolutionary Drive: The Effect of Microscopic Diversity, Error Making & Noise,* Foundation of Physics, Vol. 17, No. 7, July, pp. 723-728 (1987).
7. J. Maynard-Smith, Game theory and the evolution of behaviour, *Proc. R. Soc.* London B. 205: pp. 475-488 (1979).
8. S.J. Gould and E.S. Vrba, Exaptation – a Missing Term in the Science of Form, *Paleobiology* 8(1): pp. 4-15 (1982).
9. S.J. Gould, *The Structure of Evolutionary Theory*. Cambridge, MA: Belknap (2002).
10. P.M. Allen, Evolution, Population Dynamics and Stability, *Proc. Nat Acad Sci*, USA, Vol..73, No. 3, pp. 665-668 (1976).
11. P.M. Allen, Ecology, Thermodynamics and Self-Organisation, in *Ecosystem Theory for Biological Oceanography.* Eds Ulanowicz and Platt, Canadian Bulletin of Fisheries and Aquatic Sciences, 213. pp. 3-26 (1984).
12. D. Lack, *Darwin's finches.* Cambridge University Press; 1953. Sci. Am. 188(4): pp. 66-71 (1947).
13. R.M. May, *Stability and Complexity in Model Ecosystems* (Prinecton University Press, Princeton (1973).
14. R. Bowman, *Morphological differentiation and adaptation in the Galapagos finches.* Charles Darwin Foundation for the Galapagos Islands - Contribution 1; University of California Publications in Zoology, Vol. 58 (1961).

15. P.M. Allen and W. Ebeling, Evolution and the Stochastic Description of Simple Ecosystems, *Biosystems*, 16, Elsevier Scientific Publications pp. 113-126.
16. P.M.Allen, What is the Science of Complexity? In *Emergence*, Volume 3, Issue 1, pp. 24-42 (2001).
17. H. Tsoukas and R. Chia, *On organizational becoming: Rethinking organizational change, Organization Science*. Linthicum: Sep/Oct Vol. 13, Iss. 5; p. 567, 18 pgs (2002).
18. P.M. Allen and J.M. McGlade, Modelling Complex Human Systems: A Fisheries Example, *European Journal of Operational Research* 30, pp. 147-167 (1987).
19. P.M. Allen, Evolving Complexity in Social Science, in *Systems- New Paradigms for the Human Sciences*, Eds Altman and Koch, Walter de Gruyter, Berlin, New York (1998).
20. P.M. Allen, Of Fish and Fishermen: Models of Complexity, in *Mathematics Today*, pp. 18-24, Vol. 36 No. 1 (2000).
21. P.M. Allen, *Knowledge, Ignorance and the Evolution of Complex Systems*, in Frontiers of Evolutionary Economics: Competition, Self-Organisation and Innovation Policy, Eds J. Foster and S. Metcalfe, Edward Elgar, Cheltenham (2001).
22. P.M. Allen, *A Complex Systems Approach to Learning, Adaptive Networks,* International Journal of Innovation Management, Vol. 5 No. 2 (June 2001) pp. 149-180 (2001).
23. P.M. Allen, M. Strathern and J.S. Baldwin, Complexity and the Limits of Learning in *Journal of Evolutionary Economics*. Vol. 17 pp. 401-431 (2007).
24. J.S. Metcalfe, J.S., 1998, "Evolutionary Economics and Creative Destruction", Routledge, London.
25. J. Foster and J.S. Metcalfe, *Modern evolutionary economic perspectives: an overview,* in Frontiers of Evolutionary Economics: Competition, Self-Organization and Innovation Policy, Eds J. Foster and S. Metcalfe, Edward Elgar, Cheltenham, UK.
26. B. McKelvey, *Organizational Systematics,* University of California Press CA (1982).
27. B. McKelvey, Evolution and Organizational Science in Baum J, Singh J, *Evolutionary Dynamics of Organizations*, Oxford University Press, pp. 314-326 (1994).
28. I. McCarthy, Manufacturing Classifications: lessons from organisational systematics and biological taxonomy, *Journal of Manufacturing and Technology Management- Integrated Manufacturing systems*, Vol. 6, No. 6, pp. 37-49 (1995).
29. I. McCarthy, M. Leseure, K. Ridgeway and N. Fieller, Building a Manufacturing Cladogram, *International Journal of Technology Management*, Vol. 13, No. 3 (1997).

30. R. Ramlogan, A. Mina, G. Tampubolon, and J.S. Metcalfe, Networks of knowledge: The distributed nature of medical innovation, Scientometrics, Vol. 70, No. 2. pp. 459-489, Feb (2007).

CHAPTER 11

ORDINATE LOGICS OF LIVING SYSTEMS

Jerry LR Chandler

Krasnow Institute for Advanced Studies, George Mason University, Fairfax Virginia, USA

Washington Evolutionary Systems Society, 837 Centrillion Drive, McLean Virginia, 22102

E-mail: Jerry_LR_Chandler@mac.com

The nesting of chemical concepts is used to categorize the dynamics of the emergence of life and reproductive developments within living systems. The nesting ground is the list of chemical elements. The nexus of the parts of atoms serve to ordinate the natural sciences in terms of the natural electrical progression. The order of the atomic numbers creates the natural order of the electrical progression of the perplex number system, a new form of diagrammatic logic (Chandler, in press).[1] The abstract nest of relations is constructed from ur-symbols by creating scientific grammars (synthetic symbol systems) to relate matter to time and space. The ordinate logics, that is, logics based on scientific ordering relations, are defined in terms of the nexus of rule-based ur-symbol systems. A hierarchical nesting scheme is used to visualize the emergence of biological dynamics from predicate, subordinate, ordinate and co-ordinate logics. Ordinate logic expresses the natural relations of chemical sciences as a diagrammatic logic of matter in space. Motion in space is coordinated by the real number system. The Binary Principle of Mutational Action, a source of composition of the nested dynamics of living systems, follows from the co-ordination of perplex and real number systems.

1. Introduction

A dictionary may list two definitions of a nest. A nest as a biological residence is a common meaning. Abstractly, a nest as a gradation of objects, each fitting within a parent, is a second definition. In mathematics, the nesting of parentheses is used in the sense of separating the terms of logical operations, such that an unambiguous order of arithmetic operation is implied. Metaphorically, appropriately used, the parentheses are peeled away by consecutive operations, much as the layers of an onion enclose one another.

The concept of nesting is somewhat akin to the notion of scale or material hierarchy.[2,3] A physical example is the gradation of sizes of a set of Russian dolls where every larger doll spatially encloses a smaller doll, except for the smallest doll of the set. A natural biological example of a three dimensional nesting is that of an onion. The anatomy of the human body can be described, not as a nesting, but rather as a material partitioning into spatial subsystems, organ systems, organs, tissues, cells, and sub-cellular parts. Physiology, in contrast to anatomy, is grounded in chemical dynamics. The nesting of biochemical dynamics in generating biological functional capacities, roughly speaking, rests on metabolism and distributed control processes that co-ordinate chemical changes with temporal and spatial changes.[4] For example, the disciplines of biochemistry, genetics, neurobiology, pharmacology and toxicology study relations between chemical structures and the integrated nesting of dynamical structures in living systems. Like the concrete examples of onions and Russian dolls, biochemical dynamics starts with a smallest initial object.

Logical analysis of biological dynamics reveals two deep conundrums. The conundrums are highly abstract and go to the heart of our notions of mathematics and logic. Yet these conundrums are well known and simple to state. The atomic numbers correspond exactly with the list of chemical elements and the list of integers, but, amazingly, arithmetic operations are forbidden on the atomic numbers! The metaphor of hierarchical nesting is frequently invoked in describing the dynamics of living systems, yet the a priori associations of numerical operations with geometries of anatomy are not possible! Consequently, the hierarchical organization of evolution as the emergence of order is based on the properties of matter that neither invoke nor provoke these deep scientific conundrums. The spatial organization of living systems as well as the biochemical, hormonal and neural cycles that generate functional capacities must originate from other sources of order.[5]

The natural electrical progression of the list of chemical elements was conceived of as a concrete signification of a new abstract number system, the perplex number system[1] (The polygonal number system and the pyramidal number systems of the Greeks are ancient examples of non-real number systems.[6]) The elements of mathematical objects of the relational algebra are termed "labeled bipartite graphs" and stand in one to one correspondence with the chemical elements. Logic operations on the electrical relations describe the bindings of particles to one another. (For example, a transpose operation changes a relation of a unit from an integer number to another.) The atomic numbers serve as the primordial reference for the perplex categories. Consequently, the logical possibilities for the nesting of chemical relations within dynamical systems are several, depending on the logical priority given to material, temporal and spatial relations within a nest. As both the real number system and the perplex number system use many common symbols, we start by defining the ur-symbols to ensure that each domain of discourse, each logical space, originates in a specified system of meaning. A preliminary communication of this paper has appeared.[7]

2. Ordinate Logics

Quantities, a common domain of scientific discourse, will be separated on the basis of order, that is, ordinate relationships. By the term ordinate, I mean to associate order with the concept of logic. By the term order, in this context, I mean to convey the regular and irregular association of logical terms. For a simple example, the popular "Sudoku" puzzles are solved by identifying the exact order for each of the 81 terms in the nine partitions of the square array. Finally, by the term ordinate logic, I mean to convey the association of terms so as to correspond with causal processes, the grammars of nature. The mental analysis required to find one sequence of paths that establish the solution of a Sudoku puzzle serves as a simple conceptual example of natural causality. The principal goal is to define ordinate logics such that the structures and functions of living systems can be exactly symbolized in terms of component macromolecules (DNA, RNA, proteins) and other material objects (see[8]). Aristotelian syllogisms and the recently defined meso-syllogisms, as reductive and synthetic associative term logics, respectively, are useful perspectives in implementing ordinate logics.

Three ordinate logics will be described here, ordinate, subordinate, and co-ordinate. All ordinate logics share a common property with order relations as a listing of symbols. Each ordinate logic is grounded by a collation of unique symbols. An ordinate logic is distinguished by the grammar, which specifies the rules of association of symbols to generate a message. A nexus of symbols is necessary for symbolic communications. In other words, the rules of association of symbols create the meanings of the logical messages communicated by the ordination. The form, the style or the iconicity of an individual symbol may be necessary for communication but it is not a necessity for an ordinate logic. Thus, I assume that all ordinate logics are imposed intentionally on a symbol system by applying the grammar such that symbols cluster into a pattern. Thus, a grammar serves to arrange the order of symbols or to change the existing nexus of symbols. Each ordinate logic differs from other ordinate logics by the reference symbols admitted into the synthetic system's repertoire, and by the repertoire of synthetic grammatical rules for forming clusters of symbols. Each ordinate logic differs from other ordinate logics by

(i) the reference symbols admitted into the synthetic system's repertoire, and
(ii) by the repertoire of synthetic grammatical rules for forming clusters of symbols and
(iii) by the nature of conjunctive and disjunctive operations used for the syndeton.

From these repertoires, a meaningful message can be created that can be interpreted as an order relation along a line, in a plane or in space. For example, mathematical "knots" emerge naturally as properties of ordinate logics of matter in space.[9]

The motivation for forming these new categories of logic arises from the tension among the numerical relationships among the natural sciences. For example, the chemical sciences use the atomic numbers as a source for calculations of the irregular valence of chemical elements. The mathematical operations of multiplication and division are not applicable to the atomic numbers. Universal terms are attributed to the properties of matter by the International System of Units (SI). The natural units of chemical elements associate with the discrete units and integers of electrical properties (valences). Particular qualities are attributed to each unique chemical entity, resulting in a unique graph grammar for naming chemical substances that is formalized by the International Union of Pure and Applied Chemistry (IUPAC).

The ordinate logics form a system of logics all originating in the same set of symbols. Meaning is assigned to the symbols based on the grammar of the individual ordinate logic in order to accommodate both the perplex and real number symbols. The conceptual distinctions among ordinate logics can be symbolized with the four icons shown in Figure 1. The circle is symbolic of the closure over the choice of symbols. The line symbolizes the causal processes defined by the grammar for pairing symbols to form a nexus. The position of the line in the circle distinguishes arbitrarily the different causal processes associated with the pairing of the symbols.

Fig. 1. Separation of ordinate logics on the basis of composition of distinctive marks into distinctive patterns of relations is signified by the individual objects. Each icon represents a different method of interpretation of ur-symbols.

3. Composing Ordinate Logics

Order can be imposed on quantities. The individuality of an ordinate logic originates in three choices: the choice of its terms, the choice of its relations and the choice of its operations. Correspondingly, any two different ordinate logics can be distinguished from one another by the differences of terms, the differences of relations or the differences of operations. Consequently, many ordinate logics are possible within this syntactical framework and, indeed, many are necessary to describe the calculations of living systems.

Ur-symbols are the common basis of all the ordinate logics. The prefix ur- is of Germanic origin and implies primordial or primitive origins. This term "ur-symbol" was introduced by Brian Rotman, in *Mathematics as Sign* in reference to the symbols for zero and one.[10] Here, the term ur-symbol is used in the sense of the complete set of artificial symbols used in human communication.

The composition of ordinate logics starts with the concept of a list of unique marks, or ur-symbols, as with a number system.[11] All symbols used by ordinate logics will be chosen exclusively from this list and are essential to the naming of the ordinate logic, the method of calculation. Consequently, an ordinate logic is grounded as a sub-list of the ur-symbols. Such a list of marks assigns a unique order to every member of the list. In this context, a mark initiates the listing and a different mark terminates the list. The list is closed.[12] All other marks are preceded by a unique antecedent mark and followed by a different mark. Thus, each member of a list is in a unique position and can convey a unique meaning within a symbol system. Such a sub-list of ur-symbols corresponds with the terms used to designate the natural integers. In practice, such a list consists of a few hundred or, perhaps in some cases, a few thousand symbols.

The well-defined list of marks serves to inchoate each ordinate logic within a specified context of ur-symbols. The same ur-symbol may appear in two or more ordinate logics but the meaning assigned to each symbol is determined by the particular ordinate grammar in which it appears. The encoding of meaning assigned to any symbol, any member of a particular ordinate logic, is arbitrary and such meanings may vary from one ordinate logic to another ordinate logic. For example, as noted above, the symbols for integers encode one set of meanings in the ordinate logic of matter and another set of operational meanings in real number system. In summary, the list of unique ur-symbols serves as the common basis for constructing synthetic symbol systems that use different codes to construct different logical messages using the same symbols.

A mark may be used repeatedly, such as with the concept of a "point" as marks of position on a line. If the marks themselves are indistinguishable, such as is the case with points, they are termed logical units. The position of the mark (the unit) in the list, can be used to distinguish one unit from another, even though the marks themselves are indistinguishable, as, for example, on the real number line.

If the marks themselves are to be distinguishable, a unique sign must be assigned to each of the listed marks. In this case, each mark of the list is distinguished by two qualities, the unit position in the list and the sort of the distinguishable symbols of the list. The sortal symbols will correspond exactly with the order of the natural language terms, the integers. Thus, two qualities of the list of marks, the units and the denotations of individual sorts (integers) are separate and distinct. Hence, logical operations can be assigned to either qualities of a mark. (Parenthetically, I note that these two qualities generate self-reflexive quantities that correspond exactly with the electrical parity quantities of atomic numbers.[1])

4. Ordinate Logics for the Natural Sciences

An ordinate logic, as a synthetic symbol system, composes a set of symbols, a listing of the symbols and a grammar (rules, axioms) into intentional messages, nexus (arrangements, relations) of symbols. The concepts of order, matter, time, space and causality will be treated as separate and distinct scientific terms, each term intended to communicate separate and distinct information.

First, a sketch is given of the intimate interrelationships between the systems of ordinate logics and scientific concepts that motivate this categorization. The order of conceptual development will be, first, ordinate logic of matter, second, the sub-ordinate logic of units and thirdly, co-ordinate logic as a combination of the logics of matter and units. The ordinate logic for calculations on matter (the atomic numbers) is the starting point for scientific realism. It is based on the electrical progression of the perplex number system. In the electrical progression, units and integers are separate and distinct sorts of objects, separate and distinct sources of relations. An alternative concept of units will become the basis of a sub-ordinate logic that serves to calculate size and / or distance as defined in the International System of Units.

The ordinate logic for matter is founded on the perplex number system, a diagrammatic logic abstracted from the electrical properties of matter. It is a term logic that respects the individual sorts of matter and their nomination. The system of diagrammatic logic is based on the use of a referential set of ur-symbols, each ur-symbol consisting of two qualities, units and integers. The nexus of symbols is in space. Pairing of symbols forms logical relations. A line marks each intentional pairing. The line implies a logical conjunction of the pair of adjacent terms. Unique species are formed by creating specific adjacencies of pairing relations among collations of units and integers. The formation of ions and molecules from atoms are described by the logical operations of transpositions and synductions. Each unique nexus is assigned a unique name to distinguish the specific sort of entity. This is the origin of scientific nominalism. The formal names of such diagrams are labeled bipartite graphs. The reader is referred to the website of the International Union of Pure and Applied Chemistry for the rules of this system of naming individual objects. Arithmetic operations are not defined in the ordinate logic of matter as the operations multiplication and exponentiation violate the conservation of matter principle. The reader is referred to my recent paper[1] for details of this ordinate logic.

The structures of other ordinate logics (sub-ordinate, co-ordinate and possibly others) can now be developed from the perplex number system by either removing qualities or adding qualities to it. The common source of other ordinations is the listing of ur-symbols. The semantic meaning (relational capacities) assigned to the ur-symbols may change to accommodate the logic relations associated with partitions and compositions of matter or with purely abstract reasoning of mathematics. The order associated with the perplex number system is preserved in the listing of symbols, even though the logical operations on the symbols differ from those of the perplex number system. Removing one of the qualities from the perplex logic reduces it to a sub-ordinate logic, either simple or complex. Adding a subordinate logic to the perplex logic generates co-ordinate logic. Higher order extensions follow.

A sub-ordinate logic can be formed from the set of homogeneous ur-symbols, points, arranged along a line or in a plane. The rules for logic operations are the usual arithmetic ones. Calculations are defined to be consistent with a conceptualization of distance. Such a system of reasoning may include one or several "variables". For the organization structures of sub-ordinate logics, see the diagrams in Mac Lane[11].

Co-ordinate logic can be composed from an ordinate and a sub-ordinate logic. As the two grammars differ, such a co-ordinate logic interrelates two systems of calculations. A co-ordinate logic is necessary to calculate the relations between chemical reactions and thermodynamic variables. Chemical structures are expressed in terms of labeled bipartite graphs of the ordinate logic. Thermodynamic variables are expressed in terms of sub-ordinate logic. The Duality Principle of Mutational Action can be used to associate objects defined in terms of ordinate logic with objects defined in terms of sub-ordinate logic.

Two equations, arbitrarily labeled C (P) and C (R) in the diagram below, are necessary for the expression of mutational action. The first equation, C(P) is written in symbols of the perplex number system. This equation expresses the empirical observation that the same number of perplex particles are on the left and right sides of the arrows representing chemical equilibrium, even though the binding relations among the particles differ. The second equation expresses the ratiocination of the equilibrium constant from the amounts of each chemical identity present at equilibrium. (These two equations necessitate a re-statement of the traditional "Law of Mass Action", a principal linkage between discrete chemical theory and continuous mathematics of thermodynamics.)

The following non-commutative diagram (Figure 2) illustrates the non-commutative mappings of the Binary Principle of Mutational Action.

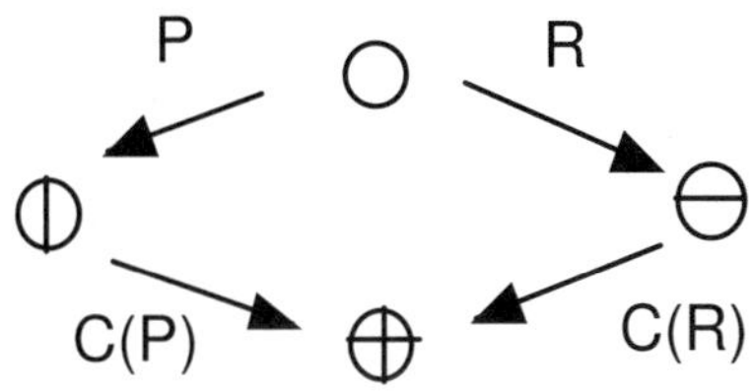

Fig. 2. The non-commutative diagram of the relations between ur-symbols and the Binary Principle of Mutational Action.

This example of co-ordinate logic is one of the many bridges between the material structures of life and the dynamical functions of living systems. The conjunction of measurements expressed in both subordinate and ordinate logics are necessary for the exact calculation of the equilibrium constant. Metabolic reactions normally refer only to the perplex logic of chemical structures. For example, the well-known triadic relations of DNA, RNA and protein do not express the subordinate relations implicit to the chemical equilibrium constants of co-ordinate logic. (This topic will be developed later.)

The relations between natural language and ordinate logics are of critical importance for exact scientific communication. Historically, the order of the ur-symbols in normal English sentences is arbitrarily defined in terms of the precedence. The nexus is a sequential arrangement of symbols into words, sentences (propositions), paragraphs, chapters, ... The resulting term logic is based on judgment of semantics rather than the syntax of ordered lists. Predicate logic, which emerged from Aristotelian syllogisms, is categorized as a non-subordinate logic. In certain instances, a weak ranking of relations emerges from the grammar of nouns, adjectives and adverbs. For example, according to Jespersen, English grammars can impose ranks on the order of terms in sentences.[13] The classic example of two sentences with equal ranking of words illustrates the intertwined relationships of phrase and sentence rankings.

"The furiously barking dog..."
"The dog barks furiously."

Such ranking relations may be inconsistent with fixed order relations. The ordinate logics, including traditional number theory, avoid the paradoxes of set theory and hence are suitable for scientific work.

5. Discussion

The natural sciences are grounded in order relations. An immediate objective of this paper is to create logical spaces for communications among the natural sciences. Ur-symbols become the basis for choices of terms, operations and relations for the logics of scientific communications. The development of the capacity for scientific communication depends on creating synthetic symbol systems that can be recorded and interpreted by the community of scientists. The functional capacity for such symbolic communication depends on encoding and decoding of messages that correspond with observations and theories. Rapid progress in the natural sciences has lead to a plethora of specialized symbol systems, each justified by a localized interpretation of distributive and associative concepts. Communication among technical disciplines is extremely difficult to realize in practice. The concept of ur-symbols creates a common basis for the ordinate logics and the ordinate logics provide a common basis for integrating mathematical, physical and chemical reasoning. Consequently, a common plait of scientific reasoning over multiple disciplines can be woven by the ordering of specific terms, operations and relations within these ordinations. Ideally, esoteric expressions can become exoteric by reducing the complexity barriers to logical communications.

Ur-symbols, as signs, can represent either an object in the external world or a mental construct of the external world. (These separate notions of signification were already described by the Modistae in the 13th Century in terms of "modi essendi and modi intelligendi" within Speculative Grammar[14]). Thus ur-symbols are not logically encumbered by the undesirable assumption such that only the logical scientific operations on mathematical variables are valid. Rather, I presuppose that multiple encodings and decodings of symbols play a crucial role in quantitative thinking.[3,15] For example, the information encoded in atomic numbers is functionally separate and distinct from the information encoded in the real numbers. The concept of ur-symbols facilitates multiple interpretations. Consequently, correspondence between observations and scientific calculations depends on the choice of meaning for symbols used in the calculations.

The ordinate logics separate the mathematics of order into at least two principal forms. One sub-list of ur-symbols creates correspondence relations with diagrammatic networks of graphs of relations of categorical terms. Another sub-list of ur-symbols corresponds with rules for arithmetic operations and continuous functions and necessitates augmentation with syncategormatic terms. This separation of forms is necessary to account for the different methods of pairings used for scientific calculations by the perplex and complex number systems, respectively. The grammar of symbolic pairings is separate and distinct for these two forms of scientific communication.

The conjecture that co-ordinate logics can be extended to living systems is being explored. Molecular biology and genomics may provide the empirical basis essential to extend co-ordinations to bio-logic ordinations, thereby contributing to the quantitative foundation for the emergence of system biologies and clinical therapies.

References

1. J. LR Chandler, *Introduction to the Perplex Number System.* Discrete Applied Mathematics, 2008, (in press).
2. J. LR Chandler, *Complexity: A phenomenologic and semantic analysis of dynamic classes of natural systems.* WESScomm, **1**: p. 34 (1991).
3. J. LR Chandler, *Semiotics of Complex Systems and Emergence within a Single Cell.* Semiotica, **125**(1/3): p. 87 (1999).
4. J. P. Vanbremeersch, A. E. Ehresmann, J. LR Chandler, *Are Interactions between Different time scales Characteristic of Complexity?* in *Actes du Symposium ECHO,* Universite de Picardie: Amiens, France. p. 162 (1996).
5. J. LR Chandler, *A New Mechanistic Model For Risk Assessment.* Fundamental and Applied Toxicology, **5**: p. 634 (1985).
6. J. Conway, Guy, Richard K., *The Book of Numbers.* New York: Springer Verlag. 310 (2006).
7. Chandler, J.L., *Logics of Synthetic Symbol Systems*, in *Systems Research and Cognition*, M. Malatesta, Editor. ISSS: Tecumseh, Ontario. p. 11 (2007).

8. M. Alavanja, J. Aron, D. Brown, and J. LR Chandler, *Biochemical Epidemiology: Potential contributions to cancer risk assessment.* J. Nat'l Cancer Inst., **78**: p. 633 (1987).
9. E. Flapan, *When Topology Meets Chemistry*. Cambridge University Press. 241 (2000).
10. B. Rotman, *Mathematics as Sign, Writing, Imaging, Counting*. Stanford University press. 170 (2000).
11. S. Mac Lane, Mathematics*: Form and Function*. Berlin: Springer Verlag, 476 (1986).
12. J. LR Chandler, G.Van de Vijer, Eds. *Closure*. Vol. 901. New York Academy of Sciences: New York. 352 (2000).
13. O. Jespersen, *The Philosophy of Grammar*. New York: Norton. 348 (1965).
14. J. Pinborg, *Later Medieval Philosophy*. Cambridge: Cambridge University Press. 1055 (1982).
15. J. LR Chandler, *Complexity VII: Composing Natural Science from Natural Numbers, Natural Kinds and Natural Affinities*, in *Tempos in Science and Nature: Structures, Relations and Complexity*. New York Academy of Sciences. p. 75 (1999).

CHAPTER 12

SKILL LEARNING, BRAIN ENGAGEMENT, CONTEXT AND THE ARTS

Martin F. Gardiner

Center for the Study of Human Development, Brown University
Correspondence to: 4 Kimberly Ann Drive, Greenville, R.I. 02828
E-mail: Martin_Gardiner@brown.edu

I discuss the thesis that the learning of each type of skill requires the brain to assemble and develop a processing ensemble whose integrated activity allows the individual to engage behaviorally to the extent of observed capability. If capability improves, the opportunities of the brain processing ensemble for engagement must improve as well. Three ways are discussed in which context during skill learning can influence the development of capability. These include context of use perceived by the learner, context of outside authority related to the learning, and context of relationship of processing to similar processing resources of use to other skill applications. I propose that learning connecting arts skill to other skill learning may be especially useful to the skill learner by fostering incorporation of forms of processing especially favored within brain evolution, and that the use of processing favored for other reasons by evolution can help to explain the appearance of the arts within human culture.

1. Introduction

Everything we experience happens in context. The context is often essential to understanding the experience.

The context of Pavlov's famous experiment with the salivating dog was that the dog was living in a laboratory and was being fed there, and a bell during the learning trials always preceded the presentation of the food. The dog became conditioned to expect the food soon after the sound and began to salivate in expectation that food would soon follow. Would the same sound, once learned in this situational context still elicit the same or a similar response in a different context? How do differences and similarities in context affect the conditioned response? Pavlov already showed that context had an important role in what has come to be called Classical Conditioning, and more recent literature has continued to show this[1,2]. If we humans hear a special sound which became important previously in a particular context, we often remember not only our reaction to the sound, but also something of the context in which it became important to us.

In this paper, I will discuss three ways in which context can influence skill learning. I will review evidence that the way a person's brain engages with a particular area of skill affects the nature of a type of capability that they can achieve, and argue how context can influence the extent of capability learned.

2. Brain Engagement: Evidence from the Electroencephalogram (EEG)

By the terminology "Brain Engagement", I refer to what the brain is doing to produce every aspect of behavior in which we engage[3]. I include not only brain activity related to what reaches conscious awareness, but also other activity on which our behavior depends but which does not reach our conscious awareness[4].

My doctoral research provided some of the evidence showing that even extremely subtle changes in behavioral engagement are associated with changes in activity of the brain[5]. I compared the very weak electrical signals reaching the scalp from the brain comprising the "electroencephalogram" (EEG) as subjects paid attention to the same auditory stimuli in two different ways. In one condition, subjects had to classify the pitch of each stimulus they heard, in the other, its loudness. The stimuli actually varied both in pitch and in loudness from presentation to presentation, but the presentations were so brief (50 ms), and the differences to be distinguished so small, that one had to prepare for the judgment required before the stimulus appeared. Average electrical activity coming from the brain during the time when the mental judgment was formed (the "Average Evoked Potential" (AEP), or in more recent literature termed the "Event-related potential" (ERP)) was found to be identical for both conditions for the first 250 ms following the stimulus, but then to differ significantly in form during the next 250 ms, depending on what aspect of the stimulus was being attended to.

However generated, the EEG must somehow reflect behaviors of large numbers of brain neurons in order to reach signal strength sufficient to carry through the skull to the cortical electrodes. Jerry Lettvin, whose extraordinary work with colleagues on visual information processing of the frog[6] based on recordings from single neurons I will discuss shortly, once suggested to me that EEG should be thought of as like noise from a boiler room that could tell you roughly what was happening, but little more. On the other hand, my teacher Ted Bullock, specialist in encyclopedic study comparing brain designs throughout the animal kingdom[7,8], remarked to me that EEG must be telling us something important about brain design and the human brain, since it appears only in vertebrates, and increases in complexity within vertebrate brains of greater complexity, and reaches its highest development in the human brain. Research in which I participated at that time, and other related research, was already showing that computer analysis of EEG from humans and of related signals from animals could detect changes consistent with, but also helping to clarify evidence from other research approaches regarding organization and changes in organization of brain activity during conscious active behavior[9,10,11,12]. More recent work involving EEG and other forms of brain imaging is too enormous to review here, but more recent work has only reinforced the evidence that even very subtle changes in brain processing can often involve sufficiently large scale changes in brain activity that they can be detected in signals such as EEG.

EEG changes imply changes in organized brain activity involving many neurons. We all know that we can shift virtually instantly from engaging in one type of behavior to another. EEG studies imply that, at least in the human and other higher vertebrates, such changes can often involve changes in activity within not just a few, but within large numbers of brain neurons.

3. Brain Engagement Processing Ensembles

Not only EEG studies as just discussed, but other forms of brain imaging imply that large ensembles of neurons typically participate in active behavior. My own opportunities to study activity from implanted electrodes from a few patients undergoing studies prior to brain surgery always found evidence of participation of brain processing not only in the cortex, but in every subcortical location where electrodes had been placed, including within the hippocampus and amygdala, when these patients were engaging actively in the mental tasks I had used in my doctoral dissertation[5,13,14].

Let us call the ensemble of neurons performing brain engagement of a particular kind the "Processing Ensemble" responsible for the observed behavior and behavioral capability. The evidence from EEG and other imaging shows that activation can shift rapidly from one processing ensemble to another as our behavior changes. When we are talking, our processing ensemble for talking is being used. When we are doing math, our ensemble for math is being used. When the subjects in my doctoral experiment changed from attending to the pitch to attending to the loudness of acoustical stimuli, they somehow changed either the ensemble of neurons they were using, or the way they were using these neurons, or both to a sufficient degree that this change was detectable in as gross a sign of brain activity as the EEG.

4. Complex Brain Engagement, and Skill Learning

Whatever we are able to do is based on engaging with our brains in some way. Learning is certainly partly about acquiring, remembering and then retrieving information. I need to know the name of the city where I live. But learning is not only about information, but also about becoming able to develop capabilities of many kinds. Every type of skill development appears to pass through stages[15,16]. This implies that the brain engagement processing responsible for capability at every stage must change if behavioral capability is to improve. Accomplished chess players, for example, have been shown to actually perceive the arrangement of chess pieces differently than do beginning players[17,18], seeing immediately patterns of pieces and sensing from these dangers or opportunities, whereas beginners see each piece individually, and must analyze from that basis.

Learning that requires the subject to memorize and recall information can be distinguished from learning of how to do things[19,20,21,22], but many areas of advanced capability seem difficult to separate in this way. As I am reading, for example, I must recognize every letter, but within the context of capability I have learned to structure this information into syllables, the syllables into words, the words into phrases, the phrases into sentences, the sentences into still higher forms, and from all this processing I must derive meaning. Many types of information storage and retrieval are clearly involved, but within a context of complex processing involved in reading. My brain must be carrying out very complex but also very well coordinated processing to support this capability. The result is capability somewhat akin to what we have seen in the advanced chess player. I am principally aware of developing meaning as it forms as I read. When I have become accomplished at reading, all that supports this awareness takes place beneath my awareness but is critical to it.

Mathematical skill again integrates information and procedure. A beginning student must be able to identify each number and each operational symbol, but from the beginning, when thinking of these symbols mathematically, must engage with them mentally, i.e. with their brains, completely differently than if they were being engaged with under the context of reading[23,24,25]. Numbers are symbols that are organized sequentially. The meaning of "3" and how it is to be used mathematically, at least in arithmetic, depends on its position in the counting sequence as coming after "2" and "before" 4. Thinking of its mathematical meaning in any way other than this leads to difficulties. The letter "e", by contrast, in the context of reading takes no meaning from its position within the alphabetical sequence as coming after "d" and before "f". Rather in English, "e" is a vowel, and takes on particular properties as such, while "d" is a consonant, with different properties from those of a vowel; "f" is also a consonant, but has properties that differentiate it not only from vowels, but also from the consonant "d". The properties I refer to here concern connections of the letters to the utterances of spoken language. The connections to verbal meaning are still more complex, but entirely different from those involved in mathematical engagement.

Unless a teacher is careful to prepare students to think differently when they are engaging with symbols mathematically and in reading, there are opportunities for confusion. To add to the possibility of confusion, beginning students are usually expected to begin to learn the numbers by learning to name them, thus associating a verbal gesture to each numbers. These names must be used for counting out loud, and thus take on some of the properties of other verbal names, but not others, and have special properties of their own. If at the same time the student is expected to memorize the letters of the alphabet in order, there is again opportunity for confusion, since the order can be helpful for memorizing, but has no further significance of the type exhibited by number symbols.

The very beginning student learning reading and math, then, must learn to understand something seemingly quite advanced philosophically: that the meaning of a symbol depends on how it is to be used. This apparent complexity is not an illusion. The skills young children are expected to learn are all complicated. By the time they arrive at school, children must have already learned at least two extremely complicated capabilities – to walk and to talk. Some will also have come with initial skills involving numbers. They are now expected to add at least three further complex capabilities, reading, writing and math, the first two related to talking, the last to numbers. In addition, they must develop individually and socially, and improve those general mental skills that intersect with their more specific learning. This is an impressive amount of learning to be handled, a great challenge both to the students and also to their teachers.

5. Context of Purpose within Skill Learning

To treat symbols used in reading, writing, and math properly requires the child to understand from the outset how these symbols are to be used. This focus on use must continue as each type of capability develops. The reader must learn how letters as part of syllables, syllables as part of words, words as part of phrases, phrases as part of sentences, sentences as part of paragraphs, and paragraphs as part of still larger forms work together to convey meaning. The meanings of words and other verbal structures must be stored in some organized way in memory, so that they can be retrieved and integrated into processing as needed. Spelled out in this way, the task sounds almost hopelessly difficult, and yet this is what every reader must accomplish to become a good reader. Learning how to talk and understand spoken language is of comparable complexity, and has already been learned, showing that such complexity can be mastered. But now the child must learn how to build something further that connects to this earlier achievement.

Writing is similar in many ways to reading but now connects to speaking rather than the listening to spoken language.

In mathematics, a seemingly completely different way of thinking from that in talking, reading and writing must be developed to accommodate first addition and subtraction, then multiplication and division, then fractions and decimals. Introduction to geometry presents further challenge. Again, to learn this capability fully, the student must learn how the foundation of meaning of numbers relate to how they are used in each operation, and how the operations in turn can be used profitably by the student.

How every bit of what is to be learned is to be used must somehow be taken into account within the learning if the final product of capability is to become an integrated whole. Everything that contributes to the whole capability must come to have one or more roles, one or more values, within the whole capability.

Potential for use is critical to skill. Consider my capability to extract useful information about the world I observe with my eyes by processing the information passing from my eyes into my brain. I am aware of intensities of light or colors in what I see only in relationship to and helping to inform me about the location, arrangement and characteristics of forms such as books or walls from which the light reaching my eyes is reflected. Somehow genes and my learning collaborate to build this capability to extract very useful information from vision. A case reported by Sachs shows that if vision is unable to develop normally, a process of constructing visual perceptional skills that begins only in adulthood can be lengthy and difficult[26], capability to perceive forms coming only gradually, and in this case study, quite incompletely.

Information of the organization of forms we perceive in our immediate vicinity is not presented to awareness as might be readily measured in terms of Euclidian geometry, but in a non-Euclidian geometry. Parallel lines do not actually get closer together as they approach the horizon, but are perceived as if they do. Renaissance painters made painting seem closer to our own perceptual perspective, and this made us feel that they were more "realistic". Why do we present information from our eyes to ourselves in this way? I would propose that this treatment of visualized space references direction of all that is observed in relation to our own capabilities for movement. We need to be able to escape from or approach towards things shown to us by our vision. Our vision locates objects we are observing in a contextual geometry especially useful to us. That two parallel lines seem to join at a great distance helps us to realize that from our location we would have to start to move exactly in the same direction to reach a position on either of the lines where they appear to join.

Matúrana, Lettvin, McCulloch and Pitts demonstrated that a frog, by contrast, gets much more limited information from the eyes than do we[6]. In the frog, information from the eyes is processed to emphasize location and movement of small moving dark dots comparable in size to flies. Since these frogs eat flies, this information enables them to capture their meal, the visual information processed in the context of the question "is there a meal out there, and if so, where?"

Human capability for walking is highly utilitarian. Full capability includes not only means to move forward smoothly at various speeds and stop and restart, while maintaining balance, but also to adjust to contours and terrain, and adjust path to avoid obstacles.

A good bicycle rider again must be able to maintain balance, now while moving smoothly at considerable speed, and during turns by leaning properly, must learn to change speeds, perhaps to shift gears to maintain momentum. Bikers who can go beyond the confines of roads must have integrated further capability to adjust to variations in terrain.

Learning verbal language appears to occur especially favorably in young children at moments when its value is most obvious to the learner[27]. Tomasello and colleagues report, for example, that a mother who initiates a lesson in naming by pointing to something whose name she wishes the child to learn, waits for the child to look at it, and then tells the name, is less effective than one who watches the child, notes what the child is gazing at, and then names what the child has chosen to gaze at. In the less effective procedure, the infant has been interrupted from something else they are doing and asked to switch gears to learning desired by the mother. In the more effective procedure, the infant is already showing interest in something, and in a period of life when they are busy learning names for things, this appears to be an especially favorable time to present the name to the child.

In all these examples, then, purpose and value of the capability to the learner as achieved are evident. Every aspect of observed behavior can be related to the use to which that behavior is put. I would go further and propose that contribution to purpose runs through the choice and engineering the brain applies as it builds integrated capability. This must be what achieves the integration in which every part operates within the context of the goals of the whole ensemble. When integrated in terms of contribution to the whole, the whole performance becomes not the sum of the activities of the parts, but rather the result of the activity of the parts[28,29,30].

How does learning typically take place when such integrated capability is achieved? Considering learning to walk as an example, the learner before starting is aware of current capability, and then, wants strongly to improve on it. An infant not yet walking knows that he or she can crawl, that he or she can command his or her arms and legs to move, and may have seen others walking, but does not yet know how to do this. But he or she wants to become able to walk. Now sufficiently motivated, the learner engages actively in trying approximations of successful capability and then trying to improve his or her capability through repeated attempts. The infant makes attempts to walk, perhaps copying behavior he or she has seen. The learner does not necessarily expect to succeed instantly or even quickly (but will of course be delighted if he or she does), but is able to receive feedback rapidly from his or her own experience, both from failures and from sense of improvement, which can help him or her know when he or she has achieved capability and often, very usefully, when he or she is making progress towards capability. As component skills are learned, they are practiced, over and over, until they can be carried out more and more predictably, and more and more automatically. All these characteristics are well known. They must all be contributing to the building of a total capability in which sense of purpose is central to performance that is achieved. We can say that all the learning takes place within a context of purpose. At every stage the learner is able to sense to what extent he or she is achieving what he or she strives for, and to continue striving when he or she feels that further improvements are needed.

A constricted context of purpose, can, in turn, I propose, diminish the extent of capability that is achieved. In some work with middle school students, as an example, I discovered that all were able to carry out the methods of multiplication and division that they had memorized years earlier, but only one among them could provide a single example of an application to which either of these operations could be applied. The students were able to solve some algebraic equations, but again almost none could provide examples of how to use these capabilities. They had apparently been taught how to do these operations, but whatever applications if any had been learned, beyond the need to be able to reproduce the calculations, perhaps for a testing situation, had largely not stuck. This could suggest that broader application of skill beyond being able to reproduce calculations had for whatever reason not been strongly integrated into the capability that had been developed.

Standardized testing of elementary grade students consistently shows far better performance on ability to carry out mathematical procedures than on ability to apply these procedures to word problems. The weakest performance within reading testing typically involves understanding what is read.

Teachers are of course aware of these testing outcomes and spend class time and homework time drilling students on mathematical word problems and on procedures for understanding what they read. But there can be reasons why a student may not profit adequately from such opportunities. I will now turn to two such reasons, involving further aspects of context.

6. Context of Authority within Skill Learning

When a child learns to walk, he or she knows at all times the extent of capability, strives to improve it, knows that improvements are related to the striving, and takes satisfaction in the gains achieved. The same is true of learning to talk, and ride a bicycle.

In an important sense, such learners feel ownership of and personal connection to such capability. They know that it is they who now have learned how to walk, or ride a bicycle. They may be aware of help they have received, but they know that this help alone was not sufficient – that they themselves have contributed to the learning through their attempts at improvement, their practicing. Indeed, they may well realize that neither they nor their helpers can fully explain how they are accomplishing what they have somehow become able to accomplish. This quality of being able to do even while not being able to fully explain how is true of walking and bicycle riding, of talking, playing an instrument or singing beautifully, of reading, writing, composing music, playing jazz, boxing, skiing. This is a quality which appears to be present in many complex capabilities that we particularly admire. This quality of achieving beyond explanation can contribute strongly to the sense of ownership the learner feels regarding what he or she has been able to learn.

The sense of ownership, in turn, can allow the learner to consider the usefulness of the capability in relationship to goals and initiate further learning actions to further develop capability when that seems useful.

Consider now the many cases of learning where what is learned comes less from personal experience, and more from information supplied by an authoritative outside source. The elementary student is typically told and shown by the teacher how to add, subtract, multiply and divide numbers, how to solve word problems, how to pronounce words, what words mean, how to analyze what they are reading. A student receiving such information from the teacher realizes that the teacher knows more about these topics than does the student, and may feel that his or her major job in learning is to remember facts or how to carry out procedures as he or she has been told. In later stages of schooling, information from authoritative sources continues to often be important. A student is told how to do and use algebra, how to do geometric proofs, how to carry out statistical procedures. Medical students learn from authority the parts of the body, the characteristics of diseases, the typical treatments of diseases.

Human progress depends critically on developing and learning to apply information about the world, and a principal purpose of schooling is to make knowledge of facts and procedures available to a learner. But the way the learner treats authoritative information in relation to learning can affect the development of capability related to that authoritative information.

The work of Kohlberg[31,32] presents an extreme example of what is at issue here. In studies of moral development, Kohlberg and colleagues compared the ways in which subjects reached decisions regarding moral behavior. One group of subjects analyzed only in terms of what a source of authority such as a parent had taught them concerning right and wrong, and referred only to the authoritative source to justify each decision. By contrast, other individuals analyzing the same problems demonstrated greater personal responsibility for their moral decisions, analyzing, considering, and trying to relate moral decisions to considerations of what they believed moral behavior should be, and articulating reasons rather than only authority for their decisions. Both groups had developed capability at making moral judgments, but made the decisions in very different ways depending on how they related their choices to authority.

Obviously it is important to take advantage of authoritative information during skill learning. But when we choose a doctor to advise us regarding a disease, we hope to find one who keeps in contact with new information as it appears, can bring to bear personal experience with the illness, and, perhaps most importantly, shows in the way they discuss the disease with each patient that they have thought and continue to think about what they are learning about the disease, and adjust their capability at advising regarding the disease accordingly. Such a doctor displays ownership of and sense of responsibility for their capability even though some of their learning may be coming from information from outside sources rather than their personal experience. Integrating authoritative information into skill may include not only evaluating the sources, but also considering the relationship of new information to what is already learned.

A learner largely focusing on authoritative outside information may feel no need to further develop capability once authoritative information judged adequate has been gained. A doctor who does not update skill as new research is published is an example, but so also can be a student who limits learning in a particular area of capability only to what has been covered in class or in homework assignments. Inevitably situations will arise that have not been covered by the authoritative source.

Learning based on remembering authoritative information alone must face the limitations in our capability to store and retrieve information. Of course print publications, today supplemented by computer sources, makes much more information available to us than we are able to readily memorize and then retrieve. But our brains address skills by combining and integrating storage and retrieval of information with other forms of processing. I will now turn to evidence from the arts that I propose can help us better understand the role of other aspects of processing within skill and also help us understand how to use the arts to help develop such components of skill and apply them to developing capabilities beyond the arts alone.

7. Context of Brain Processing Interrelationships within Skill Learning

There is increasing evidence I now wish to discuss that though skills may often be developed independently from one another, leading to great differences in capabilities for different skills in the same individual, this need not necessarily always be the case. Sometimes skill development within an individual in one area of skill seems to aid development in another area of skill, and it is even possible for both areas of skill development to be helping one another[25,33,34].

The concept of useful interrelationships within skill developments is not a new one[35,36,37]. Here I would like to focus attention on interrelationships that can appear even when the goals of the behavior seem completely different, for example, interrelationships between capability at singing on pitch, and at doing addition and subtraction within arithmetic[25,33].

There are different ways to engage with the brain in order to achieve some degree of capability. Different forms of brain engagement may be equally effective in the short run, but not so in the long run. I propose here that the most effective engagement in the long run, in the sense both of most broadly supporting application of capability and also its further development, is often engagement which captures similarity to processing used effectively in seemingly different applications of skill development. I will first present an example from our research to illustrate, then present examples from other research and discuss more generally what these examples can imply.

Our research I wish to review was published in Nature in 1996, and then in more complete form several years later[25,33]. Approximately half of the 1st grade students in two Pawtucket, Rhode Island schools received a "test arts" program of two half hour lessons in skill training in musical singing by the Kodaly method[38] and 40 minutes of visual arts training each week. The control students in these schools received the standard arts curriculum for the district, 40 minutes of music or of visual arts in alternate weeks focusing on appreciation rather than skill development. In other regards, test arts and control curricula were identical, except that test arts students had an hour less per week than the control subjects for other academic subjects. Students who received the test arts skill training in music and visual arts were selected by the schools and on analysis were found when entering from Kindergarten to be weaker than controls on standardized achievement testing. Nevertheless, they accelerated in their learning both in reading and in math compared to control peers in the same schools, more strongly in math than in reading, catching up to their peers in reading and actually significantly surpassing them in math by the end of 1st grade. Ending 2nd grade, those receiving two years of the test arts had the highest performance in math, those with one year either in 1st or 2nd grade had the next highest and those with no test arts had the lowest math performance. Reading did not show this same statistically significant trend, but students in test arts who had caught up with peers in 1st grade did not fall back in 2nd. In a follow up study, in which only musical skill was again developed by the same Kodaly method, the capability reached in math was found to be significantly correlated with singing capability developed regarding musical pitch[25,39].

What similarity in brain engagement processing might account for interrelationship between pitch skill and math skill learning in 1st and 2nd graders? Case has shown[23,24] that most effective math learning in addition and subtraction at this stage must promote development of thinking related to the number line,. That is, the students must come to think of the symbols in sequence, and also come to think of addition and subtraction as operations moving up or down this sequence[33]. Rather than having to memorize the sums of integers, a child engaging mentally with addition in this way can solve 3 + 2 = ? by starting at 3 moving two steps up the sequence, and reaching 5. If a child who has only been memorizing the sums of integers is asked why the sum of 3 plus 2 gives five, he may be able to recall the answer 5 he has been taught, but can provide no further justification why this is the case. The same child if he has learned to think of the numbers in sequence, and of addition as an operation on this sequence can explain that 5 must be the answer if you start from 3 and move 2 steps further, which the + sign has told you to do.

In music, pitches used in a song can be memorized for each song, and most children begin learning songs in this way. But when many songs using the same and related pitches are to be learned, it becomes increasingly useful to the learner to mentally organize the pitches to be used. Within most music, and all the music the students in these studies learned to sing, the creators of the songs that were learned thought of the pitches in terms of their organization within musical "scales". Pitch is a perception related to the frequency of vibration of the source of the sound. Pitches used in melodies are chosen from highly restricted choices, made by considering relationships of pitches within a scale to one another that are then exploited within the music. Compared to the total range of hearing, pitches used within melodies sit relatively near one another. When arranged in musical scales, musical pitches are organized sequentially from lower to higher pitch, and the well known pitch names "do", "re" mi", "fa" "sol". "la" and "ti" name the pitches in the relative order moving up the scale from lower to higher. The Kodaly method uses many procedures that encourage the students to organize pitches in their minds according to musical scales. These methods include the use of the "do" "re" "mi" names that refer to location of pitch within the scale, the use of hand gestures that convey changes in relative pitch position within the scale, the teaching of many songs built from the same scale which encourages the students to organize pitches in terms of scales, and the teaching and practicing of ability to shift melodies to new pitches where the relative frequency relationships determined by the musical scale are nevertheless preserved. The learning of pitch begins from two note scales and then, as organization of these pitches is learned, moves to scales with further added notes.

What is common to the pitch and math mental engagement being learned, then? Both forms of mental engagement clearly build on application of sequential representation, even if in entirely different ways. Why should the pitch learning and math learning help one another? Because, I propose[15], the value of sequential representation to pitch engagement being learned encourages the development of sequential representation within mathematical engagement as well, to the profit of mathematical engagement[40]. Effects in the opposite direction, from math to music have been observed as well[15].

Our ongoing research is implying that rhythm skills in music, by contrast, are less strongly related to mathematical capability in these young students, and more strongly related to progress in reading and writing[15]. As our research is continuing we are continuing to find evidence of many connections between particular aspects of musical skill as it develops and progress not only in academic skills but also in personal and social skills[15].

A variety of related studies by other groups have also been appearing. One of the most widely discussed demonstrated positive effect in preschoolers beginning to learn piano playing on their ability to fit together simple puzzle pieces properly into a composite figure[41]. Another important series of recent and continuing studies are using brain imaging to show evidence of connections between musical and verbal skill developments[42,43].

As connections between musical and other types of skill development continue to be documented, each is typically discussed in relation to specific hypotheses that can account for the observations. For example, I have been relating the connections between math and musical pitch learning we observed to uses of sequential representation[15]. Rauscher and colleagues have proposed that their observations concern what they term improvements in visuo-spatial processing[41]. The recent studies relating musical and language development appear to involve improvements in capability for processing rapidly changing signals[42,43].

I propose a more general implication of such observations beyond the specific hypotheses concerning brain processing and learning that they address. The human brain is able to support a staggeringly large number of different capabilities, all different from one another in important respects. But similarities in particular details of processing are also now being found, even when overall goals and uses of the related capabilities are clearly different. I propose this as reflecting a human brain that has evolved capability for developing a variety of processing strategies and also of finding a way to adapt these strategies to many different ends. What can be signaled by connections between such seemingly different capabilities as learning to sing many songs with proper pitch and learning how to do adding and subtracting especially effectively can be the value of the aspect of brain processing that unites them. The brain strategy or strategies for brain engagement that unites piano playing and puzzle assembly capability may also be more generally valuable to other applications of skill learning, and what it is that the brain is calling upon that improves speed for processing both rapidly changing musical and speech signals warrants continued examination. If these are useful skill developing tricks that the brain has evolved, where are they carried out in the brain, when are they used and when not and how are they integrated into specific skills? Such connections as are being demonstrated by connections between arts and other learning can be opening a potentially useful further window on the means the brain uses to develop our most honored capabilities.

The growing evidence I am discussing, then, suggests that interactions with music and other arts learning are not only enjoyable and valuable for their own sake, and that these values and pleasures that they give can motivate students to develop capabilities in the arts that help them more broadly in skill learning as well[15,33], but that the arts may also be nudging the students in especially useful directions, helping students to incorporate processing into their brain engagement which proves especially useful to them. Why might that be? Continuing from earlier discussion[34,44], I propose for further investigation an answer that can help us to see better why arts have evolved and developed so consistently within human culture. I propose that the benefits of the arts, the pleasures they provided, the opportunities to touch emotions in ways that were not available from other means, the opportunities to provide new types of group experiences, to approach religious concerns, to provide fresh opportunity to create, all of these benefits became apparent as the arts developed, and helped to motivate their further development. But I doubt that however important these benefits became that they were sufficiently strong to provide evolutionary advantage in and of themselves. Rather, I propose that the arts came into being by building on mental capabilities that were evolutionarily beneficial for reasons other than their application to the arts. Perhaps at first by chance, and then by purposeful development, we found ways to build on what our brain had found useful to us for other reasons. If my analysis is correct, the arts then provide the basis for learning useful skills for the very reason that they are constructed from useful components of skill.

This proposal implies that it can be possible to find many useful connections between arts skill and other aspects of learning. In fact, examination of the elementary school curriculum is revealing potentially useful connections between aspects of musical or other arts skill and many major components of skill development in reading, writing and math[45], and such exploration is being continued into upper grades.

References

1. I. Pavlov and G.V. Fol'bort, *Die höchste Nerventätigkeit (das Verhalten) von Tieren*, J. F. Bergmann, Munich (1926).
2. J.H. Maes, *Context and classical conditioning*. Nijmeegs Institut voor Cognitie en Informatie (NICI), Nijmegen (1993).
3. M.F. Gardiner, in *Handbook for arts for critical thinking: Part I (3rd edition)*, Thinking in Music, Inc., Phoenix (2007).
4. J. Julian, *The origins of consciousness in the breakdown of the bicameral mind,* Houghton Mifflin, Boston (2000).

5. M.F. Gardiner and D.O. Walter, in E. Donchin and D.B. Lindsley (Eds), *Average Evoked Potentials: Methods, Results & Evaluations*, NASA, Sp-191, Washington (1969).
6. H.R. Maturana, J.Y. Lettvin, W.S. McCulloch and W.H. Pitts, J. Gen Physiol., 43(6), Suppl. 129 (1960).
7. T.H. Bullock and G.A. *Horridge Structure and function in the nervous systems of invertebrates*, Freeman, San Francisco (1965).
8. T.H. Bullock, R. Orkand and A. Grinell, A. *Introduction to nervous systems, Freeman*, San Francisco (1977).
9. D.O. Walter and M.F. Gardiner, in C.C. Pfeiffer and J.R. Smythies, (Eds.) *International Review of Neurobiology*, Vol. 13, Academic Press, New York, p. 343 (1970).
10. M.F. Gardiner and D.O. Walter, in S. Harnad, R.W. Doty, L. Goldstein, J. Jaynes and K. Krauthmer (Eds.), *Lateralization in the Nervous System,* Academic Press, New York, p. 481 (1979).
11. F.D. Abraham, H. Bryant, M. Mettler, W. Bergerson, F. Moore, J. Maderdrut, M. Gardiner, D. Walter, and R. Jennrich, *Brain Research* 49: 349 (1973).
12. E. Donchin and D.B. Lindsley (Eds), *Average Evoked Potentials:Methods, Results & Evaluations*, NASA, Sp-191, Washington (1969).
13. M.F. Gardiner, A. Dymond and A. Tymchuk. Presented at 1st Annual Meeting of Society of Neuroscience, Washington (1971).
14. M.F. Gardiner, A. Dymond, D.O. Walter and A. Tymchuk. *Elect. Clin. Neurophysiol.*, 37, 427 (1974).
15. K.W. Fischer, *Psych. Rev.* 87 (6); 477 (1980).
16. K.W. Fischer and T.R. Bidell, in Damon, W.,and R.M. Lerner, R.M. *Handbook of Child Psychology, Vol. I: Theoretical Models of Human Development*, 467 (1998).
17. DeGroot, A.D. *Thought and choice in chess,* Moulton, The Hague (1965).
18. W.G. Chase and H.A. Simon. *Cognitive Psychology*, 1, 33 (1973).
19. D.H. Kasler, *Learning and memory in normal aging*, Academic Press, San Diego, (1994).
20. M.J. Nissen, in LR. Squire and N. Butters (Eds). *Neuropsychology of memory* (2nd ed), Guildford Press, New York (1992).
21. J.R. Star, in B. Fishman and J.O'Connor-Divelbiss (Eds). *4th International Conference of the Learning Sciences,* Erlbaum, Nahwah, N.J., p. 80 (2000).
22. E.A. Quintero-Gallego, C.M. Gomez., E.V. Cesare, J. Marquez and Perez-Santamarias. *Behavior and Brain Function,* 2:9 (2006).
23. R. Case, *The mind's staircase: Exploring the conceptual underpinnings of children's thought and knowledge*, Lawrence Erlbaum Associates, Hillsdale, N.Y. (1992).
24. R. Case and Okamoto, Y., *Monographs-of-the-Society-for-Research-In-Child-Development,* 61: (1-2), 283 (1996).

25. M.F. Gardiner, *Journal for Learning Through Music*, New England Conservatory of Music, Boston, MA., 72 (2000).
26 O. Sachs, An anthropologist on mars: seven paradoxical tale, Vintage (1996).
27. M. Tomasello, *Constructing a language: a usage based theory of language acquisition,* Harvard University Press, Cambridge (2003).
28. E. Morin, (translated from French by Carlos Gershenson), Colloquium *Intelligence de la complexite: epistonologie et pragmatique*, Ceriy-La Salle, France (2005).
29. E. Morin, (translated from French by Nidra Poller), *Seven complex lessons in education for the future,* United Nations Educational, Scientific and Cultural Organization, Paris (1999).
30. S. Geman, D.F. Potter and Z. Chi., *Quarterly of applied mathematics*, LX, 707 (2002).
31. L. Kohlberg, in D. Goslin (Ed.), *Handbook of socialization theory and research*, Rand-McNally, New York, p. 347 (1969).
32. L. Kohlberg, in T. Likona (Ed.), *Moral development and behavior*, Holt, Rinehart and Winston, New York, p. 31 (1976).
33. M.F.,Gardiner, A. Fox, F. Knowles, and D. Jeffrey, *Nature*, 381:284 (1996).
34. M.F. Gardiner, in R. M. Lerner, L.B. Schaimberg, P.M. Anderson, and J. Miller (Eds). *Encyclopedia of Human Ecology*, ABC-Clio (2003).
35. E.L. Thorndike and Woodworth, R.S. *Psych. Rev*. 8, 247-261 (1901).
36. E.L.Thorndike, *Educational psychology*, (Vols. 1 and 2), Columbia University Press, New York (1913).
37. H.J. Kalusmeister, *Educational Psychology* (5th ed.), Harper and Row, New York (1985).
38. L. Chosky, *The Kodaly concept: creating an environment for musical learning*, Prentice-Hall, Englewood Cliffs (1981).
39. M.F. Gardiner and Olson, E. *Presented at Fowler Symposium*, University of Maryland (1999).
40. M.F. Gardiner, Presented at Annual Meeting of American Association for the Advancement of Science (AAAS), Philadelphia (1998).
41. F.H. Rauscher, G.L. Shaw, L.J. Levine, E. I. Wright, W.R. Dennis, and R.L. Newcomb, *Neurological Research* (1997).
42. Gaab, N., Tallal, P., Kim, H., Lakshnariyann, Glover, G.H. and Gabrielli, J.D.E. *Ann. N.Y. Acad. Sci*, 1060, 82 (2004).
43. P. Tallal and N. Gaab, *Trends in Neurosciences*, 29 (7), 382 (2006).
44. M.F. Gardiner, unpublished paper (2004).
45. M.F. Gardiner, *Handbook for arts for critical thinking: Part I* (3rd edition), Thinking in Music, Inc., Phoenix (2007).

CHAPTER 13

UTILIZING FRACTAL TIME

Terry Marks-Tarlow

The Institute for Fractal Research
Goethestrasse 66, 34119 Kassel, Germany
E-mail: markstarlow@hotmail.com

Linear concepts of time plus the modern capacity to track history emerged out of circular conceptions characteristic of ancient and traditional cultures. A fractal concept of time lies implicitly within the analog clock, where each moment is treated as unique. With fractal geometry the best descriptor of nature, qualities of self-similarity and scale invariance easily model her endless variety and recursive patterning, both in time and across space. To better manage temporal aspects of our lives, a fractal concept of time is non-reductive, based more on the fullness of being than on fragments of doing. By using a fractal concept of time, each activity or dimension of life is multiply and vertically nested. Each nested cycle remains simultaneously present, operating according to intrinsic dynamics and time scales. By adding the vertical axis of simultaneity to the horizontal axis of length, time is already full and never needs to be filled. To attend to time's vertical dimension is to tap into the imaginary potential for infinite depth. To switch from linear to fractal time allows us to relax into each moment while keeping in mind the whole.

1. Introduction: Circles to Arrows to Spirals of Time

With the invention of the clock came a mechanical, "clockwork universe" plus a linear concept of time. Prior to this mechanistic model characteristic of modern industrializing societies, was a much more organic conception of time based on weather cycles of the sun and seasons, relevant to agriculture. In *Myth of the Eternal Return*, anthropologist Mircea Eliade examined cyclic time still characteristic of traditional cultures[1]. In such societies the particular is merged with the universal, the individual with the archetypal, the sacred with the profane, as each moment relates back to the origin of all moments. Without the written word to record history, the present moment becomes amplified to the timeless and eternal, and there is a unity to all of life.

With the introduction of the clock came the arrow of time plus time lines on which to peg the course of events in collective memory. Launched by time's unidirectional arrow, Eliade's perfect circles flew forward as spirals. In conjunction with the written word, recorded history was born. The invention of the clock advanced modern societies by helping to coordinate people. Simultaneity was a means to align activities, communications and meetings in distant locations. A higher level of organization was achieved, especially in cities, where events proceeded more by social needs than by conditions of weather or nature. With the power to tag each moment and track history, collective awareness has grown of a unique trajectory from past to present and into the future. With increased recognition of the uniqueness of each moment came new opportunities for individualism, with reflective awareness centered on the uniqueness of each self moving through time.

Within this formal recognition of unique moments lie seeds for nonlinear conceptions implicitly enfolded within the linear arrow of time. To acknowledge each moment as unique is to touch upon the nonlinear emergence of each moment from the whole of things, its complete interdependence with everything else, plus the sensitive dependence of constituent elements upon their own precise history. Enter the field of chaos theory and that one of its pioneers, Ilya Prigogine, uses nonlinear models of chaos in order to establish the irreversibility, that is unidirectionality, of time's arrow[2]. Along with nonlinear models come new possibilities for advancing beyond mechanistic, mechanical models for understanding time, allowing us to return to more organic ones, based on the self-reflexive complexity of modern society.

2. Time and Subjective Space

Even as nonlinear aspects of time begin to infuse into popular culture, most of us are stuck with linear models, at great cost to the psyche. When time is viewed as a line, it becomes a precious commodity that all too often either runs out or runs ahead of us, leaving us sprinting to catch up. When time is parceled into discrete packets, we are tempted to shift into a scarcity model. The accompanying "not enough" mentality can leave us frantic and depleted. On the flip side is the opposite stance of gorging by greedily trying to stuff more and more activities into smaller and smaller parcels of time.

Jungian psychologist James Hillman once told author Michael Ventura that the hardest thing to treat during psychotherapy is a person's schedule[3]. We so easily lose ourselves in the business and busyness of our schedules. When this happens, our spontaneity and creativity get lost along with us. There is danger that we become as robotic and mechanical as our conception of the clock time now running us. And along with this danger comes that for multigenerational reverberations, as time pressured parents drive children with their hyper-aroused physiology, often with the very best intentions to stimulate and enrich the lives of their children. I find it both sad and ironic that out of efforts to gain control over our days through the clock's coordination, there is an epidemic of loss of control over lives, now so ruled by the clock that we have lost touch with natural rhythms present in the moment.

3. A Nonlinear Model

There is an old adage, "If you want to get something done, give it to a busy man." Gender bias aside, this comment, one of my father's favorites, has rolled around in my head for years, its irony most delightful. I can see others unconsciously operating according to this nonlinear wisdom. For example, a savvy parent in a Los Angeles area public elementary school recently rectified a nonproductive parent enrichment board by loading it up with busy working parents.

Perhaps the paradoxical nature of such pieces of folk wisdom has primed my current interest in nonlinear dynamics. When I came across an arcane piece of research conducted by mathematician and nonlinear scientist, Manfred Schroeder, I certainly took note[4]. Schroeder discovered that the likelihood of completing an unfinished task actually decreases, rather than increases, over time. From a linear perspective, at best this finding is counter-intuitive; at worst it makes no sense. When time is conceived of empty baskets that are hung on a line, then the more time available, the more baskets there are to catch the task, and the greater likelihood one has of completing it. But just as linearity is a tiny, mostly human-made corner of a fundamentally nonlinear universe, this mechanistic model of clock time turns out to be a far cry from the "real" time of embodied living, which is nonlinear and fractal.

4. What is Fractal Time?

Fractals, a new form of geometry discovered by Benoit Mandelbrot[5], comprise complex shapes that possess detail on multiple scales. Fractals are inherently counter-intuitive in that they operate more like Merlin's bag of magic than like an ordinary handbag or basket. This is because the detail of a mathematical fractal is infinitely deep. Further, what we perceive during observation depends intimately upon the size and scale of our measuring stick. The smaller the measuring device, the larger the measurement, until at the extreme of zero length, we wind up with an infinitely long curve. What this translates to is this: the closer we look at a mathematical fractal, the more there is to see. In this way, fractals resemble self-reflection within the psyche, where there too, the closer we look, the more there is to see.

Fractals display a property called self-similarity, by which the pattern of the whole repeats itself within a fractal object's parts. Fractals also display a related property called scale invariance, by which the identical or similar pattern repeats on multiple size or time scales. In the case of a physical fractal, such as a mountain range or coastline, the quality of ruggedness of the range or jaggedness of the shoreline repeats, such that similar bends or curves appear approximately or statistically on smaller and smaller size scales. Because fractal form emerges out of temporal processes related to their construction, fractals are dynamic process-structures by which time gets etched into form.

Biological researchers such as Iannaccone and Khokha[6], Mac Cormac and Stamenov[7] and West[8] have established that physical fractals abound in human physiology. Self-similar branches exist on multiple size scales within our lungs, circulatory system and neuronal circuitry of the brain. We concretely see time etched into fractal form on the wrinkled faces of the elderly, whose self-similar folds within folds of skin reveal the history of their emotional states and relationships that have animated the muscles on their faces through the course of their lives. Bruce West believes Western medicine missed the boat by not recognizing the importance of fractal scaling within human physiological functioning. In contrast to the simple forms dictated by reductionistic methods, homeostatic models and linear statistics, healthy bodies depend upon fractal variability, including detail at multiple levels of scale.

Whereas physical fractals, like neural branching and wrinkles reveal time etched into form, temporal fractals enable us to detect the reverse – the evolution across real time of spatially extended dynamical systems. In the social and medical sciences, linear statistics, like the Bell curve, that flatten out variability towards central tendencies are giving way to recognition that inverse power law distributions are more ubiquitous, forming wherever nature self-organizes at the edge of chaos[9]. When a power law is present, events display frequency spectrums with self-similar fluctuations across time scales. Power laws serve as the persistence of a kind of proto-memory in nature. More and more events that otherwise appear random and uncorrelated now reveal short- and long-range temporal correlations in the form of these fractal dynamics. Fractals help us better to understand how nature weaves randomness into deterministic forms using chance. In this way the present can be strictly determined by the past, in such a way that the future nonetheless remains unpredictable.

Whether in river flows, foreign exchange markets, patterns of epidemics, or functioning of physiological systems, when fractals are present, the macroscopic pattern functions as a unified whole that both emerges from complex interactions of component parts, as well as constrains the component parts. To better understand how fractal variability supports healthy life, consider this recent controversy. As white water rafting has gotten more and more popular in the United

States, a number of companies have developed means of creating artificial waves in rivers to extend their rafting season and commercial capabilities. Because these machines mechanically create waves, i.e., turbulence, on only one size scale, the life cycle of fish in these rivers, especially salmon, is becoming seriously disrupted. The problem stems directly out of the loss of complexity in the water. Because there are no longer little eddies within eddies, the fish are robbed of a variety of places to rest, spawn, feed and travel safely.

Fractal patterns often merge form and function within spatial and temporal aspects of nature. Consider global cloud cover. On the one hand, the fluffs within fluffs of a cloud represent a physical fractal, with this property of scaling explaining the great danger pilots faced when enveloped by a cloud before the invention of radar. Because near and far shapes appear identical across scale, pilots had no way to gain distance perspective without using an outside object as a point of reference. Meanwhile, global cloud cover is not only a spatial event but also a temporal one revealing self-similar fluctuations of meteorological parameters on multiple nested time scales, from seconds to years.

In 1987 nonlinear scientist Per Bak unified the relation between spatial and temporal power law behavior with his notion of "self-organized criticality"[9]. His idea was that fractal complexity arises spontaneously in nature in far from equilibrium processes. Self-organized variability occurs right at the critical point where avalanches of change become asymmetrically poised to occur on every size scale. Epitomized by earthquake dynamics, a power law appears in the form of a log-log relationship: a high probability exists for lots of tiny earthquakes on tiny size scales, a medium probably exists for medium sized earthquakes, and very low probability exists for "the big one" to strike, although given enough time, even the largest of earthquakes is bound to occur.

5. Fractal Time and Subjectivity

Given that fractal temporal processes comprise much of nature, both outside and inside the human body, from the movement of the earth's magnetic plates to the motion and inter-coordination of much of our internal physiology, how do we understand and use fractal time subjectively?

Simply put, fractal time consists of multiply nested temporal cycles. Just as the linear is a subset of the nonlinear, nested temporal cycles are already present implicitly within analog clocks (the old fashioned time keeping device with second, minute and hour hands). Here is one way to visualize the motion of the clock dynamically as a fractal flower. Keep the hour hand in its traditional place in the center of the clock. Move the inner most part of the minute hand to the outermost end of the hour hand; then move the innermost part of the second hand to the outermost end of the minute hand. With all three hands connected as forced oscillators, let the ticking begin and continuously record the motion of all three clock hands over time and in space. With each temporal unit freed to form a spiral instead of a circle, at the end of a 12 hour period, we wind up with two layers of spirals within spirals and a beautiful fractal flower adorning the inner circle created by the hour hand. The outer face of the clock will be trimmed with a medium sized spiral looping around each of the clock's numbers, and this medium circle of spirals will be trimmed with a smaller circle of spirals looping around its perimeter.

This fractal flower image of the clock can help us to visualize how our bodies comprise an even more complex fractal flower. Imagine circadian and ultradian rhythms of each of the body's organs and physiological processes in terms of loosely coupled nested time cycles, each with its own intrinsic dynamics[10]. The beating of the heart, the pulsing of the breath, the turn of the digestive system, the action of the liver, the rhythms of the sleep/awake cycle, each comprises its own separate spiral, some more strongly, others loosely coupled in a panoply of internal rhythms.

Moving from the body back to the mind, Susie Vrobel devised a purely phenomenological application of fractal time in a book introduced by Otto Rössler[11]. Vrobel complexified the ordinary time line by making a distinction between time's qualities of length, depth and density.

- Length = incompatible facts indicating "before-after relations";
- Depth = nested, compatible facts indicating "during relations";

- Density = relation between length and depth, for comparing subjective duration and features of time series.

In my own idea of fractal time, independently derived and outlined below, I also draw upon a depth element.

6. Infinite Moments

Fractal geometry is based on nonlinear equations, some as simple as $z \rightarrow x^2 + c$, iterated for every point on the complex number plane, which generates the Mandelbrot set. The complex number plane consists of two axes – a horizontal axis for the real numbers (the rational combined with irrational numbers) plus a vertical axis for the imaginary numbers (based on *i* or the square root of -1). Consider an ordinary one-dimensional time line as the real number axis, to which we add the depth dimension by expanding our graph into a two-dimensional Cartesian grid. The quantitative dimension of length is represented on the x-axis by the real number line, while the imaginative dimension of depth is represented on the y-axis by the imaginary number line.

By tapping into the complex number plane and adding the vertical dimension of depth, time is broadened from discrete, finite baskets that hang on a one-way timeline. Any given now – or point in time – becomes richly and potentially infinitely textured, moving into qualitative aspects through Vrobel's concept of density. By tapping into imaginary numbers' capacity to model hidden or implicit dimensions, we gain flexibility to stack an infinite number of cycles for any given point along time's vertical dimension.

Before explaining how to utilize fractal time in everyday life, here is some personal background. Perhaps motivated by my father's emphasis on the efficiency of busy people, I have felt like an oddball among my friends by nearly always having more than enough time for anything. I enjoy a cordial relationship to time, rarely battling my schedule. When all goes well, I sustain high productivity in a relaxed atmosphere. If I do feel rushed, it is because all too often I hurry to be early, or because I tend to express my own neuroses through being counter-phobic towards deadlines. That is, I feel compelled to hand papers or assignments in early because I have difficulty trusting I can pull anything off under pressure. When emotionally balanced, however, I often operate in Csikszentmihalyi's timeless realm of "flow[12]," where I luxuriate in the spacious illusion of more than enough time. These characteristics of mine have gotten more extreme with age. As I slow down naturally, I appreciate the moment more, while ironically continuing to get more productive.

What is my secret to a relaxed, organic, yet fertile relationship with time? I consider the various projects and dimensions of my life as vertically nested cycles. Each operates according its own time scale within a separate level of description. Each possesses unique features and internal dynamics. Each nested cycle is present all the time, each a part of my identity, whether or not I do an activity related to it. To stack projects and facets of myself along the vertical dimension corresponds subjectively to a rich sense of life, while unleashing unlimited possibilities to pack time, without time pressure. Each project is always present and keeps moving, however slow its time scale might be.

In order to allow time to bloom into a fractal flower, certain conditions must be fulfilled:

(i) Projects need to be ongoing and cannot be overly constrained by deadlines.

(ii) If deadlines exist, one must work well inside them so that projects can become structured and paced in a self-organized fashion;

(iii) All vertical cycles need to be continually monitored, and occasionally adjusted or nudged at critical points. Procrastination does not work here, but winds up unnecessary anyway. From a nonlinear, dynamical vantage point, when operating in far from equilibrium conditions, systems are always moving, rather than static entities that require effort to move them. There is no need to overcome inertia, but only to maintain momentum. In the case of group activities, keeping momentum often requires task delegation.

(iv) Utilizing fractal time also necessitates large chunks of unstructured time, along with other chunks of variable lengths. This way, time gets parceled according to the intrinsic dynamics of one's schedule and day. In my case, some of my days are tightly packed, like the three days every week I dedicate to my psychotherapy private practice. Here I pack in 8 to 12 hours of clinical work, while each of these days is punctuated by dance classes or yoga breaks midday. My four other days are relatively unstructured, filled with attention to my family and friends, writing projects, drawing inspirations, rock climbing, skiing, etc. I tend to add projects on that have slow time scales. Once a year I curate a children's art show in a local gallery; four times a year I teach a yoga workshop; once a month I hold a live model figure drawing session.

On fractal time, what one does depends upon how much time one has plus how one is feeling. With a whole or half day free, I might immerse myself in working on a book or going on an outing all day. With an hour free, I might choose to do bills. With 15 minutes to spare, I might check email or return telephone calls. Rarely do I make lists. An advantage to operating on fractal time is that it is organically organized, spacious, depth permits intuition and spontaneity. I always keep a drawing or two laying around, a paper or book in process, a project around the house to tweak when the urge strikes. Conversely days or moments when I am anxious, foggy, uninspired, or preoccupied, I plunge instead into chores, like house cleaning, laundry, food shopping, or sticking my hands in the dirt to weed or clear leaves.

7. Fractal Flowers versus Too Many Balls in the Air

Within a fractal system of keeping time, the simultaneity of nested cycles is not the same thing as multi-tasking. Multi-tasking, which means doing multiple tasks simultaneously, is an adept person's solution to packing the time line by juggling as many activities as possible all at once. Three screens might be opened on the computer; the television turned on; the person talking on the phone, while trying to catch a bite to eat. Multi-tasking still involves a frantic pace to fill into time's limited baskets.

In his way, multi-tasking is quite the opposite of fractal time. Ideally, when on fractal time, one focuses on a single activity at a time, even though all activities are proceeding simultaneously under the surface, however invisibly. Whereas multi-tasking leads to fractured focus, fractal time helps to pinpoint and sharpen focus. Whereas multitasking is often done with speed, utilizing fractal time includes activities that easily fit into available time slots, so one can arrive to events early, slow down immensely, enjoy the moment and still have time for spontaneous deviations from the plan when unexpected opportunities or invitations arise. Because all projects are accomplished well-before their deadlines, there is more flexibility with a schedule based on fractal time than on one regimented by a day timer.

8. Psychological Complexity

From a nonlinear vantage point, I join with psychiatrist and educator Daniel Siegel to define mental health in terms of psychological complexity poised on the edge of chaos, between extremes of rigid order and utter disorder[13]. Here there is lots of variability, maximum access to a wide repertoire, plus the responsive capability to shift on a dime and flow with current circumstances. I believe that to understand and use fractal time is an important aspect of psychological complexity constituting enabling greater integration and alignment with our bodies' natural rhythms. In contrast to left brain, linear rationality, a fractal conception of time is corresponds with right-brain holistic, intuitive processes that implicitly register the complexity of the whole. To understand and use fractal time is to attend to grand scale and continuous picture of our full being and how this whole lines up with discrete, finite states of doing.

I believe our bodies already operate on fractal time. With a more conscious understanding of what our bodies already know, we can become more self-reflective about our unconscious wisdom. By conscientiously implementing a fractal conception of time and by teaching our children how to do the same, we can help motivate our selves along with the next generation to stop wasting time and tuning out in order to recover. Instead we can feel refreshed by tuning into the richness of our own imaginations and longings. Psychologically we will need less to escape from our poor time-pressured aching psyches, leaving us better able to take advantage of our privilege and wealth, both inner and outer.

References

1. M. Eliade, *The Myth of the Eternal Return: Cosmos and History*. Princeton University Press, Princeton, NJ (1954/2005).
2. I. Progogine, *The End of Certainty: Time, Chaos, and the New Laws of Nature*. The Free Press, New York (1997).
3. J. Hillman and M. Ventura, *We've Had a Hundred Years of Psychotherapy – And the World's Getting Worse*. HarperCollins, New York (1992).
4. M. Schroeder, *Fractals, Chaos, Power Laws*. W.H. Freeman, New York (1991).
5. B. Mandelbrot, *The Fractal Geometry of Nature*. W.H. Freeman, New York (1977).
6. P. Iannaccone and M. Khokha (Eds.), *Fractal Geometry in Biological Systems*. CRC Press, New York (1996).
7. E. Mac Cormac and M. Stamenov (Eds.), *Fractals of Brain, Fractals of Mind*, John Benjamin, Amsterdam (1996).
8. B. West, *Where Medicine Went Wrong*. World Scientific, Singapore (2006).
9. P. Bak, *How Nature Works: The Science of Self-Organized Criticality*. Springer –Verlag, New York (1996).
10. D. Lloyd and E. Rossi (Eds.), *Ultradian Rhythms in Life Processes*. Springer-Verlag, New York (1994).
11. S. Vrobel, *Fractal Time*. Houston: The Institute for Advanced Interdisciplinary Research, Houston (1998).
12. M. Csikszentmihalyi, *Flow: The Psychology of Optimal Experience*. Harper & Row, New York (1990).
13. D. Siegel, *The Developing Mind: How Relationships and the Brain Interact to Shape Who We Are*. Guilford, New York (1999).

Part III

Disentangling Temporal Simultaneous Contrasts

CHAPTER 14

RELATIVITY OF SCALES: APPLICATION TO AN ENDO-PERSPECTIVE OF TEMPORAL STRUCTURES

Laurent Nottale[1], Pierre Timar[2]

[1]*LUTH, CNRS, Observatoire de Paris et Université Paris-Diderot*
5 place Janssen, 92190 Meudon, France
E-mail: laurent.nottale@obspm.fr
[2]*262 rue Saint-Honoré, 75001, Paris,France*
E-mail: ptimar@magic.fr

The theory of scale relativity is an extension of the principle of relativity to scale transformations of the reference system, in a fractal geometry framework where coordinates become explicitly dependent on resolutions. Applied to an observer perspective, it means that the scales of length and of time, usually attributed to the observed object as being intrinsic to it, have actually no existence by themselves, since only the ratio between an external scale and an internal scale, which serves as unit, is meaningful. Oliver Sacks' observations on patients suffering from temporal and spatial distortions in Parkinson's and encephalitis lethargica disease offer a particularly relevant field of application for such a scale-relativistic view.

1. Introduction

The theory of scale relativity is founded on a new extension of the principle of relativity. This principle can be expressed by stating that the various properties of the physical world to which it applies do not exist in an absolute way, but only relatively to some reference system.

Up to now, physical theories of relativity have been applied to position (in space and time), orientation and motion (as measured by velocity and acceleration), and as a consequence, to gravitation.

The theory of scale relativity is a new generalization, in which the principle of relativity is also applied to scales (as measured in particular by the resolution of the measurement apparatus), and, as a direct consequence, to quantum mechanical systems and gauge fields. It has a potential wide domain of applications, because scale transformations play a central role, not only in physics, but also in many other sciences, in particular in biology, medicine and human sciences. Here we give an example of tentative application to the understanding of certain manifestations of Parkinson's disease.

2. Relativity of Position, Orientation, Motion and Gravitation

Concerning position, relativity means that the position of an object cannot be defined by itself, in an absolute way. This is clear if one imagines a single point-mass in a space which would be totally empty, except for it. The very concept of position of this point would have no meaning, and it could even not be defined. Therefore, there are no positions, only inter-positions. One needs, in order to define the position of an object, to refer it to other objects, which serve as reference system. This is true of any object, and then in particular of the reference system itself, whose origin can be defined only relatively to another reference. It would be of no use to try to go on and on, to a third reference system, etc…, in the hope to finally find an absolute reference. Indeed, the essence of relativity is that the property called "position of an object" is actually not an intrinsic property of the object. It is falsely attributed to the object in its own, while its true nature is that of a relation.

The same is true of orientation, with the difference that the changes of position correspond to the "symmetry" called translation, while the changes of orientation correspond to rotations. With rotations, another essential trait of relativity emerges: namely, the projective character of the physical "reality". The apparent length, width and height of an object changes as a function of its orientation relatively to the observer. The same change can be observed, if we turn the object in front of us, or if we ourselves turn around the object at the same angle.

But it was with its application to motion at the end of the XVIth century that the principle of relativity was officially born. The relativity of inertial motion was discovered in Western science by Galileo ("[...] for all things that take part equally in it, it does not act, it is as if it were not. [...] The motion is as nothing"), who elevated it to a principle: "Let us therefore set as a principle that, whatever be the motion that one attributes to the Earth, it is necessary that, for us who [...] partake of it, it remains perfectly imperceptible and as not being"[1].

It is nevertheless remarkable that this statement of Galileo's was anticipated by the principle of relativity-emptiness of Buddhist philosophy, according to which the various phenomena are empty of proper (absolute) existence and appear only in a relative and interdependent way. For example, one finds in Nagarjuna's 2000-year-old writings[2] an analysis of the relativity of motion (among other phenomena) which is almost word for word identical to Galileo's analysis.

Einstein's first insight about the relativity of gravitation, which he characterized as "the most fortunate of all his life" was of a similar nature. Namely, he realized in 1907 that if a man moves in free fall in a gravitational field, he no longer feels his own weight[3]. Einstein has translated this statement into the principle of equivalence, according to which a gravitational field and a uniform acceleration field are locally equivalent, so that a gravitational field can be locally canceled (or created) by the choice of an accelerating reference system.

The key point raised by Galileo and Einstein is, both for motion and gravitation, their absence of absolute existence. The velocity of a body, its acceleration, and the gravitational force felt by a body are not intrinsic or local physical properties of the body, but relative properties that depend on the reference system. They are not individual properties of objects, but inter-properties between objects, which have therefore the status of relations. By changing the reference system, the property changes (this is relativity), but there is one reference system, namely, the proper one, in which the property vanishes (this is 'emptiness').

A final general remark about the nature of the principle of relativity concerns its connection with space-time theories. It is also remarkable that all successive theories of relativity (Galileo's relativity, Einstein[4]-Poincaré[5]'s special relativity and Einstein[6]'s general relativity) are naturally achieved in terms of progress in the description of the geometry of space-time (respectively Euclidean space, then Minkowskian, then Riemannian space-times). This is easily understandable if one realizes that space-time is defined as an inter-relational level of description between objects. This is clear in the metric of general relativity which is a generalization of Pythagoras' relation. The account of more and more general transformations between coordinate systems therefore naturally leads to more and more general geometries of space-time.

3. Scale Relativity

In the theory of scale relativity,[7] the observation scale (in particular, the resolution at which a system is observed or experimented) is considered as characterizing the state of the reference system, on the same footing as its origin, its orientation and its motion. It is a long-known experimental fact that the scale of a system can be defined only in a relative way, since only scale ratios do have a physical meaning, never absolute scales. The principle of relativity is therefore extended so as to apply also to the scale transformations of the reference systems. In this approach, the resolutions are considered, not only as a property of the measuring device and of the measured system, but more generally as a property that is intrinsic to the geometry of space-time: in other words, space-time is considered to be fractal, i.e., explicitly scale-dependent.[8,9]

More generally, one gives up the hypothesis of differentiability of the space-time continuum, i.e., one works in a generalized geometry of space-time which contains both the differentiable and nondifferentiable cases. In such a framework, one therefore recovers standard differentiable physics, but new physics is also derived from the properties of nondifferentiable continua (which can be proved to be fractal).

The development of the theory involves

(i) the construction of scale laws coming under the principle of scale relativity[10] that describe the explicit scale dependence of a fractal space-time;

(ii) the construction of the equations of motion in a nondifferentiable space-time:[7] particles are identified with the space-time geodesics, and the main consequence of nondifferentiability and fractality is the transformation of classical mechanics into a quantum-type mechanics (in its standard microphysics form, but also in a generalized form that may apply to some fractal and partly nondifferentiable macroscopic systems[7,15]) ; the various mathematical quantum tools (complex wave functions, spinors, bi-spinors) can be built as manifestations of the non-differentiable geometry, while the Schrödinger[7,11], Klein-Gordon[11,12], Pauli[13] and Dirac[11] equations are successively derived as integrals of the geodesics equation;

(iii) the construction of gauge fields as manifestations of the fractality of space-time,[12,14] in analogy with gravitation being the manifestation of its curvature in Einstein's general motion relativity.

4. Log-Periodic Laws

One of the applications of the scale relativity approach concerns the log-periodic and discrete scale-invariant laws which have been found to give a good description of many spatial and temporal structures.[16] Such laws can be easily obtained as solutions of second order scale-covariant differential equations acting on the scale variables.[10] These equations have actually the form of a wave equation in the scale space, naturally leading to periodicity of the logarithm of a scale variable.

Let us consider in particular the case when such an approach is applied to discrete evolutionary time sequences.[17,18] The scale variable can be identified in that case with the time interval between two evolutionary events, $L_n = T_{n+1} - T_n$.

The scale relativity view leads to the idea that, instead of an invariance of these intervals (which would lead to the periodic behavior which is typical of the kind of phenomena we use to define clocks and time units), it is the *ratio* of these intervals which is physically meaningful. The simplest possible situation is therefore a case of invariance of these ratios, namely:

$$(T_n - T_{n-1}) / (T_{n+1} - T_n) = g = cst.$$

Such a simple geometric progression converges toward some critical time T_c, since the sum of the intervals is the well-known convergent series $(1 + g^{-1} + g^{-2} + ...) = g / (g - 1)$. The sequence of event chronology can therefore be described in terms of the log-periodic discrete law:[17,18,19]

$$T_n = T_c + (T_0 - T_c)\, g^{-n}.$$

This simple law, which is typical of the chronology of, e.g., earthquakes,[24] describes either an acceleration toward the critical time (precursor of the crisis), or a deceleration from the critical time (replicas of the crisis). When approaching the critical time, the intervals become shorter and shorter, in such a way that an infinite number of events is needed to reach it (in real situations, threshold effects imply a finite number): namely, the critical time is sent to infinity in terms of the scale variable $log|T\text{-}T_c|$, and it therefore becomes a *horizon*. When applied to space intervals, it has been known since Leonardo da Vinci to describe the discrete scale invariance of the lengths of branchs, sub-branchs, etc., of vegetal trees, rivers and various hierarchical arborescent systems.[17]

Such a law has been proved to yield satisfactory (and statistically significant) models of punctuated evolutionary sequences, such as those encountered in many lineages of species evolution,[17,18,19] of economic and social evolution,[17] of embryogenesis,[20] and of other evolutionary phenomena in human sciences.[21] The critical time obtained for the crisis / non-crisis rhythm of Western civilization on time scales of several thousand years (T_c = 2050-2080)[17] has since been confirmed by Sornette and Johansen on several independent R&D, economic, market and demographic indices on time scales of several hundred years.[22] It is remarkable that in all these cases, the values of the scale ratio g seem to cluster around well-defined values, in particular $g = 1.73$ (close to $\sqrt{3}$) and its square-root $g = 1.32$. The understanding of this universality is still a matter of conjecture.[17,23] (Note however that going from a value to its square-root corresponds to a doubling of the number of events considered, which may be a consequence, e.g., of changing the threshold in the choice of the events: it has been shown on several sequences that such a change does not affect the value of T_c, within error bars).[19]

It is also worth remarking that this law is a simple generalization of usual periodic laws. Indeed, one recovers the standard periodic clock as the special case $T_c \to \pm\infty$. One may therefore consider this law, not only as being specific to the description of the chronology of events in time crises (as it appears near the critical time), but also as a description of the spontaneous behavior of all natural clocks, which are known to drift with respect to each other if they are not regularly synchronized.

5. Application to Parkinson's Disease

Parkinson's Disease (PD) is a degenerative disorder of the central nervous system that belongs to a group of conditions called motor system disorders, which are the result of the loss of dopamine-producing brain cells. The four primary symptoms of PD are tremor, or trembling in hands, arms, legs, jaw, and face; rigidity, or stiffness of the limbs and trunk; bradykinesia, or slowness of movement; and postural instability, or impaired balance and coordination. We have learned that patients who survived Encephalitis Lethargica have PD-like symptoms.

The works of Dr Oliver Sacks[27] ('Awakenings' as a book, film and documentary) have led us to see the disease in a way that allows us to think we could use the scale relativity approach and principle to help understand it at a fundamental level.

Especially a symptom called festination – that is, involuntary shortening of stride and quickening of gait that occurs in such diseases as PD – is also used to describe speech. For example, the expression of words can accelerate while talking, and the space between words will become shorter and shorter. This happens to such a degree that it can be difficult to understand the individual who has festinating speech. The same thing can be observed with the "three claps test". The patient will clap every second then unconsciously accelerate the tempo and meanwhile the space between his hands will become shorter until he seems to freeze.

We think that the loss of the internal device that synchronizes us to external space-time would be expressed by chaotic-like perception of dilation and contraction. In forthcoming works, we will analyse these kinds of observations in a quantitative way, comparing them in particular to the log-periodic law described above, which has exactly the same kind of properties of time acceleration and deceleration (that may go up to freezing) and of space dilations and contractions, and which emerges as a natural manifestation of scale relativity.

We shall only give here, as a first step, various elements supporting such a view. The behavior of PD patients indeed seems to definitely come under the principle of scale, which concerns in the first place the essence of any measurement (and therefore, by extension, our own senses and sensory organs). Because of the universality of this principle, our perception system is itself subjected to it, in particular as regards our location in space and time. This location can be effective only through a measure (even if it is only implicit), i.e., through a comparison between a reference system (which defines units) and the measured 'object'. Therefore, the estimates of length and time intervals made by the perception system of any sentient being (whether Parkinsonian or not) necessarily follow from the comparison between spatial and temporal objects with internal rules and clocks. The main point consists of understanding that, concerning time, only the ratio between the measured time interval and the period of the reference clock has a meaning, as do concerning space, the ratio between the size of the measured object and the units defined on the reference rule. But a space or time interval, by itself, from its own side, without any reference, has no physical meaning, and actually no existence at all.

The consequence of this fundamental relativity of scales is that it is totally impossible to distinguish a dilation of the object (seen or measured) from an equivalent contraction of the unit of reference (either spatial or temporal). This impossibility is not a simple difficulty that a more complex experiment could hope to overcome. It is fundamental, since it involves the very mode of being of phenomena, which never exist in an absolute way, but only relatively to a reference system. So a distinction between a dilation of the object and a contraction of the unit would have actually no meaning. If $Y = B/A$, multiplying B by 10 or dividing A by 10 gives exactly the same result, $Y' = (10\ B)\ /\ A = B\ /\ (A/10)$. The result of the perception is Y then Y' itself, but neither of the two terms that contribute to it.

The consequence as regards Parkinsonian and post-encephalitic patients is that their deformed perceptions are no more or less 'illusory' ("What you call 'festination' is no more than a normal reaction to an abnormal perception. We Parkinsonians suffer from illusions"[27], as do unaffected subjects. In both cases, what is perceived is only the ratio between external and internal scales, so that a deformation of the internal units is by essence indistinguishable from a deformation of the external (spatial and temporal) objects. Since it is a general and natural tendency for our perceptions to be projected outside (as proved, e.g., by our experience in the dream state of consciousness), such an internal deformation naturally appears as projected outside. The difference between the unaffected and the affected subject is therefore only a question of synchronization (or of desynchronization) of internal rules and clocks relative to the external ones.

Let us indeed quote Sachs:[27] "Frances D [...] has often depicted for me the strange and deeply paradoxical world in which she lives. These patients describe a fantastical-mathematical world remarkably similar to that which faced Alice. Miss D. lays stress on the fundamental distortions of Parkinsonian *space,* on her peculiar difficulties with angles, circles, sets, and limits. She once said of her 'freezing': 'It's not as simple as it looks. I don't just come to a halt, I am still going, but I *have run out of space to move in...*You see, *my* space, *our* space, is nothing like *your* space: our space gets bigger and smaller, it bounces back on itself, and it loops itself round till it runs into itself.'"[27]

Sacks observed that patients may experience illusions or misleading conjectures about space-time when they are hurried or festinant. Equally, they may have illusions when dawdling or procrastinant: "Now let us examine Parkinsonian behaviour in this light, concerning ourselves especially with *illusions of scale* [emphasized by Sacks]. I have had letters from Frances D., and other Parkinsonian patients, which showed singular (and often comic) disparities of scale: I remember one such letter from Frances D., of which the first page was in a perfectly formed but microscopic hand (so small I needed a magnifying-glass to decipher it), while at the start of the second page (which was in normally sized script) she had written: 'I see that what I wrote yesterday was far too small, although I didn't see this at the time. Today I have borrowed a ruler and ruled lines on this page, and I will use the lines to guide my writing, so that I don't inadvertently make it tiny again.'"[27] Sacks recalls that other letters Frances D. and certain other patients produced were, although perfectly shaped and formed, sometimes marked by enormous writing, a fact these patients were oblivious to.

Several aspects of the set of symptoms and reactions of patients[27] may then be accounted for in this framework, such as freezing, distortions, dilations, and contractions, and, as a consequence, the necessity of external stimuli:

- rhythmic, but also moving music, as regards time: it amounts to replacing the distorted or destroyed internal clock by an external clock which is *likely to be internalized*: "Edith T., a former music teacher, [...] said that she had become 'graceless' with the onset of Parkinsonism, that her movements had become 'wooden, mechanical-like a robot or doll,' that she had lost her former 'naturalness' and 'musicalness' of movement, that - in a word - she had been 'unmusicked.'"[27] Edith T. added that, fortunately, the disease was accompanied by its own cure: "'Music,' she said, 'as I am unmusicked, I must be remusicked.' Often, she said, she would find herself 'frozen,' utterly motionless, deprived of the power, the impulse, the *thought* of any motion; she felt at such times 'like a still photo, a frozen frame' – a mere optical flat, without substance or life. In this state, this statelessness, this timeless irreality, she would remain, motionless-helpless, *until music came*",[27] Sacks reports that he is often asked what music can serve to awaken such patients, and what exactly is going on at such times. It appears that rhythmic impetus must be present, but that this rhythm must be 'embedded' in a melody: "Raw or overpowering rhythm, which cannot be so embedded, causes a pathological jerking; it coerces instead of freeing the patient, and thus has an anti-musical effect. Shapeless scrooning ('slush,' Miss D. calls this), without sufficient rhythmic/motor power, fails to move her – either emotionally or motorically – at all. One is reminded here of Nietzsche's definitions regarding the pathology of music: here he sees, first and foremost, 'degeneration of the sense of rhythm.' 'Degenerate' music sickens and forces, 'healthy' music heals and frees. This was precisely Miss D.'s experience; she could never abide 'banging' or 'slush,' and required a firm but 'shapely' music."[27] The question arised whether any kind of firm and shapely music would suffice to get Frances D. going in the right way. This turned out not to be the case: "The only music which affected her in the right way was music she could *enjoy;* only music which moved her 'soul' had this power to move her body. *She was only moved by music which moved her.* The 'movement' was simultaneously emotional and motoric, and essentially autonomous (thus distinguishing it from passive jerkings and other pathology)."[27]

- Regular spacings, rules, staircases, help from someone who walks in a normal way, etc., as regards space: it amounts to replacing the distorted or destroyed internal rules by external rules, also *likely to be internalized.* ("So, in the most general terms, we find that ametric-dysmetric Parkinsonian patients can be activated and regulated, ordered and organised, by either sort of measure: by regular marks in a conventional, formal (and linear) space-time, e.g. stairs, steps painted on the ground, clocks, metronomes, and devices that count in a simple, regular, and orderly manner; or by co-action and co-ordination with a concrete, living activity or agent. Thus, in the case of Miss D., the chalking of regular lines on the ground enabled her to walk stably, but like a puppet or robot; but taking her arm, and strolling with her, enabled her to stroll like a normal human being."[27]

The keypoint here seems to be the need for introjection by the patient of the clocks and rules. Purely mechanical clocks and rules are not sufficient, as many quotations of patients by Sacks shows.[27] This may allow one to understand (but also to suggest improvements to) the various strategies that have been proposed to help PD patients.[25,26]

6. The Will of the Ball

A scene in the movie shows Dr "Sayer" with a patient, an old lady paralysed in her wheelchair. He turns his back a while to write some clinical items but when he looks at her again she is bent forward in such a way as to catch in her right hand her glasses that have slipped from her nose. Then she maintains the position. This observation gave him the paradigm that was expressed as the "will of the ball". As patients are deprived of their own willpower, they seem to borrow it from an external provider. Next scene: the same patient is shown to the senior neurologists, catching on the volley, in an amazingly single gesture, a ball that is thrown to her.

What is the "will" of the ball? One may think that even an old woman cannot dislike baseball if she is American, and assume that the will of the object is the internalisation of the very movement of a recognized and somewhat liked (or, rather, "sufficiently liked") object. That is what moves us. The object can be anything that we encounter in the (external) world. It may be a thing, a music, a rhythm, an order spoken or not spoken, a touch. It works on us in an unconscious way. This means we don't have to think or consider our wish (which is much too long a process to catch anything) but it is not only a simple catching reaction. Dr Sayer said that a simple reaction would have been to deflect the ball.

For a psychoanalyst, an object is a psychologically invested object. Otherwise, it would not exist for us. A psychiatrist is used to learning from pathological manifestations to understand regular psychic processing and vice versa. A perfectly common process is called "introjection". Freud wrote: "In so far as the ego is auto-erotic, it has no need of the external world, but, in consequence of experiences undergone by the instincts of self-preservation, it acquires objects from that world, and, in spite of everything, it cannot avoid feeling internal instinctual stimuli for a time as unpleasurable. Under the dominance of the pleasure principle a further development now takes place in the ego. In so far as the objects which are presented to it are sources of pleasure, it takes them into itself, 'introjects' them (to use Ferenczi's term); and, on the other hand, it expels whatever within itself becomes a cause of unpleasure."[28] It seems that introjection is a very appropriate concept to "send the ball" between the level of neurological observation and that of psychological understanding. Why is it working? Because the damaged cerebral zone is one part of the structures that support a function we can describe as a synchronization device. It shows that the use of some external device seems to bypass the damage and dramatically allows the function. In a non-invasive way, it seems to work like electro-stimulation does (the electrodes are introduced downstream).

Dr Sacks shows that, for one person, the way to find a proper external device is the result of intuitive finding, of playing capability. The movie shows, for example, the drawing of a hopscotch game giving Dr Sayer the solution for Lucie's walking inhibition: she absolutely needed a continuous guideline on the floor to walk from her hell to her heaven. Her desire was not enough to support action. Testimonies of the vanishing of the very ability to think of moving led Dr Sacks to write that a frozen Parkinsonian does not exist in the regular space time. That gives an important demonstration of the need to distinguish between the desire and the actual move. The move needs a well enough working body and welcoming context. The desire is possibly an aid to finding/creating (we use Winnicott's terms) ways to make things good enough.[29,30,31]

Acknowledgments

We would like to gratefully thank Prof. Otto Rössler and Susanne Vrobel for their kind invitation to contribute to this festschrift.

References

1. Galileo Galilei, *Dialogo sopra i massimi sistemi*, Torino (1975) (Dialogue sur les deux grands systèmes du monde, 1630).
2. Nagarjuna, *The Philosophy of the Middle Way*, (Kalupahana D.J. trad., State University Press of New York, 1986).
3. A. Pais, *Subtle Is the Lord: the Science and Life of Albert Einstein*, (Oxford University Press, New-York, 1982).
4. A. Einstein, *Annalen der Physik* **17**, 891 (1905).
5. H. Poincaré, *C. R. Acad. Sci. Paris* **140**, 1504 (1905).
6. A. Einstein, *Annalen der Physik* **49**, 769 (1916), translated in *The Principle of Relativity,* p. 111, Dover (1923, 1952).
7. L. Nottale, *Fractal Space-Time and Microphysics: Towards a Theory of Scale Relativity*, (World Scientific, Singapore, 1993).
8. L. Nottale and J. Schneider, *J. Math. Phys.* **25**, 1296 (1984).
9. L. Nottale, *Int. J. Mod. Phys.* **A4**, 5047 (1989).
10. L. Nottale, in *Scale invariance and beyond*, Proceedings of Les Houches school, B. Dubrulle, F. Graner & D. Sornette eds., (EDP Sciences/Springer,

1997), p. 249.
11. M.N. Célérier and L. Nottale, *J. Phys. A: Math. Gen.* **37**, 931 (2004).
12. L. Nottale, in *Relativity in General*, (Spanish Relativity Meeting 1993), J. Diaz Alonso & M. Lorente Paramo, Eds., p. 121, Editions Frontières, Paris (1994).
13. M.N. Célérier and L. Nottale, *J. Phys. A: Math. Gen.* **39**, 12565 (2006).
14. L. Nottale, M.N. Célérier and T. Lehner, *J. Math. Phys.* **47**, 032303 (2006).
15. L. Nottale, *Astron. Astrophys.* **327**, 867 (1997).
16. D. Sornette, *Phys. Rep.* **297**, 239 (1998).
17. L. Nottale, J. Chaline and P. Grou, 2000, *Les arbres de l'évolution: Univers, Vie, Sociétés*, Hachette, Paris (2000).
18. J. Chaline, L. Nottale and P. Grou, *C.R. Acad. Sci. Paris*, **328**, 717 (1999).
19. L. Nottale, J. Chaline and P. Grou, in *Fractals in Biology and Medicine*, Vol. III, Proceedings of Fractal 2000 Third International Symposium, G. Losa, D. Merlini, T. Nonnenmacher and E. Weibel, Eds., Birkhäuser Verlag p. 247 (2002).
20. R. Cash, J. Chaline, L. Nottale and P. Grou, *C.R. Biologies* **325**, 585 (2002).
21. I. Brissaud, *Math. Sci. Hum., Mathematics and Social Sciences*, **178**, 41 (2007).
22. A. Johansen and D. Sornette, *Physica* **A 294**, 465 (2001).
23. D. Queiros-Condé, *C. R. Acad. Sci. Paris* **330**, 445 (2000).
24. D. Sornette and C.G. Sammis, *J. Phys. I France* **5**, 607 (1995).
25. T.A. Kaminsky, B.J. Dudgeon, F.F. Billingsley, P.H. Mitchell, S.J. Weghorst, *Journal of Rehabilitation Research and Development*, **44**, 437 (2007).
26. Rescue Project, http://www.rescueproject.org/
27. O. Sacks, *Awakenings*, pp. 60, 61, 347, 339, 340 354 (Picador, 1973-1990).
28. S. Freud, *Instincts and their vicissitudes*, Complete Works, Ivan Smith, p. 2971 (2000).
29. L. Nottale, *Être et ne pas être,* in *Penser les limites. Ecrits en "honneur d'André Green*, César Botella Ed., Delachiaux et Niestlé, Paris, p. 157 (2002).
30. P. Timar, *La transitionnalité de l'espace-temps psychique*, Société Psychanalytique de Paris, Science et Culture, http://www.spp.asso.fr/Main/PsychanalyseCulture/SciencesDeLaComplexite/Items/3.htm
31. L. Nottale and P. Timar, *De l'objet à l'espace psychique*, *Psychanalyse et Psychose* **6**, 195 (2006).

CHAPTER 15

UNPACKING SIMULTANEITY FOR DIFFERING OBSERVER PERSPECTIVES AND QUALITIES OF ENVIRONMENT

Bill Seaman

Chair and Graduate Program Director, Digital+Media
Rhode Island School of Design, 2 College St., Providence, R.I. 02888
E-mail: bseaman@risd.edu

We take for granted that we know what an observer is. In actuality, there are many different kinds of observation that we employ, related to differing qualities of environment. These modes of observation include direct human observation; human observation that is augmented by differing tools and/or machines; computational models functioning as machinic observers, in which humans become super-observers; chemical observers where particular changes register particular qualities; and potential future observers - Neosentient machines. For each of these different kinds of observers, the notion of simultaneity is problematic in some manner. The definition is clear enough: simultaneous - existing, occurring, or operating at the same time; concurrent: simultaneous movements; simultaneous translation. Yet, notions of time are different on the micro and macro scale. Humans also have been observed to include up to a one half-second time buffer, adjusting for the body's distributed sensing functionality and potentially its embodied relation to memory and past thought. These facts throw off our concept of something happening at "the same time." Thus, time, spatial relations, the body, differing machines and chemical modalities each can be framed linguistically as having different perspectives in relation to the notion of simultaneity and/or the limits of registering simultaneous events.

1. Introduction

Our knowledge gained from direct observation is further informed by the knowledge that we have obtained through augmented forms of observation, new forms of machinic measurement, computer simulation, social and cultural learning, science as a set of fields, different forms of logic, linguistic framing, self reflection, and memory. The definition of the observer must in each case be carefully qualified in relation to the particular instance of study as well as the modality of that particular study. As we begin to work in teams of researchers that cross traditional boundaries, the definition of 'the observer' will need to be unpacked according to the particular domain of 'experience' and/or of 'experiment' that is being undertaken, as well as the perspectives that become enfolded in such an undertaking.

2. The Human Observer

Each individual is the center of his or her own particular sense of *nowness*. Physics has shown how entities registering events in slightly different locations perceive the timing of those events in subtly differing ways. These differences of locality are amplified by distance to the event as well as by motion in relation to that event. The event's positionality and spatial movement are also factors that come into play. As we unpack notions surrounding definitions of the observer and their 'relative' natures, we discover a series of difficulties related to our conceptions of simultaneity.

The human observer has many interesting characteristics. They represent a physical volume in space that is continuously in motion across multiple scale domains from the micro, to the macro. Human observers are dynamically nested in an environment that enables them to consciously make observations. They are examples of *far from equilibrium* [a] dissipative structures, and are thus dynamic systems.[1] They become structurally coupled with their environment to maintain their

[a] Scott Kelso has articulated a dynamic systems approach to cognition in his book Dynamic Patterns (cf. S. Kelso 1988).

living status, to make first-hand observations, as well as to communicate about those observations.

A definition of simultaneity is as follows: simultaneous - existing, occurring, or operating at the same time; concurrent: simultaneous movements; simultaneous translation.[2] Thus we must first ask what in particular is to be meant when two events arise simultaneously? What are the specific relational qualities which contribute to the notions surrounding the simultaneously registered?

We are interested here in events that are in some manner observable to the body without a machinic apparatus (in this human-observer-centered definition) via multiple modalities of human sensing.

Events have a particular set of qualities as do the sense modalities of the observer that register them:

(i) The event inhabits a particular volume (or changing volume) that is observable by the human observer who has a volume;
(ii) The event exhibits a spatial relation to the observer;
(iii) The event exists over a particular duration of time;
(iv) The event takes place in a particular environment that may have qualities that alter the observation of that event.

We must also agree that the observer has the intention to observe for simultaneity in regard to the particular event under scrutiny. In general, we could say that the observer has an intention to observe, and seeks to be aware of the observation. The observer socially, culturally and linguistically frames the observation as part of this process of observation. This framing may occur with more or less self-reflection in terms of its social, cultural and linguistic spheres.

In *A Dynamic Systems Approach to the Development of Cognition and Action*, Thelen and Smith point out the following: "...a category is created in a context in a trajectory of internal activity in time. The trajectory *always* is a complex product of the immediate context, the just prior internal activity, and the history of reentrant mappings between the heterogeneous processes that make up the system."[3] They talk about this process as being time dependent. "By 'time dependent' we mean that each event in brain and body has not only a here and now but also a history and an effect on the future."[3] Categorization as a linguistic framing mechanism is also always implicated as a human tool in articulating simultaneity.

One problematic notion of defining simultaneity for a particular human observer is the following: the human becomes aware of his or her environment at a different speed consciously than unconsciously. Benjamin Libet in his book *Mind Time* has experimentally shown that conscious awareness has a measurable lag. He states: "Our experiment showed that people thought they were aware of a sensory stimulus about 500 msec before they could possibly become aware of the stimulus. This discrepancy became known to us objectively; it is no longer a theoretical speculation. We called this phenomenon 'subjective referral of consciousness sensory awareness backward in time'."[4] If we are registering something to be classified as happening "at the same time" via one or more of our sense modalities, we therefore encounter a problem. If our consciously observed awareness of an event is actually approximately 500 msec after the event, which "same time" are we talking about – the actual time at which the event has occurred, or our delayed conscious registering of that event in time? In this case, we are considering our observer to be conscious of the event that he or she is seeking to observe. Yet, this consciousness registers a false reading of timing.

Libet also points out other qualities specific to our perceptual apparatus which may affect our understanding of "simultaneous" events. Given the distributed nature of the body and the time it takes for signals to reach the brain, the body brings these signals together to make a coherent understanding. Libet states: "The primary evoked potentials can have a latency between about 5 and 40 msec, depending on the location and the modality of the stimulus. If all simultaneous stimuli are perceived subjectively by different sense modalities as being synchronous, we would have to assume that the brain does not "consider" this range of variability in latencies as subjectively significant."[3] Thus, our notion of the "synchronous" or happening "at the same time", here must also include a human duration of acceptable latency or difference.

It is interesting to note that our enfolding of past experience also can potentially alter our observations. Libet argues the following:

> "Our evidence indicates that a substantial period of neural activity (500 msec of 'on' time) is in fact required to elicit awareness of the sensory event. That delay provides a sufficient psychological opportunity during which unconscious brain patterns can alter the content of the experience before awareness of it appears. Indeed, the experimental phenomenon of subjective referral of a conscious sensory experience backward in time provides relatively direct evidence for one kind of modulatory distortion of the subjective experience. The delayed experience is subjectively timed as if it were not delayed at all."[3]

The projective nature of our ongoing observations, enfolding our embodied history, is central to a deep knowledge of current context. Our 'meaning summing' enfolds a series of thought processes during the delay period, informing our understanding of our 'conscious' now. Thus the body intermingles multiple kinds of knowledge to inform our sense of the present context as it unfolds.

Yet Libet points to other kinds of distortions. He states: "Any modulations of modifications of the developing experience would be unique to the person involved. It would reflect the person's own history of experiences and his (or her, Seaman's insertion) emotional and moral make up. But the modulations are made unconsciously. Consequently, one may say that the unique nature of a given person can express itself in unconscious processes."[3] Thus, our enfolding of past experience also contributes to our understanding of the current context under observation arising from the subconscious domain. This may also *taint* our ability to register events as happening simultaneously or shift our linguistic framing of a particular interpretation related to the timing of an event. Such thought bias can even shape our ability to register an event in terms of a particular contextual classification or "category."

On a deeper level, we must state that time is a continuum functioning in ongoing relation to each 'observer.' We can humanly parse time to differing degrees, yet in actuality between any two moments is an infinity of potential fractional divisions. Humans have their biological limits in terms of parsing this infinity using their senses and self-awareness. The closer we are in bodily proximity to an event, the less evident timing differences become. Our "now" appears to us with events happening from many different distances. The experiential qualities of these many events become folded into our individual self-centric sense of the now. The speed of light and sound predicate a different *frame of observation* for an event relative to each observer. The observer can at times register a discrepancy in the witnessing of an event's simultaneity by differing senses, even at quite a short distance away from it. How often have we watched someone hammer something in the distance? The visual stimulus reaches us long before the sound, making an odd impression of the hammer making the sound as it has long since left the surface that it has hit. Thus, each sense also enfolds a particular set of qualities in terms of registering particular events.

Think of the light of a distant planet that just now reaches us – our human-centric now. We ask how much time has passed since that perturbation was originated? Given that particular time frame and distance, we do not know the current state of the source of emission, only our time-centric perception of it.

We may seek to define an "objective" delay time with which to adjust our approximation of the simultaneous. Unfortunately, due to the biological differences inherent to each observer[3], we can not accurately define a precise offset with which to "adjust" our measurements. Thus, human observers may only register a vague, shared sense of simultaneity based on location and motion of both observer and event and the particular properties of the sense modalities employed as well as the nature of the environment that the event is being registered within.

The human observer must accept what Brian Massumi calls the "smudging" of the now, where our limits and qualities of perception become the "now" as we know it, lived within the parameters of our human framing. In his work "Parables for The Virtual", Massumi, in the chapter entitled *Strange Horizon: Buildings, Biograms, and the Body Topologic,* states:

> "Experience smudges. You get a thirdness: a supplemental effect not reducible to the two stimuli's respective durations considered

> separately. You get a supernumerary difference, a qualitative difference arising from the interrelation of recursive durations. To put it bluntly, you get a relational time-smudge. A kind of hypertime. Think about it. Since any lapse of time is infinitely divisible, and at every instant there must be some kind of stimulus arriving through one sense channel or another, if you try to fill in what happens in the half-second lapses of awareness, things get downright hallucinogenic. Say at .01 seconds a second loop begins; at .02 seconds another begins ... but at .015 seconds there will have been an intervening beginning, and also at .0125. You're left with an infinite multiplication of recursively durational emergent awarenesses, madly smudging each other. You get an exponentially self-complicating relational mess."[5]

Accepting the spatial biases of bodies in motion and the subtle differences of bio-timing may well mean accepting a range of problematic notions surrounding simultaneity and the human observer. Another avenue to explore is the augmentation of the human observer with machinic affordances that may better register the qualities of events in time and space.

3. Human Observation of Measurements that are Augmented by Differing Tools and/or Machines

It seems a natural human desire to attempt to transcend the limitations of the body as regards time, space and simultaneity, and so we devise tools to augment our perceptions. Given the limitations of the human observer as discussed above, we seek to find other ways to measure or demarcate simultaneous events that do not become distorted. We do this by decoupling the measuring instruments from the human observer and their human-centric time frame. We devise tools and machines that amplify and/or transduce particular qualities (bringing us metaphorically "closer" to the seeming understanding of an event outside of the body's non-machinic capabilities); we elucidate systems of data visualization and sonification (various event transduction systems and/or simulation systems) to point at particular qualities of difference over time; and we seek to parse time to a finer degree with machines that can recognize a far greater number of divisions per unit time.

We must always take into account the time that a transduction takes to be registered by our observational systems, as well as the nature of the environment in which the event is being measured. In humans, this "difference" in biological time-frames, based on specific qualities of our physiology, becomes a problem as articulated above. One wonders to what degree we can also attribute a related set of problems to machines? Do machines always take the same time to register an event perturbation? Can we always subtract this duration (or make it equivalent across a series of measurements) to point at simultaneity? To what degree do the environmental circumstances also alter such a measurement? We must say that for every newly authored machine, a new set of qualities related to observation become enfolded into our human, observer-centric understanding of the world. We must in each case intellectually adjust and extend our linguistic frame appropriately in order to share our experiences of derived "transduced" and/or "simulated" data. Gordon Pask in *An Approach to Cybernetics* states: "Nearly always we can tell the machine something, and surely, we should tell it, as part of its programme. But we should not prevent it indulging its own breed of recognition or expect logical nicety amongst the attributes it selects. We do not really recognize signatures in terms of neat geometrical attributes - they 'remind us of faces' or seem more or less "wiggly" and we must tolerate just as unruly attributes in an automaton."[6]

Language as a framing and articulating device must also be seen as technology on an equal footing with the machinic data that it helps to articulate. Our mathematical systems and computer codes are expanded examples of these linguistic framing devices, yet as Pask elucidates, new qualities of interpretation enter the domain of observation engendered by the nature and qualities inherent to our tools. When we examine reality closely, we observe a series of cases in which what had seemed to be a scientific "certainty" proves to be of no certainty at all. On subsequent examination (or through additional study and/or advanced technological means), these "certainties" have proven to be false. The following litany, provided by Hamilton and Bonk, points to some of these cases:

> "Heisenberg's Uncertainty Principle, Von Neumann's Monte Carlo Method, Gödel's Incompleteness Theorem, The 2nd Law of Thermodynamics and statistical description of matter, Wittgenstein's verdict on the sayable and the sentence. These are all acknowledgements of the limitations of an hermetic cognitive system."[7]

In terms of the event qualities discussed above, we must in each case ask how a tool or augmenting machine alters our understanding of the observed:

- Any machine or tool will always have a degree of inaccuracy as the human does;
- The machinic observer, like the human, is a volume and represents a set of specific embodied qualities and a spatial locality as an observer-centric system;
- A machine or tool may be able to highly specify spatial location of an event over time as may a mathematical or computational system. Each of these modes of observation also has an 'acceptable' degree of accuracy with which we may (re)define notions of simultaneity.
- An event exhibits a spatial relation to the machinic observer or tool over time;
- Machines may be capable of far higher time resolution than human observers;
- Linguistic framing will continue to pose problems regarding how we define particular terms – e.g. the "observer" is defined differently in different experiments.

Machinic amplification and/or transduction systems have opened up entirely new worlds of observable domains – think of the scanning-tunneling microscope as one example. Yet we must always remember that we are in this case observing a mediated experience that has attributes unique to its environment, qualities specific to the mode of transduction, and qualities specific to the mode of legibility (visualization and sonification). Again, we ironically decouple the human observer from a time-centric perspective of direct observation in order to more accurately register and parse time via particular tools and machines that take on the observer-centric time frame for us.

We must always consider how our tools and linguistic framing bias our understanding of events. Can we objectively minimize the time it takes a tool to register an event, or measure aspects of it to an adequate degree of accuracy so as to better know the time/space qualities of the event in itself? Does the ambiguity of language also pose a set of hidden problems? When we truncate a particular number in a problem due to the limits of some computational system, how does that approximation affect future measurements?

Each event has an infinite set of parameters that we can never get at – an infinite depth of time/space qualities that fall outside of the human and/or machine potential of registering (see Von Neumann below). We devise systems and machines that enable us to better approach that depth, or delimit a piece of an event to study it. Yet, again invoking Zeno's paradox, we can only go part way there (given that time is a continuum), never reaching the exactitude of knowing the time of an event in and of itself. In our considerations of simultaneity we are always up against the limits of the tools, machines and framing languages in terms of articulating finite qualities of accuracy. This is not to say that we have not made extremely accurate devices. Additionally, we must account for the fact that human observers will subsequently interpret and frame the data that such machines and tools provide.

4. Micro vs Macro Time Scales

We have spoken of limitations in both human and machines observation systems in the registration of time/space phenomena. We have also discussed the limitations of framing languages including text, mathematical languages and computational code. When we shift scales and seek to study micro events, we again come up against a problem in our ability to demarcate the simultaneity of events. In his book, *Endophysics, The World As Interface*, Rössler states: "The observer is the sum total of the dynamical processes which go on inside him or her. There is a macro dynamics (the coarse-grained responses of the dissipative structure called "observer"), and there is an underlying, much faster microdynamics. Even the most rapid macro change in the observer lasts several orders of magnitude longer than a micro time slice does. The micro time slices therefore are necessarily 'integrated over' from the macro point of view. An analog to 'flicker fusion' – but much faster – can thus be predicted to hold good for micro time slices."[8] In terms of our ability to articulate simultaneity, we must point to an interesting fact, that time functions differently on the micro scale. Rössler states: "Norman Campbell proposed in 1921 that all the phenomena characteristic of the quantum domain might be explicable by a single monistic assumption; that time ceases to be well-defined in the microrealm. More specifically, he said: "[Time is] a statistical conception, significant only with respect to large aggregates of atoms [so] that it is as meaningless to speak of the time interval between atomic events as of the temperature of an isolated molecule."[9] In this light, we further point to problematics surrounding our subject. We again approach a limit that we must try to transcend. Alternately, it must also be clearly stated that the human observer, even if de-coupled, at this time, must always be enfolded in terms of linguistic interpretation of machinic results. Systems that include other classes of observer must at this time always include the human observer as the final linguistic interpreter of the results of the computation, even if the computer is contributing to results on a deeply informed manner through human programming. This situation may change when Neosentient machines participate as self-aware observers (see below, section 7).

5. Computational Models Functioning as Machinic Observers, in Which Humans Become Super-Observers

More and more regularly, we find ourselves using computer simulations to help provide insights into the nature of reality. By devising a highly specific model of the physics that characterizes our world on the scales of the micro and/or macro domains, we can begin to elaborate a system that houses both the (computational) observer and the environment within a single system. Peter Weibel provides this articulation of

Endophysics, named by David Finkelstein[b] and later elucidated at length by Otto Rossler:

> "Endophysics shows us to what extent objective reality is necessarily dependent on the observer. Ever since the introduction of perspective during the Renaissance and of group theory in the 19th century, we have known that the appearances of the world depend in a lawful manner on the localization of the observer ("codistortion"). Only if one is completely outside a complex universe is a complete description of the latter possible (cf. Gödel). According to endophysics, it is only in a model that this position on the outside of a complex universe is possible, but *not* in reality itself. Endophysics hence provides an approach to a general model and simulation theory (and also to the "virtual realities" of the computer age). It is an outgrowth of chaos theory, to which Otto Rössler has contributed since 1975."[8]

The human observer in the future will run such a computational simulation and gain insight into situations that might normally be inaccessible due to the nature of measurement problems and the nature of time at the micro scale. Rössler discusses another problematic aspect of discerning simultaneity as defined by Von Neumann:

> "A chaos-generated cellularity may, if it is the finest systematic feature of a macroscopic system like you and me, leave some indelible mark on everything one touches or tries to touch … The first who apparently saw this was John von Neumann, when he formulated: "The result of the measurement is indeterminate because the state of the observer before the measurement is not known exactly. It is conceivable that such a mechanism might function because the state of information of the observer

[b] Finkelstein designed a program for a "holistic" physics in the spirit of Bohr, but discrete. He hypothetically attributed both the quantum limit and the relativistic limit to the fact that the whole is not accessible to us. Later, he indicated an explicit example of a dissipative finite automaton (computer) whose internally evaluated state is different from the objectively existing one. Shortly thereafter, he endorsed the two notions "physics from without" and "physics from within," by proposing the antonyms "exophysics" and "endophysics" as more attractive terms. The name "endophysics" is his creation.

> regarding his own state could have absolute limitations, by the laws of nature."[10]

Rössler has posited the notion of giving "assignment conditions" to be authored for the virtual observer as part of computer code (or in equations) within a highly exact simulation environment as a means to approach this problem. Where "initial conditions" cannot be known, one might "specify" the observer "microscopically.":

> "Newton divided the world into laws and initial conditions. This ingenious reduction of the whole world to comprehendible machinery left out one element – assignment conditions. The spin theory of Michael Conrad and O.E.R. offers itself as a possible starting point. This macro-assignment needs to be differentiated from micro-assignment."[11]

Elsewhere, Rössler further elucidates the concept of "assignment conditions":

> "Besides the two types of agency introduced by Newton (the "laws" and the, once-and-for-all specified, "initial conditions"), there is a new type of agency: the "assignment conditions." Assignment is a deliberate accident, so to speak. This can be understood as follows. The same universe run in a computer, so to say, acquires many faces (interfaces or cuts) depending for which particular, microscopically specified, subsystem (called observer) the rest of the universe is put on screen. A microscopic change in the observer-world assignment can radically change the interface. "[12]

Thus, given the limits of computability, and the limits of reflecting the complexity of experience, such systems hold great promise in helping to elucidate the world in a manner that transcends the use of current tools and/or observational systems. Again, we point to the limits and qualities inherent to computation, mathematics and linguistic framing surrounding the articulation of simultaneity. Yet, pragmatically we must live within these limits and come to devise new technologies and linguistic framings to inform us about the world with increasing accuracy.

6. Chemical and/or Electrochemical Observers where Particular Changes Register Particular Qualities

Another class of observer might be called the chemical and/or electrochemical observer. Yet, we must also admit that humans are members of this class – functioning as sentient electrochemical observers. Chemical observers function in relation to an event and reflect state changes within particular chemical or electrochemical reactions. The diffuse nature of chemical reactions also make the notion of simultaneity difficult to discern – part of a chemical reaction may have transpired while other parts of the reaction are still in process (see 2. The Human Observer above). We might ask 'simultaneous with what part of the reaction process'? Chemical observation may be used in conjunction with other observational systems mentioned above, forming hybrid observational approaches. Again, we here speak of limitations of the observing system in terms of acceptable accuracy, and the qualities and nature inherent to the observing system be it singular or hybrid in nature.

7. Neosentient Machines - Potential Future Observers

Rössler and I have discussed the potential of creating an intelligent situated computer and related robotic system. This presents a new class of observer. In an operational definition, we consider a *Neosentient* robotic entity to be a system that could exhibit the following functionalities: it learns; it intelligently navigates; it interacts via natural language[13]; it generates simulations of behavior (it 'thinks' about potential behaviors) before acting in physical space; it is creative in some manner; it comes to have a deep situated knowledge of context through multi-modal sensing; it displays mirror competence.[14,15] We have entitled this entity *The Benevolence Engine.* We believe that the inter-functionality of this system is potentially complex enough to operationally mimic human sentience.[16] Such an observer might be able to dynamically function as a measuring system; function as a computational simulation system; parse time on a unique scale; have electrochemical computing devices participate in the process[16]; and have introspection about these observational potentials. Again, like the other systems, this system would also have a set of limitations, yet it may assert some interesting new characteristics in terms of being a first-hand observer to measurements, given that the system has introspective capabilities. It thus does not need to be de-coupled from the event in the same way a human observer does.

8. Conclusions and Outlook

I have shown our relation to simultaneity to be problematic as it pertains to different classes of observer and the relative perspectives that arise through their observatory natures. In particular each class of observer has a set of limits and qualities inherent to its processes of observation in relation to both the environment and scale on which the event occurs; as well as the nature and characteristics of the event itself. I have discussed a series of limitations that pertain to differing approaches to observation including the limits of language, the limits of mathematical systems (as an example of a linguistic system), the limits of computational systems (also linguistic in nature) and the limits of human observation related to sensing and cognition. In this light, one must carefully frame the acceptable limits that apply for a given experiment whenever articulating simultaneous events. Given the infinite nature and complexity of experience, we can never know the world in and of itself. We can only tend toward coming to know an event at the greatest "depth" that science and our framing methodologies allow. As Wittgenstein suggested… we can only point. Casti, speaking about Wittgenstein and complexity, articulates this relation:

> "The main claim of Wittgenstein's picture theory is that there must be a link between the logical structure of a given language and the logical structure of a real-world fact that a statement in that language asserts. Since the link is itself a relationship in the real world, it's reasonable to suppose that there is some way to express the character and properties of this link using the grammatical rules of the language. But after years of struggling with exactly how to do this, Wittgenstein came to the conclusion that the link between the real world and its expression in language cannot be 'said' at all using language; rather it must be

> 'shown.' We can't express everything about language using language itself; somehow we must transcend the boundaries of language. Thus Wittgenstein says that we cannot really speak about the world, but only 'point.'"[17]

As new technologies provide us with ever more exacting capabilities for "objective" measuring, it will be increasingly important that we remain mindful of the deep complexities surrounding conceptions of simultaneity.

Acknowledgments

I would like to acknowledge discussions with Otto Rössler, Brian Massumi and Daniel Howe.

References

1. S. Kelso, Dynamic Patterns, *The Self-Organization of Brain and Behavior*, A Bradford Book, MIT Press, 4 (1988).
2. http://dictionary.reference.com/browse/simultaneity
3. E. Thelen and L. Smith, *Dynamic Systems Approach to the Development of Cognition and Action.* MIT Press, Cambridge, 70, 121, 122, 128, 182, 183 (1994).
4. B. Libet, Mind Time, The Temporal Factor in Consciousness, 113 (2004).
5. http://64.233.169.104/search?q=cache:RcDtSlm8TUJ:www.radicalempiricism.org/textes/Strange%2520Horizon.pdf+massumi+Strange+Horizon&hl=en&ct=clnk&cd=1&gl=us&client=safari
   See also B. Massumi, *Parables for The Virtual,* Durham [N.C.] , Duke University Press, (2002).
6. G. Pask, *An Approach to Cybernetics*, with a preface by Warren S. McCullock, Harper and Brothers, MIT press 87 (1961).
7. R. HAMILTON, and E. BONK, *The Typosophic Texture.* Politics/Poetics: Das Buch Zur Documenta X. Ostfildern-Ruit: Cantz Verlag. 310 (1997).
8. O.E. Rössler, *Endophysics: The World as Interface.* Singapore: World Scientific Publishing Co. Pte. Ltd. 127 (1998).
9. N. Campbell, "Atomic structure," Nature (London) **107**, 170 (1921); "Time and chance," Phil. Mag. **I**, 1106-1117 (1926); Nature **119**, 779 (1927). As found in O.E. Rössler,. *Endophysics: The World as Interface.* Singapore: World Scientific Publishing Co. Pte. Ltd. 127 (1998).

10. J. von Neumann, *Mathematical Foundations of Quantum Mechanics,* Princeton University Press, Princeton, 439 (1955), First German edition, (233) (1932): as found in O.E. Rössler, *Endophysics: The World as Interface.* Singapore: World Scientific Publishing Co. Pte. Ltd. (18) (1998).
11. From a discussion between Seaman and Rossler related to the forthcoming book *Neosentience | The Benevolence* Engine by Seaman and Rossler.
12. http://www.lampsacus.com/documents/ROESSLERTheWorldasanAccident.pdf
13. See Steeles, Luc, http://arti.vub.ac.be/steels/publications.html
14. See K. Lorenz. *Behind the Mirror: A Search for a Natural History of Human Knowledge,* New York: Harcourt Brace Jovanovich Translated by Ronald Taylor (1977).
15. See F. de Waal, Self-Recognition in an Asian Elephant, PNAS, vol 103, no 45, 17053-17057 (2006).
16. B. Seaman and O.E. Rössler, *Toward the Creation of an Intelligent Situated Computer and Related Robotic System: An Intra-functional Network of Living Analogies*, Catalogue for Emocào Art.ficial, Itau Cultural Center, Sao Paulo, Brazil (2006), and

    B. Seaman and O.E. Rössler: Neosentience – *A New Branch of Scientific and Poetic Inquiry*, delivered by Seaman, The Planetary Collegium Montreal 2007 Summit.
17. J.L. Casti, *Complexification: Explaining an Illogical World Through the Science of Surprise.* New York: Harper Collins. 7 (1994).

CHAPTER 16

CIRCUMCISING THE VOID: (DE)CONTEXTUALISING IN COMPLEX LACANIAN PSYCHOANALYSIS

Dieter De Grave

Psychoanalytical psychotherapist
Condorlaan 24 B-2100 Deurne, Belgium
Psychiatric Center Duffel Stationsstraat 22c 2570 Duffel
Telephone: 0032+498111059
E-mail: dieter.de.grave@emmaus.be

In contemporary psychoanalysis, the true origin of the science seems to be put aside to get it in vogue with the rest of the scientific framework and psychoanalytical thinking. Although this is a defendable position from which to approach psychoanalysis, it robs it of its core. In this paper, we take the hard-core themes of psychoanalysis such as death and sex, to heart and show how they can be linked to the other sciences, such as the theory of complexity, without censoring or rephrasing the concepts or the language itself.

1. Introduction: Castrating Pretences

When I started to study philosophy after my degree in psychology, I had high hopes. I pretended that it would be possible to achieve a certain synthesis between my newly acquired psychoanalytical thinking and the brunt of philosophical thought. During my studies of psychoanalysis I had studied Schopenhauer and some other philosophers, so I believed that my enterprise would prove fruitful.

To my amazement, it was not as easy as I had imagined it to be. If you want to put psychoanalytical thought next to another framework, a lot of issues seem to conflict from the start, a very painful experience for me. During my studies of philosophy, this theme haunted me up until my dissertation on psychoanalysis and the theory of complexity. In this, I tried to put this harrowing experience to the test. Could it be that psychoanalysis really is as unpalatable to the taste to philosophy as some philosophers thought it was? And if so, why is it so? What lies at the heart of psychoanalysis which is so strange to other referential works?

For this festschrift, I would like to dive into the issue from a very personal point of view, namely my own ambivalence between a clearcut philosophical way of thinking and the nitty-gritty experience of working through psychoanalysis. In doing so, I would like to prove or at least show that a certain synergy of thinking is indeed possible, even agreeable if you are able to take a step back from the need to put *everything* into order. So, let us begin.

2. Shortage of Sex and Lack of Life as Defining Points of Psychoanalytical Thinking

Throughout psychoanalysis, a thread seems to uncoil about nothingness and want. It is Freud who first tripped over this thread of Ariadne in 1920 in his text on the pleasure principle.[1] In this text, he made a quite remarkable discovery even to his own amazement. Where he had first stated that people are always preoccupied with repressed sexual urges, he began to investigate the underlying currents for these repressions. To his own astonishment, he discovered that we are all obsessed with mortality. Even in sexuality, he discovered that the main issue is death and how to overcome it.

As we all know, only death is the true master in our existence. Everyone knows that we are going to leave our mortal coil sooner or later, for better or worse. In thinking through the sexual impulses which drive us, according to Freud, he stated that "When we, as empirics does not allow lifetime, have to believe that all the living dies through inner causes, returns to the inorganic, then we can only state: the goal of all life is death, and going back in the past, the lifeless existed before the living."[2] If his theories on sexuality were a hot issue of debate during his own existence, this death deal never quite got the response it deserves. It seemed that none of his contemporary psychoanalytical colleagues really could follow him. Freud knew this when he wrote: "I truly realize that a dualistic theory, that postulates a death-, destructive or aggression drive as an equal partner next to the Eros which manifests itself in the libido, has not received a warm reception and also has not gained much ground among psychoanalysts."[3]

Unrightly so, because, in my opinion, it is impossible to leave out the death drive if you really want to work through Freudian thought and link it to other philosophical bodies of work. The death drive is not some hogwash discovery Freud made in 1920, following the First World War and the circumstantial death of his favorite grandson. As I have shown in a previous publication, the drive towards death is noticeable even in Freud's earliest ways of thinking, namely the *Entwurf einer Psychologie* in 1895.[4]

I personally believe that Freud, albeit unconsciously, realized that his thoughts on the constancy principle of the psychophysiologist Fechner, combined with his own thinking on the drive and the *Lebensnot*, the need to live, would definitely lead him onto the wrong track. In this, I find an argument why Freud himself did not publish his *Entwurf*, and even tried to destroy it. He believed it was not for the general public.

His way of encompassing this substantial mortifying point of human existential suffering was to incorporate it in sexuality, in his works on dreams, mistakes, jokes and, of course, his clinical work.

3. À Chacun, Chacune, à Lacan, Lacune

Of all the followers of Freud, only one really understood what he was saying concerning the death drive. Although Bion is a genius when writing on the death drive, he too fell into the trap of making it into something more lively. Only Lacan in his work took the concept of a death pulsion seriously in his theoretical and clinical enterprises. The reason behind this is fairly simple. He worked out death and destruction in a new topological field, namely the symbolic.

Lacan knew that he had stumbled onto something special when he published his first grand work on the symbolic.[5] He even sought an audience with the Pope on this issue and he tried to motivate his brother, a Benedictine monk, to arrange it for him.

We can wonder why Lacan went the narcissistic mile to get the point across to such a noteable religious figurehead such as the Pope. The point is that we really have to delve into the presented text to finds the real issue on which he wanted to extrapolate. "Thus the symbol manifests itself first of all as the murder of the thing and this death constitutes in the subject the eternalization of his desire."[5] The symbol is the murder of the thing as this constitutes the unending character of desire. Brilliant as it may be, what does Lacan want to say with this sentence?

The symbol is the murder of the thing. The following sentence claims that the grave is the first symbol, which might explain my personal fascination with this theme. "The first symbol in which we recognize humanity in its vestigal traces is the sepulchre, and the intermediary of death can be recognized in every relation in which man comes to life of his history."[5]

Look back to your experience of first acquiring language, or if you are not able to do this, look at what you do when you first name something which you didn't know before. Take, for example, a chair. The first moment you name a certain object as a chair, you destroy a part of the world, that is you name something to be something. From that point on, only chairlike objects can be chairs. The rest cannot be named as chairs. You can name all objects objects, but then you are not saying much.

Now, what is a chair? Here we have the link with philosophy. What is a chair? Define it. A chair is something to sit on. A stool is too, and if I want I can sit on the floor, or on a table, a building or whatever. So, what is a chair? Something with four legs, a back to rest on? A horse also has four legs and a back I can rest on. We will leave it at this for the moment. The main issue is that we cannot define objects in a very clearcut way. There is always a certain degree of ambiguity involved. What we name a chair is a common-sense idea of what a chair should be. I have an idea of a chair, but it is very hard to exactly define what a chair is.

Now, I took a chair as an example, but you could do the same exercise with any word, any definition. The symbol, the signifier, is the murder of the thing. What does this mean? It means that language as the symbolic register sets us as human beings apart from all other living creatures.

Through language, we are able to come to grips with our own existence, and moreover, we are able to convey all our thoughts concerning this notion to other human beings. As I have shown in a previous publication, no other complex system has this capacity.[6] But nonetheless, most living systems have another quite interesting property: they are able to procreate.

And so we come to the main issue of our paper. Sex is something strange if it becomes unhinged from procreation. And this is the main discovery of Freud throughout his work and, at the same time, the grand link with his thinking concerning the death drive. Human beings as complex systems have two affiliations towards surviving the inevitable end.

Like most other sexually procreating living systems, we have a certain drive towards sexuality *qua* procreation. But, notwithstanding this, we are aware of our mortality and have devised ways of overcoming this transience though personal communication of our thoughts and feelings in our cultural interactions. Culture in and of itself is a new way of survival and procreation. It is this point which Freud and Lacan tried to fathom in their theories concerning the sexual drive. It is double: natural and cultural.

"Sexuality is established in the field of the subject by a way that is that of the lack. Two lacks overlap here. The first emerges from the central defect which is the dialectic of the advent of the subject to his own being in the relation to the Other turns-by the fact that the subject depends on the signifier and that the signifier is first of all in the field of the Other. This lack takes up the other lack, which is the real, earlier lack, to be situated at the advent of the living being, that is to say, at sexual reproduction. The real lack is what the living being loses, that part of himself qua living being, in reproducing, himself by means of sex. This lack is real because it relates to something real, namely, that the living being, by being subject to sex, has fallen under the blow of individual death."[7]

In his text 'The signification of the phallus' Lacan remarks that the male sexual organ takes up a special role in cultural exchanges. "The phallus reveals its function here. In Freudian doctrine, the phallus is not a fantasy, if by that we mean an imaginary effect. Nor is it, as such, an object (part, internal, good, bad, etc.) in the sense that this term tends to accentuate the reality pertaining in a relation. It is even less the organ, penis or clitoris, that it symbolizes. [...] For the phallus is a signifier, a signifier whose function, in the intra-subjective economy of the analysis, lifts the veil perhaps from the function it performed in the mysteries. For it is the signifier intended to designate as a whole the effects of the signified, in that the signifier conditions them by its presence as a signifier"[8] The phallus is the go-between for the natural and the cultural system, the real and the symbolic register. And why not the vagina, the female organ of sexual interaction?

Psychoanalytical thought is obsessed with monosexuality, with having a phallus or not having a phallus. And why not a double sexuality with women and men combined? Underlying this issue is what some describe as the secret behind psychoanalytical sexuality. This is the castration complex, which circulates around the theory of the phallus.

Rest assured, the psychoanalytical world does not have the chauvinistic sexist worldview it is cracked up to have. Speaking about having a phallus or not is the only way psychoanalysts like Freud and Lacan were able to combine their theories on sex and death in a culturally and naturally complex framework. They were not stating that women were inferior to men in any way, quite the contrary.

Now what does the phallus mean in relation to death and sex? The symbolic phallus, according to Lacan, is the grounding signifier. It is the signifier that is not dead, which emanates life and meaning. The symbolic phallus is the center of our symbolic network. But as Lacan stresses throughout his work, it is the only signifier which is absent in the symbolic chain. It should be there to complete the symbolic field, but it is always lacking. This is precisely the point of castration. Let us put it in other words.

The symbol which is not victim to the Thing murder is nonexistent. This means that we can never mean with words what we want to mean with them: the precise definition of what our words refer to is impossible. We all believe that our words have an exact meaning, that we can convey our thought and feelings to others in an unambiguous way. This belief is false. Our words are never precise enough to have an exact meaning. This is the reason why we speak in the first place. We speak because we are driven to make sense, even in the conviction of never fully attaining this point. Through symbolic castration, a circumcision is made of a hole in our symbolic plain of existence. Not everything is subject to symbolization. There is a workable black hole that operates like a strange attractor in the symbolic field. The same goes for the two other fields, the imaginary (body image) and the real (enjoyment) register, but going into these other two would lead us too far astray.

4. Theory of Sexual Complexity: the Problematic Self in Self Organization

In my dissertation on complexity theory and psychoanalysis, I picked out four fundamental concepts in talking about complex systems. These were:

- *Boundary* phenomena, with every complex system being subject to the idea of closure.
- *Context* phenomena, with nesting and entropic interactions as essential concepts.
- Dynamic *organization*, with neural networks, feedback mechanisms, (strange) attractors and self organization as central issues.
- Structural dynamic *hierarchies*, concerning part-whole interactions, layering, fractal structures and adaptive/anticipating systems.[9]

Using these four points of approach, I showed that psychoanalysis is fit to take up these structuring elements without doing away with its core definitions. And so I compared it with Freud's biological metaphors of the protozoa, his drive theory and its regulating principles, the distinction between conscious and unconscious processing and his topological model of the Id-Ego-SuperEgo. Psychoanalysis is a complex theoretical approach towards human suffering and its contextual effects. And, in my view, it has something crucial to add to the thinking in the theory of complexity.

Lacanian psychoanalysis provides a critique of the function of the Ego as a clinical goal in therapy or as an ethical problem-solving apparatus.[10] The Ego itself is a symptom, a mark of the human condition *par excellence*. How?

As humans, we are not only driven to procreate, we are also driven to make ourselves be heard, seen, felt and more. Freud was the first to describe an Ego drive.[11] Of course, you can say that Schopenhauer had already beaten him to it, but Freud talks about the Ego in an entirely different way.

He puts the the Ego drive in an opposition to the sexual drives.[12] The Ego drives are non-sexual and this is what defines them. From 1910 up until 1920 he struggles with this issue of Ego versus sex for the very simple reason that he could not yet find the deathly uniting aspect in both. So, again in 1920, in 'Beyond the pleasure principle' he finds the distinction between Eros and the death drive.

Sex drives are a combination of both and the Ego drives disappear from the scene from this point on to transform into the Ego function his text 'The Ego and the Id' in 1923.[13]

Only Lacan truly understood this issue from the root up. From his first publications to the very last, he stresses this point of the Ego being a problematic culmination point, a life-death drive war zone. There is nothing healthy or simple about the Ego: he even dislikes the term and speaks about the divided subject, symbolized as S. We as human subjects are always divided, never whole as Egos.

Now, within the theory of complexity, the idea of self organization is a very important one. But it is highly disputed, because no one thus far has been able to define the self concept within. And rightly so, according to psychoanalysis. The organization of the self is a symptomatic form of organization. By symptomatic, I mean that the symptom refers to an underlying 'disease'.

Disease should not be read as sickness, but as the absence of ease or rest in a certain dynamical context. The self as symptom is a way of dealing with dynamics at the edge of chaos in an underlying layer of organization. Self organization is an emergent quality, rising out of a fractal process of recombination in certain fields in touch with others. Just as fractal theory is based on self similarity and iteration (which has nice echoes in the psychoanalytical idea of repetition compulsion), so is self organization related to theories concerning contextuality, nesting and attractors. The main contribution of psychoanalysis to this field is that the emergent qualities of the self in organization circulates around an attracting void; it is not whole and this provides the stimulus for the entire enterprise. Death, lack, shortage embedded in a 'living' system make it function, this is what Lacan tried to point out. There is no possible whole or holistic theory that can do away with this crucial organizing point. Lacanian psychoanalysis shows that absence is central in structuring presence, that less produces more and so that death (lack) and sex (procreation) are intimately intertwined. This might be an interesting field of research in other form of complex thinking.

5. And Beyond: Towards a Grand Unifying Theory of Complex Psychoanalysis

Castration is the core of psychoanalytical thinking because it is important for understanding and helping others to cope with human suffering. Castration combines sexuality and mortality into one complex term. Up to now we have said practically nothing about the other famous complex, namely the Oedipus complex. So, to close off this work on complexity, some words about it.

The oedipal complex is not an incestuous theme park for frustrated Victorian gentlemen and ladies. Again, it is Lacan who forms an abstraction from this complex and talks about a first and second other in relation to a human developing subject. In this endeavor, he makes it clear that the castration complex and the oedipal complex are intertwined in an intimate relationship. You could say that, according to him, the castration complex is embedded or nested in the oedipal scene.

He describes the Oedipus as a field of interactions between three individuals and death as a structuring element.[14] The structuring function of lack rings out throughout his entire work. The death drive (castration) is a structuring attractor nested in a living network of interactions (Oedipus).

Psychoanalysis is not a philosophy because it is applied to humans and their conditions. It will prove impossible to extrapolate the psychoanalytical findings to other fields of complexity without severe translation efforts. But I hope that in this paper I have given those not familiar with or biased towards psychoanalysis another perspective on the matter. The metatheory for psychoanalysis should be the theory of complexity, as the field where interactions with other complex fields such as biology and sociology are facilitated. But I hope that this complex metatheory can become enriched with certain psychoanalytical insights such as those discussed concerning death and sex. I hope I have demonstrated that these two terms at least are more complex to interpret than many reductionist critiques (even of psychoanalytical origin) have made them.

References

1. S. Freud, *Jenseits Des Lustprinzips*, G.W.13. Imago, London pp. 1-70, (1920g).
2. S. Freud, *Jenseits Des Lustprinzips*, G.W.13. Imago, London p. 53, (1920g), (my translation from the Dutch text).
3. S. Freud, *Die endliche und die unendliche Analyse*. G.W.16. Imago, London, p. 68, (1937c), (my translation from the Dutch text).
4. D. De Grave, De doodsdrift ten aanzien van het Reële, *Psychoanalytische Perspectieven,.43/44,* pp. 33-61, (2001).
5. J. Lacan, Function and field of speech and language in psychoanalysis, in *Écrits: a Selection.,*Routledge London, p.104 (1977).
6. D. De Grave, *Time to Separate the Men From the Beasts: Symbolic Anticipation as the Typically Human Subjective Dimension*, AIP Conference Proceedings Vol. 718, American Institute of Physics, pp. 435-444 (2004).
7. J. Lacan, *The seminar of Jacques Lacan: Book XI The Four Fundamental Concepts of Psychoanalysis*, Routledge, London, pp. 204-205 (1977).
8. J. Lacan, The signification of the phallus in *Écrits: a Selection,* Routledge, London, p. 285 (1977).
9. D. De Grave *Complexiteitstheorie en Psychoanalyse,* unpublished.
10. J. Lacan, *The Seminar of Jacques Lacan II: The Ego in Freud's Theory and in the Technique of Psychoanalysis, 1954-1955*, Norton, NYC (1988).
11. S. Freud, *Die psychogene Sehstörung in psychoanalytischer Auffassung*, G.W.8. Imago, London, pp. 93-102 (1910i).
12. S. Freud, *Triebe und Triebschicksale*. G.W.10. Imago, London, pp. 209-232 (1915c).
13. S. Freud, *Das Ich und das Es*, G.W.13. Imago, London, pp. 235-290 (1923b).
14. J. Lacan, Le mythe individuel du névrosé ou Poésie et Vérité dans la névrose, *Ornicar?* pp. 17-18 (1979).

CHAPTER 17

A REVIEW OF FLICKER-NOISE SPECTROSCOPY: INFORMATION IN CHAOTIC SIGNALS

Serge F. Timashev

Karpov Institute of Physical Chemistry
10 Vorontsovo pole St., Moscow, 105064, Russia
E-mail: timashev@cc.nifhi.ac.ru

Yuriy S. Polyakov

USPolyResearch, Ashland, PA 17921, USA
E-mail: ypolyakov@uspolyresearch.com

Flicker-Noise Spectroscopy (FNS), a general approach to the extraction and parameterization of resonant and stochastic components contained in medical time series, is presented. The basic idea of FNS is to treat as main information carriers the correlation links present in sequences of different irregularities, such as spikes, "jumps", and discontinuities in derivatives of different orders, on all levels of the spatiotemporal hierarchy of the system under study. The tools for extracting and analyzing the information are power spectra and difference moments (structural functions), which complement each other's information. The stochastic component derived from the structural function is formed exclusively by "jumps" of the dynamic variable while the stochastic component derived from the power spectrum is formed by both spikes and "jumps" on every level of the hierarchy. An application of this approach to the analysis of electroencephalogram signals is discussed.

1. Introduction

Time and space series of dynamic variables that arise in studies of various natural processes and structure are often an important source of information about the state of the system and features of its evolution and structure.[1] Therefore, a reliable tool for extracting and analyzing the information contained in the series could help us to understand many processes occurring in nature, for example, the preparatory stages for a major earthquake or the development of a human disease.

A phenomenological scheme, developed by J. Nicholis[2], for representing the information stored in various complex signals provided a conceptual foundation for the practical extraction of stochastic and resonant components from time and space series. This scheme assumes that there is an infinite number of levels in the evolution hierarchy of the system under study and that there are some recurrent rules that generate information on a specific level of the hierarchy and compress it at a higher level. General ideas from the Nicholis scheme were used to develop Flicker-Noise Spectroscopy (FNS),[3-5] a phenomenological framework where the concept of the information contained in signals generated by open dissipative systems is generalized. FNS can be applied to three types of problems:

(i) determination of parameters or patterns that characterize the dynamics or structural features of open complex systems;
(ii) finding precursors of abrupt changes in the state of various open dissipative systems, based on a priori information about the dynamics of the systems; and
(iii) determination of flow dynamics in distributed systems based on the analysis of dynamic correlations in stochastic signals that are simultaneously measured at different points in space. The FNS approach has already been applied to the analysis of the structure and dynamics of various physicochemical, electrochemical, biological, medical, geophysical, and astrophysical processes.[3-5]

The purpose of this paper is to discuss the extraction and parameterization of resonant and stochastic components in time and space series using FNS. An application of this approach to the analysis of electroencephalogram signals is discussed.

2. Flicker-Noise Spectroscopy

2.1. *Basic principles*

The basic idea of FNS is to treat as the main information carriers the correlation links present in sequences of different irregularities, such as spikes, "jumps", discontinuities in derivatives of different orders, on all levels of the spatiotemporal hierarchy of the system under study. It is further assumed that, according to the Self-Organized Criticality (SOC) paradigm,[6] the stochastic dynamics of real processes is characterized by intermittency: consecutive alternation of rapid changes in the values of dynamic variables on small time intervals with small variations of the values on longer time intervals. It was demonstrated that the origins of such intermittency, which occurs on every hierarchical level of the system evolution, are associated with the occurrence of complex (multiparticle, nonlinear) interactions, dissipation, and inertia.

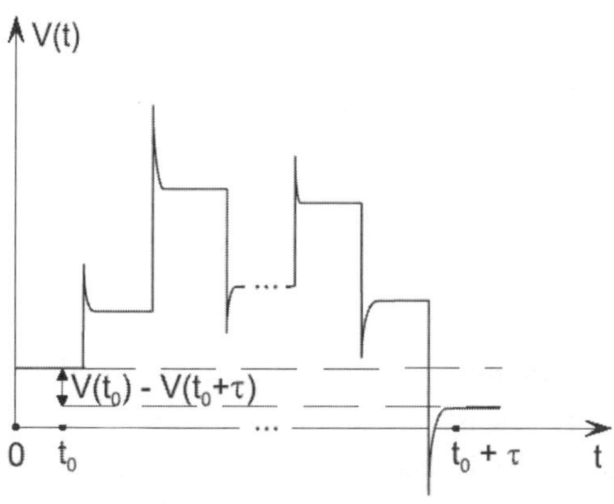

Fig. 1. Schematic of "random walk" evolution.

To illustrate the idea, consider the model process of one-dimensional "random walk" with small "kinematic viscosity" ν (Figure 1). The small value of ν implies that when the signal changes from position V_i to V_{i+1}, which are $|V_{i+1} - V_i|$ apart (in value) from each other, the system first overleaps ("overreacts") due to inertia and then "relaxes". We assume that the relaxation time is small compared to the residence time in a "fluctuation" position and the signal does not contain "enveloping" low-frequency curves, i.e., the resonant components are absent. The signal has all the main features of intermittent behavior: "laminar" phases with small variations in the dynamic variable $V(t)$ on characteristic time intervals T_l are followed by short-term spikes in the dynamic variable on characteristic time intervals τ_s $(\tau_s << T_l)$. These latter intervals accompany step-like changes in $V(t)$, which determine the value of the dynamic variable during the next "laminar" phase. In order to extract the information about the dynamics of the signal in Fig. 1, one may analyze the correlation links between the signal irregularities: step-like jumps in the dynamic variable, namely, Heaviside θ-functions, and inertial spikes, i.e., Dirac δ-functions, accompanying the step-like changes.[7]

In FNS, the information parameters are assumed to be related to the autocorrelation function, one of the basic concepts in statistical physics. This function is defined as

$$\psi(\tau)=\langle V(t)V(t+\tau)\rangle_{T-\tau}, \tag{1}$$

where τ is the time lag parameter, and the angular brackets denote averaging:

$$\langle(\ldots)\rangle_T=\frac{1}{T}\int_{-T/2}^{T/2}(\ldots)\,dt\,. \tag{2}$$

Signal $V(t)$ generally contains the "resonant" components that are specific to the evolution of the system, and the stochastic components that are related to various irregularities. The resonant modes, which may correspond to natural, external, or interferential frequencies, manifest themselves as low-frequency (slow-varying) "envelope" components of the signal. The stochastic components are characterized by wide high-frequency bands.

To extract the information contained in $\psi(\tau)$, it is convenient to analyze some transforms ("projections") of this function, for example, "power spectrum" $S(f)$:

$$S(f)=\int_{-T_M/2}^{T_M/2}\psi(\tau_1)\cos(2\pi f t_1)\,dt_1\,, \tag{3}$$

where $T_M \leq T/2$ is the part of the averaging interval T that can be used to calculate "reliable" estimates of the actual power spectrum.

This particular transform was chosen because $S(f)$ is most effective in separating out the resonances (main components of the signal) of the analyzed functions, which are represented as a set of N_r peaks characterized by positions f_{0i} and "half-widths" γ_i ($i = 1, 2, \ldots, N_r$).

The resonance contribution $S_r(f)$ to the overall power spectrum $S(f)$ can be extracted by rewriting the latter as

$$S(f) = S_c(f) + S_r(f), \tag{4}$$

where $S_c(f)$ is the continuous power-spectrum component associated with the stochastic component of dynamic variable $V(t)$. Additive representation (4) is justified because the contributions of resonant and stochastic components to dynamic variable $V(t)$ usually correspond to different time scales. In the frequency range from $1/T$ to $f_d/2$, where f_d is the sampling frequency, the resonant components mostly contribute to the low-frequency range, while all the irregularities manifest themselves in the high-frequency range.[3,4] It should be noted that the parameterization of $S_r(f)$ by finding the positions, "half-widths", and partial weights A_i of the fixed resonances can be done rather easily. By contrast, the parameterization of $S_c(f)$ is a much more difficult task.

To solve the latter problem, we also consider difference moments ("transient structural functions") $\Phi^{(p)}(\tau)$ of different orders p ($p = 2, 3, \ldots$):

$$\Phi^{(p)}(\tau) = \left\langle \left[V(t) - V(t+\tau)\right]^p \right\rangle_{T-\tau}, \tag{5}$$

where $\tau \leq T_M$.

Function (5) can also be written as a linear combination of stochastic $\Phi_c^{(p)}(\tau)$ and resonant $\Phi_r^{(p)}(\tau)$ components:[3,4]

$$\Phi^{(2)}(\tau) = \Phi_c^{(2)}(\tau) + \Phi_r^{(2)}(\tau). \tag{6}$$

Equation (6) may be used because the contribution of resonant components to function $\Phi^{(2)}(\tau)$ is mostly seen at intermediate and large values of τ. At the same time, the stochastic components contribute to the whole interval of $0 \leq \tau \leq T_M$.

The functions $\Phi_c^{(p)}(\tau)$ are formed exclusively by "jumps" of the dynamic variable while $S_c(f)$ are formed by both "spikes" and "jumps" on every level of the hierarchy, which has been formally proved by Timashev.[3] It should be emphasized that such separation of information stored in various irregularities is attributed to the intermittent character of the evolution dynamics. In other words, the information contents of $S_c(f)$ and $\Phi_c^{(2)}(\tau)$ coincide if there is no intermittency.

FNS methodology may be applied to the extraction of information from experimental time series as well as spatial series. In the latter case, when the dynamic variable (magnitude of surface roughness, space distribution in concentrations of chemical substances, codon variations in a genome, etc.) varies chaotically along the configuration coordinate x, the following substitutions need to be made:[3,8] $t \rightarrow x$ and $f \rightarrow f_x \equiv k/2\pi$, where the dimension of "space frequency" f_x and wave number k is $[x]^{-1}$.

If one analyzes time series $V(t)$, which characterize the fluctuation dynamics of a system as a whole, the dimensions of the system should not exceed its characteristic correlation length by orders of magnitude. Otherwise, the sequences of correlated fluctuations will take up only a small part of the overall system space, leading to a large number of such fluctuation sequences in the total volume. As all of these sequences are independent of each other (they are much further apart than their correlation length), the fluctuations averaged over the total volume will be described by a frequency-independent power spectrum.

2.2. *Interpolation formulas*

Let us write the interpolation expressions for stochastic components $\Phi_c^{(2)}(\tau)$ and $S_c(f)$ for stationary processes. The parameters characterizing the dynamic correlations on every level of the evolution hierarchy are assumed to be the same. Consider the simplest case in which there is only one characteristic scale in the sequences of "spikes" and "jumps". Then:[3]

$$\Phi_c^{(2)}(\tau) \approx 2\sigma^2 \cdot \left[1 - \Gamma^{-1}(H_1) \cdot \Gamma(H_1, \tau/T_1)\right]^2, \tag{7}$$

$$\Gamma(s,x) = \int_x^{\infty} \exp(-t) \cdot t^{s-1} dt, \ \Gamma(s) = \Gamma(s,0) ,$$

where $\Gamma(s)$ and $\Gamma(s, x)$ are the complete and incomplete gamma functions ($x \geq 0$ and $s > 0$), respectively. σ in Eq. (7) is the standard deviation of measured values of $V(t)$; H_1 is the Hurst constant, which describes the rate of "forgetting" by the dynamic variable of its values on time intervals that are less than the correlation time T_1. In this case, T_1 may be interpreted as the correlation time for the "jumps" in a stochastically varying $V(t)$.

The interpolating function for power spectrum component $S_{cS}(f)$ formed by the "spikes" is written as:[3]

$$S_{cS}(f) \approx \frac{S_{cS}(0)}{1+(2\pi f T_0)^{n_0}} \ . \tag{8}$$

Here, $S_{cS}(0)$ is the parameter characterizing the low-frequency limit of $S_{cS}(f)$, n_0 describes the degree of correlation loss in a sequence of the "spikes" on time interval T_0.

The interpolating function for power spectrum component $S_{cR}(f)$ formed by the "jumps" is written as: [3]

$$S_{cR}(f) \approx \frac{S_{cR}(0)}{1+(2\pi f T_1)^{2H_1+1}} , \tag{9}$$

where

$$S_{cR}(0) = 4\sigma^2 T_1 H_1 \cdot \left\{ 1 - \frac{1}{2H_1 \Gamma^2(H_1)} \int_0^{\infty} \Gamma^2(H_1,\xi) d\xi \right\} .$$

Although the contributions to the overall power spectrum $S_c(f)$ given by Eqs. (8) and (9) are similar, the parameters in those equations are different; that is, $S_{cS}(0) \neq S_{cR}(0)$, $T_1 \neq T_0$, and $2H_1 + 1 \neq n_0$. This implies that the parameters in the expressions for the power spectrum and structural function of second order generally have different information contents when one analyzes experimental time series $V(t)$. For example, characteristic times T_0 and T_1 are usually much different because they correspond to different frequency ranges of the power spectrum. The fact that the "jumps" are more regular than the "spikes" implies that the power spectrum contribution from the "jumps" will be concentrated in the lower frequency range as compared to the one for the "spikes". Thus, the high-frequency part of the power spectrum, which is often characterized by a flicker-noise dependence $S_c(f) \sim 1/f^n$, is mostly generated by the "spikes".

The interpolating function for the overall power spectrum can be expressed as:

$$S_c(f) \approx \frac{S_c(0)}{1+(2\pi f T_{01})^n}, \tag{10}$$

where $S_c(0)$, T_{01}, and n are phenomenological parameters calculated by fitting the interpolating function (10) to the power spectrum for experimental time series. As FNS introduces the dynamic multiparametric self-similarity on all levels of the hierarchy, all the phenomenological parameters, including H_1 and n, are determined by fitting the interpolations (7) and (10) to the difference moment and power spectrum calculated using the experimental series. In other words, FNS does not impose any restrictions on the values of H_1 and n, as is the case for fractal theory: $0 \le H_1 \le 1$ and $n \le 3$.

As was noted above, complex signals usually contain both "stochastic" (formed by irregularities) and resonant (system-specific) components. Therefore, our goal is to extract all parameters for both the stochastic component and resonances from time series $V(t)$.

Let us only consider the "low-frequency" resonant components that vary much more slowly than the "high-frequency" stochastic components. The overall signal $V(t)$ will then be represented as a linear superposition of the "high-frequency" stochastic component $V_c(t)$ and "slow-varying" resonant component $V_r(t)$; that is,

$$V(t) = V_c(t) + V_r(t).$$

A similar additive presentation can be used for the power spectrum and structural functions (see expressions (4) and (6), respectively).

If we consider large time intervals ($T >> \max\{T_0, T_1\}$), and the autocorrelator $\psi_r(\tau) = \langle V_r(t)V_r(t + \tau)\rangle$ averaged over time interval T depends only on the time lag parameter τ, then the functions $S_r(f)$ and $\Phi^{(2)}{}_r(\tau)$, which correspond to $V_r(t)$, become interrelated:[3]

$$S_r(f) = \int_0^\infty \cos(2\pi f\tau)\left[\Phi_r^{(2)}(\infty) - \Phi_r^{(2)}(\tau)\right]d\tau. \tag{11}$$

On the other hand, the autocorrelator resonant component $\psi_r(\tau)$ can be determined from the power spectrum resonant component $S_r(f)$ by applying the inverse Fourier transform for $T \to \infty$:

$$\psi_r(\tau) = 2\int_0^\infty S_r(f)\cos(2\pi f\tau)df. \tag{12}$$

In this case, we also have

$$\Phi_r^{(2)}(\tau) = 2\left[\psi_r(0) - \psi_r(\tau)\right]. \tag{13}$$

Stationary formulas (11) and (12) allow one to separate and completely parameterize the contributions from resonant and stochastic components of real signals. This approach can be used even when the number of resonances that are taken into account is high (hundreds or thousands).

2.3. *Parameterization algorithm*

Consider the case in which the difference moment of second order $\Phi^{(2)}(\tau)$ and power spectrum $S(f)$ are good enough to determine the necessary set of parameters.

STEP 1. Using the experimental data, evaluate the difference moment of second order $\Phi^{(2)}(\tau)$ and power spectrum $S(f)$ with discrete versions of Eqs. (3) and (5).[4] Consider only the low-frequency interval, $1/T < f \leq ¼ f_{max}$ ($f_{max} = 1/\Delta t$, Δt – interval between successive measurements), for $S(f)$ and low-lag-time interval, $0 \leq \tau \leq T/4$, for $\Phi^{(2)}(\tau)$.

STEP 2. Extract the stochastic component $S_c(f)$ from $S(f)$ using the interpolation formula (10). Here, parameter $S_c(0)$ is determined as the minimum (or 2nd, 3rd, or 4th least, depending on the problem) absolute value of $S(f)$ in the low-frequency range of the spectrum. As the absolute magnitude of the stochastic component is small, parameters n and T_{01} are found by least-square fitting to $S(f)$ in log-log scale.

STEP 3. Subtract the stochastic component $S_c(f)$ found in STEP 2 from "experimental" $S(f)$. The result represents the resonant component $S_r(f)$.

STEP 4. For this difference, evaluate the inverse Fourier transform, which according to Eq. (12) is the autocorrelator corresponding to resonant components of the signal. Using Eq. (13), calculate the resonant component of the structural function.
STEP 5. Subtract $\Phi_r^{(2)}(\tau)$ from experimental function $\Phi^{(2)}(\tau)$. Fit the interpolation function for $\Phi_c^{(2)}(\tau)$ given by Eq. (7) to this difference by the method of least squares and find unknown "jump" parameters H_1, T_1, σ.
STEP 6. Calculate $S_{cR}(f)$ from $\Phi_c^{(2)}(\tau)$ using Eq. (9). Use $S_{cS}(f) = S_c(f) - S_{cR}(f)$ to find the "spike" parameters $S_{cS}(0)$, n_0 and T_0. All the parameters associated with the stochastic components of the signal are known.

If necessary, the resonant component $S_r(f)$ can be parameterized in terms of frequency positions f_{0i}, «half-width» γ_i, and intensity A_i.

2.4. *Correlation links in the dynamics of distributed systems*

FNS also allows one to analyze different (mass, electric, magnetic) flows in distributed systems with the following two characteristics: (1) The system size is larger than the space scale of noticeable variations in the dynamic variables. (2) The evolution is "perplexed" due to complex nonlinear interactions between the components of the system. The information about the dynamics of correlation links in variables $V_i(t)$, measured at different points i, can be extracted by analyzing the temporal links of various correlators.[3,4,9]

In this review, we will limit our attention to the simplest "two-point" correlators, which are formed only by the jump irregularities and characterize the links between $V_i(t)$ and $V_j(t)$:[3]

$$q_{ij}(\tau;\theta_{ij}) = \left\langle \left[\frac{V_i(t) - V_i(t+\tau)}{\sqrt{\Phi_i^{(2)}(\tau)}} \right] \left[\frac{V_j(t+\theta_{ij}) - V_j(t+\theta_{ij}+\tau)}{\sqrt{\Phi_j^{(2)}(\tau)}} \right] \right\rangle_{T-\tau-|\theta_{ij}|} . \quad (14)$$

Here, τ is the "lag time"; θ_{ij} is the "time shift" parameter.

The dependence of correlator $q_{ij}(\tau, \theta_{ij})$ on θ_{ij} describes the cause-and-effect relation ("flow direction") between signals $V_i(t)$ and $V_j(t)$. When

$\theta_{ij} > 0$, the flow moves from point i to point j, when $\theta_{ij} < 0$, from j to i. When the distance between points i and j is fixed, the value of θ_{ij} can be used to estimate the rate of information transfer between these two points. The dependence of the value and magnitude of correlator $q_{ij}(\tau, \theta_{ij})$ on τ and θ_{ij} can be used to analyze the flow dynamics with signals $V_i(t)$ and $V_j(t)$ changing in phase ($q_{ij} > 0$) and in antiphase ($q_{ij} < 0$).

3. Analysis of Fluctuations in Electroencephalograms

The high specificity of $\Phi^{(2)}(\tau)$ and $S(f)$ patterns obtained in the analysis of complex system states may be used to diagnose diseases at their initial stages. To illustrate this capability of FNS, let us consider the analysis of the electroencephalograms (EEGs) obtained in the medical assessment of the patients in the age range from 11 to 14 who were suspected to have schizophrenia spectrum disorders.[10] The EEG signals were measured simultaneously at 16 electrodes placed on the head of a specific patient using the standard procedure. Standard electrodes with the characteristic surface of 1 sm^2 were used. The magnitudes of the electric potentials were measured with reference to the electrode placed at the left ear lobe. Sampling frequency f_0 was equal to 200 Hz.

The EEG signals represent the results of the superposition of the electric potentials for a large number (up to $5 \cdot 10^5$) of nerve cells (neurons) in the cortex. As there are some interactions in the neuron chains, the values of the electric potentials measured at different electrodes are often correlated.

The behavior of these correlations and the dynamics of the links between different cortex areas can be visually studied by plotting "two-point" FNS correlators $q_{ij}(\tau, \theta_{ij})$ given by (14). Figures 2 a, b and 3 a, b show the signals measured for approximately 38 seconds at C3 (left central area of the head) and Pz (occiput - posterior portion of the head) electrodes for two patients. The first patient supposedly belonged to the "healthy" group and was marked as S196W. The second patient was in the group with schizophrenia disorders and was marked as s-425w. The corresponding power spectrums $S(f)$ demonstrating the difference in alpha rhythms ($f \approx$ 12-13 Hz), beta-rhythms ($f \approx$ 18-20 Hz), delta rhythms

($f \approx$ 1-4 Hz), and theta rhythms ($f \approx$ 7 Hz) are shown in Figs. 2 *c*, *d* and 3 *c*, *d*.

Figures 4 and 5 demonstrate the surfaces of cross-correlators $q_{ij}(\tau, \theta_{ij})$ for the Pz and C3 electrodes calculated for two different time intervals $T_1 = (1\text{-}500)\, f_0^{-1}$ and $T_2 = (5001\text{-}5500)\, f_0^{-1}$ with the equal duration of $T = 500 \cdot f_0^{-1} = 2.5$ s. It can be seen that the functions of cross-correlators $q_{ij}(\tau, \theta_{ij})$, where index *i* stands for Pz and index *j* for C3, are dramatically different. One can see a clear signal correlation of the oscillatory type for some ranges of sequential values of $\theta_{ij} > 0$ and $\theta_{ij} < 0$. The corresponding period of the oscillations is approximately equal to the characteristic time that is inverse to the frequency of alpha rhythms ($f \approx$ 12.5 Hz), which is clearly seen in Figs. 2*a* and 3*a* for the S196W patient for both electrodes. This type of dependence $q_{ij}(\tau, \theta_{ij})$ at $\theta_{ij} > 0$ implies the correlated "retardation" of the Pz signal with respect to the C3 signal, and the correlation of $q_{ij}(\tau, \theta_{ij})$ at $\theta_{ij} < 0$ points to the reverse "retardation" of the C3 signal with respect to the Pz signal. In other words, Fig. 4 illustrates the phenomenon of "self-synchronization" between the cortex signals Pz and C3 for the S196W patient. The analysis of the cross-correlators plotted for other time intervals within the total 38 second measurement interval supported this conclusion. At the same time, the values of $q_{ij}(\tau, \theta_{ij})$ at different time intervals may vary, as seen in Fig. 4*a* and Fig. 4*b*. The cross-correlator for the s-425w patient (Figure 5), who belongs to the group with schizophrenia disorders, is characterized by a different type of behavior. Here, the effects of self-synchronization are much less pronounced and virtually disappear for most of the ranges of θ_{ij}. Of course, these differences in the behavior of the cross-correlators can be to a large degree attributed to the differences in the alpha-rhythm frequency bands of power spectrums $S(f)$ for S196W and s-425w signals (Figs. 2 *c*, *d* and 3 *c*, *d*.). At the same time, the cross-correlators provide additional information as compared to the power spectrum functions: they show the differences in the dynamics of the links between the Pz and C3 cortex areas and they allow one to analyze the process of self-synchronization between the corresponding areas for different patients. The latter may be used for diagnostic purposes.

It should be noted that the purpose of this study is not to develop a specific method to diagnose the patient's condition and determine the degree of a specific schizophrenia spectrum disorder, because the methods of diagnosis must be checked using the data for a large a number of patients. The purpose of this study is to show that FNS can be used as a tool to find significant differences in the original and visually almost identical signals and determine the features of the dynamics of the correlations between different cortex areas.

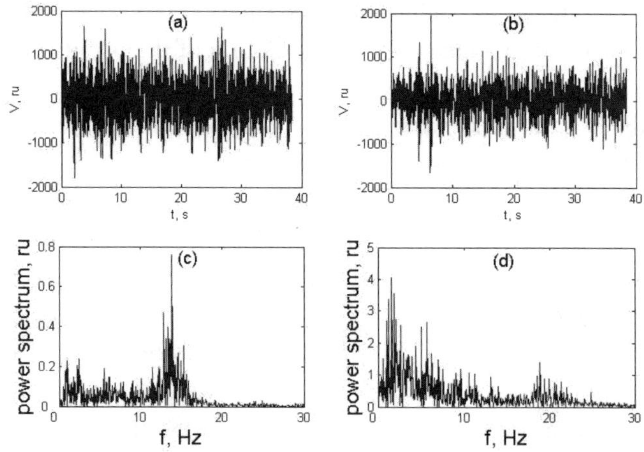

Fig. 2. EEG signals for the C3 (*a*, *b*) and corresponding low-frequency power spectrum functions (*c*, *d*) for the S196W (*a*, *c*) and s-425w (*b*, *d*) patients.

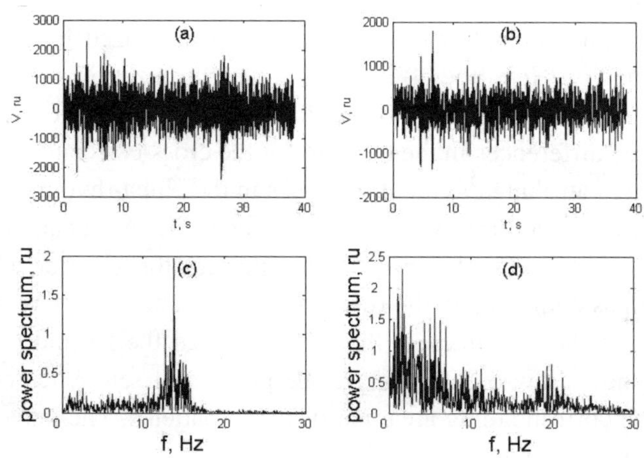

Fig. 3. EEG signals for the Pz (*a*, *b*) and corresponding low-frequency power spectrum functions (*c*, *d*) for the S196W (*a*, *c*) and s-425w (*b*, *d*) patients.

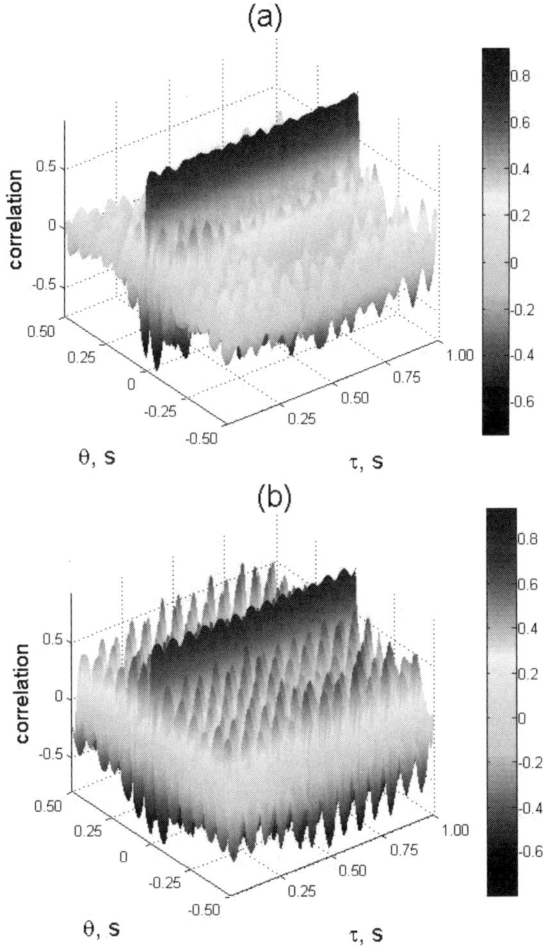

Fig. 4. C3-Pz cross-correlators (14) for the S196W patient for the intervals T = $(1 - 500)f_0^{-1}$ (*a*) and T = $(5001 - 5500)f_0^{-1}$ (*b*); f_0 = 200 Hz.

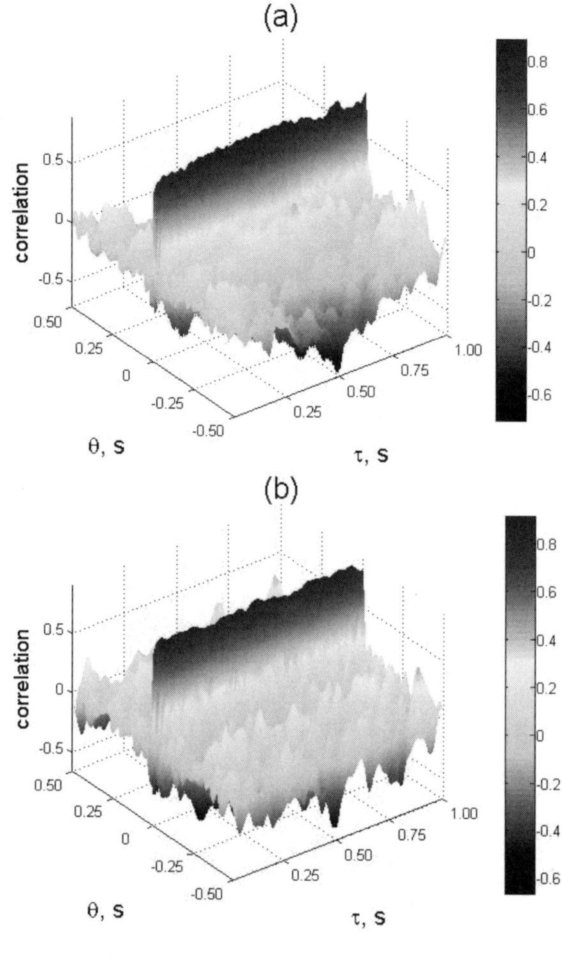

Fig. 5. C3-Pz cross-correlators (14) for the s-425w patient for the intervals $T = (1 - 500)f_0^{-1}$ (*a*) and $T = (5001 - 5500)f_0^{-1}$ (*b*); $f_0 = 200$ Hz.

Acknowledgment

This study was supported in part by the Russian Foundation for Basic Research, project no. 05-02-17079.

References

1. P. Réfrégier, *Noise Theory and Application to Physics: From Fluctuations to Information* (Springer, New York, 2004).
2. J.S. Nicolis, *Dynamics of Hierarchical Systems: An Evolutionary Approach* (Springer, Berlin, 1986).
3. S.F. Timashev, *Rus. J. Electrochem.*, **42**, 422 (2006).
4. S.F. Timashev and Yu.S. Polyakov, *Fluctuations & Noise Letters*, **7**(2), R15 (2007).
5. S. F. Timashev, V. E. Belyaev, P. S. Timashev and A. B. Solovieva, *Colloid J.*, **68**, 481 (2006).
6. P. Bak, How Nature Works. The Science of Self-Organized Criticality (Oxford University Press, Oxford, 1997).
7. H. G. Schuster, *Deterministic Chaos: An Introduction* (Physik-Verlag, Weinheim, 1984).
8. S.F. Timashev SF, A.B. Solovieva and G.V. Vstovsky, in *Advanced experimental method for noise research in nanoscale devices*, Ed. J. Sikula and M. Levinshtein (Amsterdam, Kluwer Academic Publisher, 2004), p. 177.
9. S. F. Timashev, *Rus. J. Phys. Chem.*, **75**, 1742 (2001).
10. S.F. Timashev, G.V. Vstovsky, A.Ya Kaplan and A.B. Solovieva, in *Noise and Fluctuations*, AIP Conference Proceedings, Vol. 780 (American Institute of Physics, New York, 2005), p. 579.

CHAPTER 18

HIDDEN PERSPECTIVES IN EMERGENT STRUCTURES PRODUCED BY NEURAL NETWORKS

Raymond Pavloski

Department of Psychology, Indiana University of Pennsylvania
1020 Oakland Avenue, Indiana, Pennsylvania, U.S.A. 15705-1064
E-mail: pavloski@iup.edu

This chapter summarizes recent work based on an approach to conscious observer perspectives that views the emergence of phenomenal experience as a kind of hidden biological pattern formation. Initially, models based on simple binary neuron-like elements were devised in order to contrive ways of producing hidden patterns, and these are summarized to develop the key ideas. New evidence for the natural production of hidden patterns using simulations of networks of integrate-and-fire model neurons is then presented.

1. Introduction

Perceptual completion,[1-3] multistability,[4,5] and hysteresis[6-8] imply that conscious perceptual experiences may be special cases of the cooperative phenomena that generally underlie biological pattern formation.[9,10] This observation, made famous by early Gestalt psychologists (Koffka[11]), now has a firm foundation in the methods of nonlinear dynamics.[12-15] While the perceptual phenomena provide useful methods in the search for neural correlates of consciousness,[16,17] and although computational models of them have been devised,[14,15] very little progress has been made in solving the theoretical problems involved in relating subjective, qualitative experiences to objective neural processes[18]. The approach described here assumes that conscious experience consists of *hidden* biological patterns, and uses simple neural network models and computer simulations to explore the consequences of that assumption.[19,20]

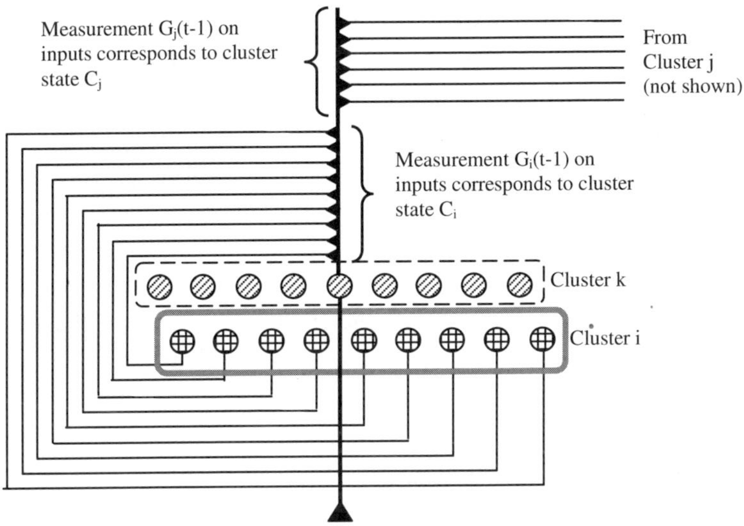

Fig. 1. Objective patterns of action potentials and a hidden pattern of cluster states. Neurons are shown as patterned circles, and axons end in filled triangles. The change in output in the center neuron of Cluster k (shown with a heavier axon line) is determined by a function of cluster states C_i and C_j, and is a component of cluster state C_k.

Figure 1 schematically contrasts objective and hidden patterns in part of a recurrent neural network. Action potential patterns occur over time on the lines that represent axons, and are embodied in a context that makes them observable. In contrast, experience is said to have a necessarily subjective ontology,[21] and to lie beyond a subjectivity barrier.[22] Such descriptions might fit self-organized patterns that are embodied completely within the dynamics of the neural network and are revealed only by considering each neuron's perspective on this network. In the models described below, meta-neural cluster states correspond to measurements that each neuron makes on clusters of neurons. Since each neuron's state partially determines one or more cluster states, and is determined by its input cluster states, cluster-cluster interactions governed by cluster states occur through the network dynamics. Given an appropriate network architecture, patterns of cluster states arise and persist. From a third-person perspective, however, such patterns are merely implicit in neural activity (much as stimuli are only implicit in neural patterns in MacKay's description of vision as the detection of optical covariation[23]). To such an observer, the interactions of neurons are merely such that *it is as though* new elements arise and interact to produce patterns of their states. In this way, cluster state patterns are hidden from outside observation.

Two recurrent neural network models developed to study hidden patterns are described below. Section 2 provides a summary of hidden pattern generation by lattice models composed of simple binary neurons. These models were developed in efforts to understand how hidden patterns might be produced. Both the model neurons and the network architecture are extremely simple and non-biological, and the states of the elements that comprise the hidden patterns are contrived. However, the simplicity of these models brings out clearly just how the hidden patterns are produced, and highlights their causal role in the network's dynamics. More recent research, discussed in Section 3, has the objective of determining if interactions among more biologically realistic model neurons can, in principle, produce hidden patterns. Rather than finding ways to produce hidden patterns, the goal is to find evidence of their existence in the dynamics of a simple recurrent network.

The final section of the chapter further considers the issue of the attributes of patterns of cluster states in light of these results. Future directions for research are also considered.

2. Lattice Models Employing Binary Model Neurons

The model summarized here has been called a latent Ising model[19], both because it was motivated by the famous Ising model of a ferromagnet that forms the basis of the Hopfield attractor neural network (e.g., Amit[24], Bar-Yam[25]), and because it possesses an energy-like function that is formally identical to that of the Ising model[19].

This model consists of N simple binary model neurons having states $s_i = \pm 1$ (presence or absence of an action potential) at each moment of discrete time. As depicted schematically in Figure 2, every element is the center element of one cluster of nine elements, and is the non-center

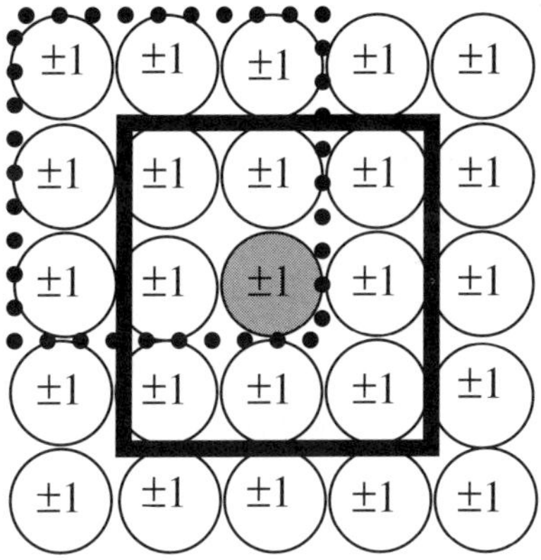

Fig. 2. Every cluster to which the highlighted element belongs is present in a small portion of a square lattice of binary neuron-like elements, with the possible states (±1) of the element on each lattice location depicted. The highlighted element is in the central position of one outlined cluster, and in the bottom right position of another.

element of eight other clusters. Periodic boundary (i.e., wrap-around) conditions[26] are employed in order to make this true for every element in the lattice; thus, some clusters contain elements on the top and bottom rows, and some contain elements in both the left-most and right-most columns.

Each neuron computes the products of the states of the neurons belonging to all N clusters of the network. If we let $c(l)$ be the set of nine elements comprising the cluster in which element l is the center element, then

$$p_l = \prod_{j \in c(l)} s_j \tag{1}$$

corresponds to a meta-neural state of this cluster. At each moment of discrete time, a model neuron updates its state by computing the products of the states of neurons for each of its nine clusters, and comparing the sum of these nine products to the sum of products for all N clusters in the lattice. Algebraically, the resulting input quantity for neuron i is

$$h_i = \frac{1}{N} \sum_{j \in c(i)} p_j \sum_{k}^{N} p_k, \tag{2}$$

in which the first summation includes the cluster state products for each of the nine clusters to which neuron i belongs, and the second summation includes all N cluster state products.

Clearly, $h_i < 0$ shows that the products of the nine clusters that include element i do not agree in general with the products formed from all clusters in the network, while $h_i > 0$ shows that they do so agree. As a result, the agreement among cluster states will increase over time if each neuron updates its state according to

$$s_i(t) \;=\; s_i(t-1)\cdot Sign\left[h_i(t-1)\right], \tag{3}$$

since the state changes only if $h_i < 0$.

The intuition captured by this update rule is formalized by introducing the energy-like function

$$E \;=\; -\left(\sum_{i,j}^{N} p_i p_j\right) \Big/ 2N. \tag{4}$$

Eq. 4 shows that E will have its minimum value when all cluster states agree. It is not difficult to show that Eqs. 2 and 3 imply that E will either remain constant or decrease on each neuron update,[19,20] so that the order parameter[24,25] of this network,

$$m \;=\; \frac{1}{N}\sum_{i}^{N} p_i \rightarrow \pm 1. \tag{5}$$

Simulations confirm this result[19]. Because 256 neuron patterns are consistent with each cluster state as measured by a product, the order that is achieved exists in the cluster product state space and not in neuron state space, providing a clear demonstration that the hidden pattern of cluster states governs the network's dynamics.[19,20]

As this brief overview shows, clusters and their states are defined by how the component neurons of the latent Ising model compute and compare inputs. A network neuron senses aspects of the state of the network that correspond to products of neuron states within clusters, and uses those sensed aspects to determine its own state, thereby contributing to aspects of the network sensed by both it and by other network neurons. In this manner, the aspects of the network that are indexed by the products are reified. The neuron clusters that they characterize interact, and those interactions are based on these reified cluster states. In effect, the architecture and dynamics (update rule) of this simple network *creates* a lattice of clusters and their states. Several generalizations of this basic model have been studied[20], differing mainly in the type and number of hidden cluster state patterns that can be generated, but all sharing the essential reification of clusters and their states.

3. Hidden Order in a Network of Integrate-and-Fire Neurons

Having a simple model that clarifies at least one way in which a recurrent network of elements produces hidden patterns is extremely valuable because it suggests how typical recurrent networks of more biologically-realistic neurons might be examined in order to determine if they naturally produce such hidden states. In order to make a start on this agenda, integrate-and-fire model neurons having recommended normalized parameter values[27] were used. These neurons function in the following way.

3.1 The Integrate-and-Fire Neuron

Each model neuron receives action potential - like signals on each synaptic site, and multiplies them by synaptic weight values. Unlike the binary neuron model considered above, an action potential has the value 1, and the absence of an action potential has the value 0. The total weighted input from a group of such sites (i.e., the scalar product $\mathbf{w} \bullet \mathbf{s}$ of a synaptic weight vector $\mathbf{w}$ and a vector of input signals $\mathbf{s}$) is used at each moment of discrete time to calculate the resulting change in conductance of sodium or chloride ions, depending on whether the synapses in question are excitatory or inhibitory, respectively. Using τ as a time constant and g as conductance, the computation can be written as

$$g(t) \quad = \quad (1-\tau)\cdot g(t-1) + \tau\, \mathbf{w} \bullet \mathbf{s}(t-1). \qquad (6)$$

This computation is performed for each collection of synaptic input sites that the modeler wishes to group together. For reasons of consistency with the earlier discussion of binary model neurons, the neurons that contribute to each set of inputs $\mathbf{s}$ will be referred to as a cluster.

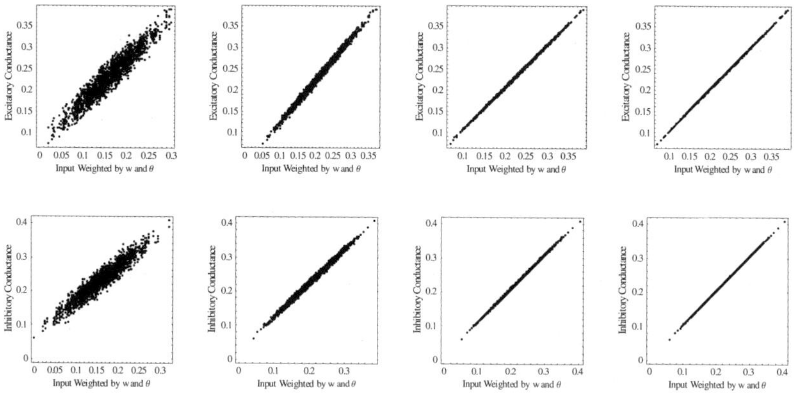

Fig. 3. Scatter plots of excitatory (top) and inhibitory (bottom) conductance, after 10 ms have passed, as a function of inputs weighted both by synaptic efficacy $\boldsymbol{w}$ and by $\boldsymbol{\theta}$. The number of columns of S and terms of θ increase from left to right.

One implication of Eq. 6 is that the value of a conductance depends not only on current input signals, but also on previous signals over some period of time, the length of which depends on τ. For example, using $g(0)$ and $s(0)$ as initial values, $g(1) = (1-\tau)\, g(0) + \tau\, \mathbf{w} \bullet \mathbf{s}(0)$, $g(2) = (1-\tau)^2\, g(0) + \tau\, (1-\tau)\, \mathbf{w} \bullet \mathbf{s}(0) + \tau\, \mathbf{w} \bullet \mathbf{s}(1)$, and so on. Since τ is typically assigned[27] the normalized value 0.7, the contribution of $g(0)$ to the value $g(t)$ declines markedly after even only a few time steps (milliseconds - ms). Suppose that after Δ such steps its contribution is negligible. Then, it is not difficult to show that

$$\begin{aligned} g(t) &= \mathbf{w}^T S\, \boldsymbol{\theta} + (1-\tau)^{\Delta}\, g(0) \\ &\approx \mathbf{w}^T S\, \boldsymbol{\theta}, \end{aligned} \tag{7}$$

where the time-decay vector

$$\boldsymbol{\theta}^T = \begin{pmatrix} \tau & \tau(1-\tau) & \tau(1-\tau)^2 & \cdots & \tau(1-\tau)^{\Delta-1} \end{pmatrix}, \tag{8}$$

entry s_{ij} of

$$S = \begin{pmatrix} s_{11} & s_{12} & \cdots & s_{1\Delta} \\ s_{21} & s_{22} & \cdots & s_{2\Delta} \\ \vdots & \vdots & \cdots & \vdots \\ s_{C1} & s_{C2} & \cdots & s_{C\Delta} \end{pmatrix} = \begin{pmatrix} \mathbf{s}_1\ \mathbf{s}_2 \cdots \mathbf{s}_{\Delta} \end{pmatrix} \tag{9}$$

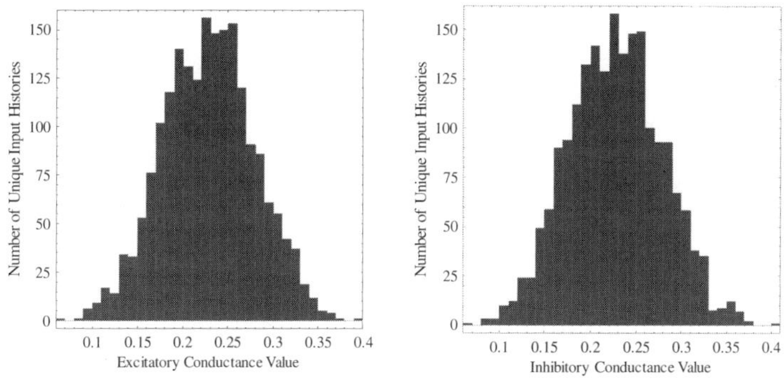

Fig. 4. Histograms showing the number of unique input histories that produce after 4 ms excitatory (left) and inhibitory (right) conductances in the ranges (0.01) specified by the abscissa of each graph.

is the state of cluster neuron i at time $t\text{-}j\text{+}1$, and $\boldsymbol{s}_j$ is a vector of all C cluster neuron states at time $t\text{-}j\text{+}1$. Thus, each conductance value is based on a history of cluster input over Δ time steps. In the simulations reported below (written in Mathematica[28]), each network neuron computed excitatory and inhibitory conductances g_e and g_i on the basis of input signals received from clusters of other network neurons, and an excitatory modulatory conductance g_m from clusters of neurons sending inputs to the network.

As an illustration, consider a set of 2000 repetitions of a simulation of a neuron receiving 11 inputs with the probability of an action potential for any input equal to 0.5, and with the center (sixth) signal having a maximum synaptic weight of 0.05, and the remaining weights falling off as a normal distribution with standard deviation (*sd*) equal to 5. Figure 3 shows graphs of excitatory (g_e - top) and inhibitory (g_i - bottom) conductances after 10 updates (ms) have passed. The abscissa of the left-most pair of graphs shows the product $\boldsymbol{w}^T\ S\ \boldsymbol{\theta}$ computed with only the first column of S and first term of $\boldsymbol{\theta}$. The abscissa of the second pair of graphs uses the first two columns of S and first two terms of $\boldsymbol{\theta}$, and so forth. As this figure makes clear, g_e and g_i both depend on a recent history of input signals S, and not just on the current pattern of inputs.

Furthermore, small ranges of g_e and g_i values are consistent with more than a single history of input signals. As shown in Figure 4, each small range of conductance values partitions the total set of possible

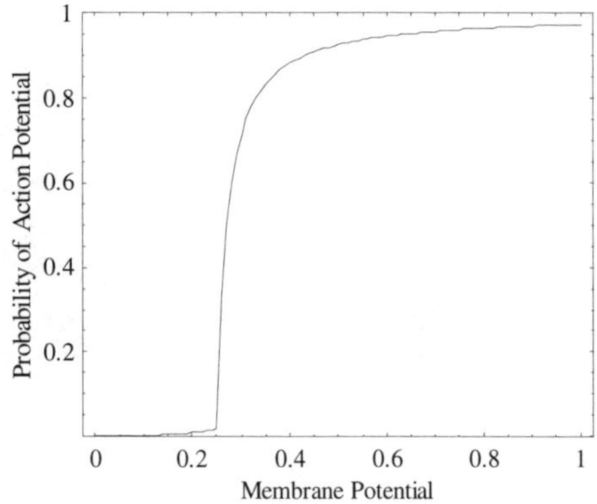

Fig. 5. The function used to determine the probability of an action potential on each update. Essentially, a hard threshold at a normalized membrane potential of 0.25 is softened by a sigmoid.[27]

input histories *{S}* into equivalence classes. Conductance values are used in this model to compute changes in membrane potential, and the probability of an action potential increases monotonically with membrane potential. Thus, patterns that fall into a given equivalence class are equally effective in changing the activity (action potential) pattern generated by the receiving neuron. Specifically, the change in membrane potential on each update (ms) is given by

$$\begin{aligned} \delta v(t) &= v(t) - v(t-1) \\ &= \tau_v \cdot \left[(g_m + g_e) \cdot \delta e_e + g_i \cdot \delta e_i + g_l \cdot \delta e_l \right], \end{aligned} \tag{10}$$

in which τ_v is the membrane time constant (set equal to 0.2)[27], g_l is a constant potassium leak conductance set equal to a normalized value of 0.1, $\delta e_e = e_e - v(t\text{-}1)$ is the difference between the excitatory sodium equilibrium potential (1.0) and the current membrane potential, $\delta e_i = e_i - v(t\text{-}1)$ is the difference between the inhibitory chloride equilibrium potential (0.15) and the current membrane potential, and $\delta e_l = e_l - v(t\text{-}1)$ is the difference between the equilibrium potassium leak potential and the current membrane potential (0.15). All conductance values are taken at time *t-1*.

Finally, a sigmoid function is used to determine the probability of an action potential as a function of membrane potential, as depicted in

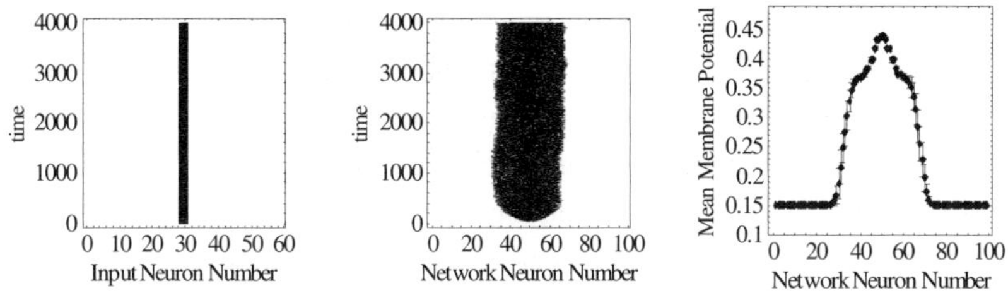

Fig. 6. Input (left) and network (middle) action potential patterns, and the mean pattern of network neuron membrane potentials achieved at steady-state (right), computed over the final 3000 updates and shown ± *sd*.

Figure 5. This function uses the recommended[27] normalized membrane potential threshold value of 0.25, and softens the resulting step function.

3.2 A Simple Recurrent Excitatory Network with Lateral Inhibition

Computer simulations have been used to explore the possibility that appropriate architectures of integrate-and-fire neurons produce hidden patterns of cluster states. As demonstrated above, cluster states, indexed by conductance values, determine each neuron's pattern of action potentials over time. Furthermore, in a recurrent network this latter pattern will be a component of excitatory and inhibitory cluster states for other neurons in the network (see below). From this perspective, the excitatory and inhibitory clusters interact by changing each other's states. In this sense, the cluster states represented by conductance values play a role analogous to the cluster states indexed by products in the binary neuron model considered above.

As pointed out by Koch & Meinhardt,[10] the concepts of local excitation and long-range inhibition play a fundamental role in the generation of biological patterns from initially homogeneous states. Therefore, initial exploratory computer simulations of integrate-and-fire neuron networks were based on these concepts. M modulatory (excitatory) inputs were distributed to the N neurons of the network, which were arranged in a line. Each network neuron also received both excitatory and inhibitory inputs from all N network neurons; however, the synaptic weights in the two excitatory cases were normally distributed with small *sd* values so that the weakest synapses made essentially no contribution. As a result, each network neuron effectively defined a modulatory (excitatory) input cluster having a state indexed by a modulatory conductance g_m, an excitatory network cluster having a state indexed by an excitatory conductance g_e, and an inhibitory network cluster consisting of all N network neurons having a state indexed by an inhibitory conductance g_i.

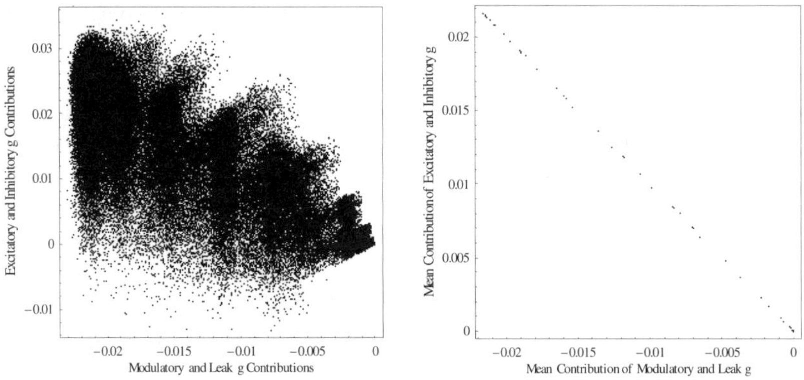

Fig. 7. The offsetting contributions of modulatory and leak conductances, and excitatory and inhibitory conductances to steady-state membrane potential are shown both for raw data (adj. $r^2 = 0.82$, $p = 9.3 \times 10^{-112170}$) and for mean values (adj. $r^2 = .99$, $p = 1.6 \times 10^{-226}$).

Simulations show that a fairly large selection of modulatory patterns, and a large range of values for the standard deviations of the three synaptic weight distributions, lead to rapidly-developing steady-state patterns of excitatory and inhibitory conductance values, and therefore to steady-state patterns of excitatory and inhibitory cluster states. The results reported below used neuron parameters equal to those reported in Sec. 3.1. The number of modulatory inputs $M = 59$, the number of network neurons $N = 99$, and the modulatory input and excitatory network neuron synaptic weight distribution $sd = 2$, while the inhibitory network weight distribution $sd = 100$. In the case of modulatory input, the largest synaptic weight value was given to the modulatory input in the relative position from 1-59 that was closest to the relative position of the network neuron from 1-99. The maximum modulatory synaptic

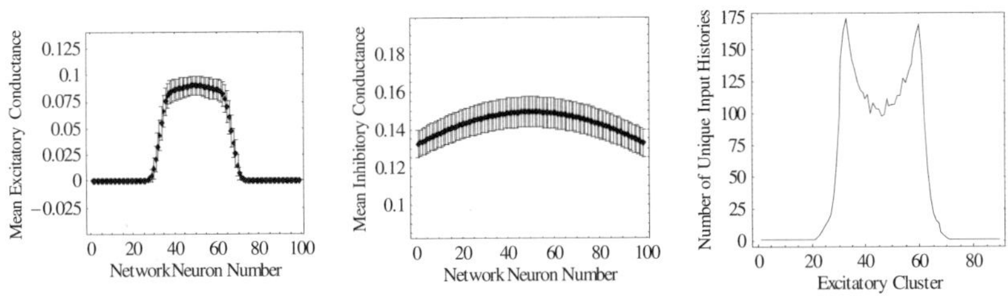

Fig. 8. The steady-state patterns of excitatory (left) and inhibitory (center) conductances are shown for the final 3000 samples of a simulation, ±*sd.* The rightmost panel shows the number of unique histories *S* of action potentials in the excitatory clusters that correspond to the conductance values in the leftmost panel.

weight was 0.15, the maximum network excitatory synaptic weight was 0.025, and the maximum inhibitory weight was 0.005. Each of the *M* input neurons fired with probability 0.95 over a simulation period of 4000 ms (updates).

Figure 6 shows typical graphs of modulatory input and network action potential patterns. This figure shows that the network achieves a steady-state fairly quickly, and that each network neuron has a stable membrane potential at steady-state. Equation 10 reveals that, when $v(t)$ is constant, there is a linear relationship between the modulatory and leak conductance contributions, and the excitatory and inhibitory conductance contributions,

$$g_e \cdot \delta e_e + g_i \cdot \delta e_i = -g_m \cdot \delta e_e - g_l \cdot \delta e_l . \qquad (11)$$

That this does in fact hold is demonstrated clearly in Figure 7, both in terms of raw data gathered over 99 x 3000 updates (left panel), and for the mean values of the conductances for each of the 99 network neurons. Thus, the network neurons adjust their excitatory and inhibitory conductances to balance the modulatory and leak conductance contributions to membrane potential, giving the steady-state patterns shown in Figure 8.

The data summarized in Figure 4 suggest that each conductance corresponds to a cluster state that is consistent with multiple input histories *S* of action potentials, at least in the case of an isolated integrate-and-fire neuron. This result also holds for the steady-state achieved by the present network simulation. The right-most panel of

Figure 8 shows the number of action potential histories that correspond to each small range of excitatory conductances given in the left-most panel.

These data suggest that simple networks of integrate-and-fire neurons can be described on two levels, in analogy to the latent Ising model. In the latter model, each neuron computes products that correspond to states of clusters of which it is a member and products that correspond to states of all clusters in the network, and the presence or absence of an action potential is determined by a relationship between these two classes of cluster state. In the present case, each integrate-and-fire neuron computes conductances that index cluster states, and the change in probability of an action potential is determined by a relationship between modulatory, leak, excitatory, and inhibitory conductances. In the latent Ising model case, an explicit energy function (Eq. 4) demonstrates that cluster states increase their alignment over time, making it clear that cluster-cluster interactions govern the dynamics of the network. No such simple expression is possible in the present case. However, it is clear that the excitatory and inhibitory cluster states change over time in order to balance the contributions of modulatory cluster states and leak conductances to membrane potential. Alignment of products is replaced by the dynamics of achieving this balance, and this in turn is a dynamics of cluster states. Once again, the dynamics is governed by cluster-cluster interactions.

4. Conclusions

The results of the investigations of these two neuron models support the hypothesis that hidden patterns can be produced by some recurrent networks. Although a careful study of the nature of hidden patterns remains to be done, a few comments regarding the properties of clusters seem particularly relevant to understanding patterns of cluster states.

First, a cluster state is an equivalence class at the level of the constituent model neurons, and it describes something about the activities of those component neurons (i.e., it is a meta-state). A product of ± 1 occurs for any of $2^8 = 256$ input neuron states, and simulations show that multiple input histories are associated with small ranges of conductance values (Figures 4 and 8). Thus, products and conductances partition the cartesian product of the set of all states of neurons within a cluster into equivalence classes.

Second, at least some properties of clusters considered as whole entities may not be intrinsic, but may arise through the organizing relations[29,30] of the network that give rise to cluster-cluster interactions. For example, equation 7 shows that each conductance value is associated with a matrix S that implements a linear transformation $\mathscr{S}$ that maps a time-decay vector $\boldsymbol{\theta}$ to a vector in the space of synaptic weights, $\boldsymbol{w}$. This suggests that each cluster has the properties of linear transformations between these spaces. The cluster state pattern is a pattern of *rules*.

Both types of model neuron employ indices of cluster states to determine their own states, which are in turn parts of cluster states. This embodiment of the cluster-cluster interactions in the dynamics of the network gives the clusters causal efficacy, and their states determine the influence that they have on the states of other clusters. It seems that future research must focus on reification (via such dual-level self-referential loops[31]) as potentially a key element in the production of hidden patterns.

Acknowledgments

This work was supported by an Academic Excellence and Innovation Award from Indiana University of Pennsylvania, and by grants from the IUP School of Graduate Studies and Research and the Pennsylvania State System of Higher Education Faculty Professional Development Council Annual Grant Program.

References

1. C. D. Bremer, J. B. Pittenger, R. Warren, and J. J. Jenkins, *Am. J. Psychology* 90(4) (1977).
2. F. A. Geldard and C. E. Sherrick, *Science* 178(4057) (1972).
3. V. S. Ramachandran and R. L. Gregory, *Nature* 350 (6320) (1991).
4. F. Attneave, *Sci. Am.* 225(6) (1971).
5. D. A. Leopold and N. K. Logothetis, *Trends in Cognitive Sciences* 3(7) (1999).

6. H. S. Hock, J. A. Kelso, and G. Schöner, *J. Exp. Hum. Percept. Perform.* 19(1) (1993).
7. T. Ditzinger & H. Haken, *Biological Cybernetics* 61(4) (1989).
8. D. Williams, G. Phillips, & R. Sekuler, *Nature* 324 (1986).
9. P. K. Maini and H. G. Othmer, *Mathematical Models for Biological Pattern Formation* (Springer, New York, 2001).
10. A. J. Koch and H. Meinhardt, *Rev. Mod. Phys.* 66(4) (1994).
11. K. Koffka, *Principles of Gestalt Psychology* (Harcourt, Brace & World, New York, 1935).
12. H. Haken, *Synergetic Computers and Cognition* (Springer, New York, 2004).
13. J. S. Kelso, *Dynamic Patterns: The Self-Organization of Brain and Behavior* 1 (MIT Press, Cambridge, MA, 1995).
14. S. Lehar, *Behavioral and Brain Science*, 26(4) (2003).
15. S. Lehar, *The World in Your Head: A Gestalt View of the Mechanism of Conscious Experience* (Lawrence Erlbaum, Mahwah, NJ, 2003).
16. F. Crick, *The Astonishing Hypothesis* (Charles Scribner's Sons, New York, 1994).
17. T. Metzinger, Ed., *Neural Correlates of Consciousness: Empirical and Conceptual Questions* (MIT Press, Cambridge, MA, 2000).
18. J. Shear, Ed., *Explaining Consciousness: The Hard Problem* (MIT Press, Cambridge, MA, 1999).
19. R. Pavloski, *Nonlinear Dynamics, Psychology & Life Sciences*, 10(3) (2006).
20. R. Pavloski, *DCDIS A Supplement, Advances in Neural Networks,* 14(S1) (2007).
21. J. Searle, *The Rediscovery of the Mind* (MIT Press, Cambridge, MA, 1992).
22. S. E. Palmer, *Behavioral and Brain Science* 22(6) (1999).
23. D. MacKay, in *Visual Neuroscience*, Ed. J. D. Pettigrew, K. J. Sanderson, and W. Levick (Cambridge University Press, New York, 1986).
24. D. Amit, *Modeling Brain Function: The World of Attractor Neural Networks* (Cambridge University Press, Cambridge, UK, 1989).
25. Y. Bar-Yam, *Dynamics of Complex Systems* (Addison-Wesley, Reading, MA, 1997).
26. R. J. Gaylord and P. R. Wellin, *Computer Simulations with Mathematica: Explorations in Complex Physical and Biological Systems* (Springer, New York, 1995).
27. R. C. O'Reilly and Y. Munakata, *Computational Explorations in Cognitive Neuroscience: Understanding the Mind by Simulating the Brain* (MIT Press, Cambridge, MA, 2000).
28. S. Wolfram, *The Mathematica Book (5th ed.)* (Wolfram Media, Champaign, IL, 2003).
29. S. Jones, *NeuroQuantology*, 3 (2004).
30. S. Jones, in *Artificial Life VIII*, Ed. R. Standish, M A. Bedau, and H. A. Abbass (MIT Press, Cambridge, MA, 2003), pp. 418-422.
31. D. Hofstadter, *I am a Strange Loop* (Basic Books, New York, 2007).

CHAPTER 19

MODELING COMMON-SENSE DECISIONS

Michail Zak

Jet Propulsion Laboratory, California Institute of Technology
Advanced Computing Algorithms and IVHM Group
Pasadena, CA 91109
E-mail: michail.zak@gmail.com

This paper presents a methodology for efficient synthesis of dynamical model simulating a common-sense decision making process. The approach is based upon the extension of the physics' First Principles that includes behavior of living systems. The new architecture consists of motor dynamics simulating actual behavior of the object, and mental dynamics representing evolution of the corresponding knowledge-base and incorporating it in the form of information flows into the motor dynamics. The autonomy of the decision making process is achieved by a feedback from mental to motor dynamics. This feedback replaces unavailable external information by an internal knowledgebase stored in the mental model in the form of probability distributions.

1. Introduction

Human common sense has always been a mystery for physicists and an obstacle for artificial intelligence. It was well understood that human behavior, and in particular, the decision making process, is governed by feedbacks from the external world, and this part of the problem was successfully simulated in the most sophisticated way by control systems. However, in addition to that, when the external world does not provide sufficient information, a human being turns for "advise" to his experience, and that is associated with common sense. The simplest representation of human experience is by *if-then rules.* However, in real world situations, the number of rules grows exponentially with the dimensionality of external factors, and implementation of this strategy is hardly realistic. One of the ways to fight such a combinatorial explosion is to represent rules in a more abstract and more generalized form by removing insignificant details and making the rules more inclusive. This procedure is usually accompanied by coercion, i.e. by reducing uncertainties about the world via forcing it into a known state regardless of the initial state. Indeed, many natural and social phenomena exhibit some degree of regularity only on a higher level of abstraction, i.e. in terms of some invariants. Within the mathematical formalism of the model of mental dynamics, this means that only variables of the highest level of abstraction are capable of classifying changes in the external world and send them to the corresponding attractor. Based upon that, the model reacts to these changes, and that can be associated with common-sense-based decision. Collaboration, competition and games are examples of those types of human activities that require common sense for predicting the intentions of a collaborator or a competitor in order to make a right decision when objective information for a rational decision in incomplete.

In this paper, by common sense we will understand a feedback from the self-image (a concept adapted from psychology), and based upon that, we will propose a physical model of common sense in connection with the decision making process. This model grew out of the model of Livings proposed and discussed in our earlier papers [1,2,3,4,5,6]. As follows from this model, the difference between behaviors of living and non-living systems is fundamental. Indeed, as has been recently recognized, the evolution of living systems is progressive in a sense that it is directed to the highest levels of complexity if the complexity is measured by an irreducible number of different parts that interact in a well-regulated fashion. Such a property is not consistent with the behavior of closed Newtonian systems that cannot increase their complexity without external forces. Indeed, the solutions to the models based upon dissipative Newtonian dynamics eventually approach attractors where the evolution stops, while these attractors dwell on the subspaces of lower dimensionality, and therefore, of the lower complexity (see Figure 1).

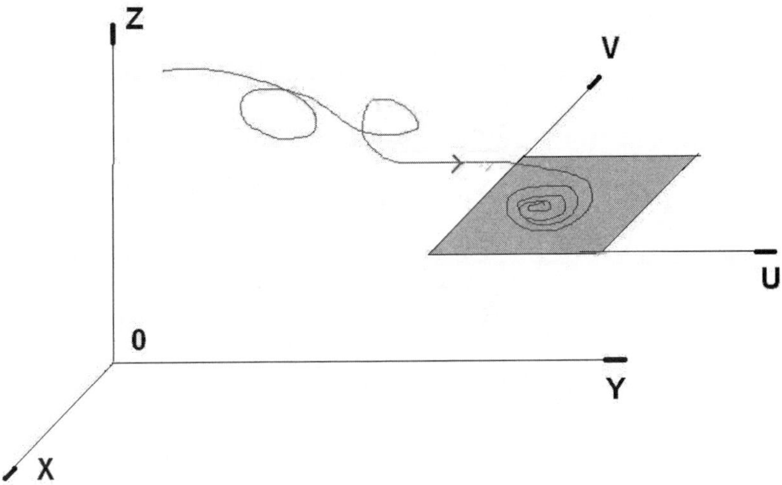

Fig. 1. Dissipative dynamics.

If thermal forces are added to mechanical ones, the Newtonian dynamics is extended to the Langevin dynamics combining both mechanics and thermodynamics effects; it is represented by stochastic differential equations that can be utilized for more advanced models in which randomness stands for multi-choice patterns of behavior typical for living systems. However, even those models do not capture the main property of living systems, i.e. their ability to evolve towards an increase in complexity without external forces. Indeed, the Langevin dynamics is complemented by the corresponding diffusion (or Fokker-Planck) equation that describes the evolution of the distribution of the probability density over the state variables; in case of an isolated system (under condition that the Newtonian forces and their gradient do not have singularities), the entropy of the probability density cannot decrease, and that expresses the second law of thermodynamics (formalized in Boltzmann's H-theorem). From the viewpoint of complexity, this means that the state variables of the underlying system eventually start behaving in a more uniform fashion with lesser distinguished features, i.e. with lower complexity, Figure 2.

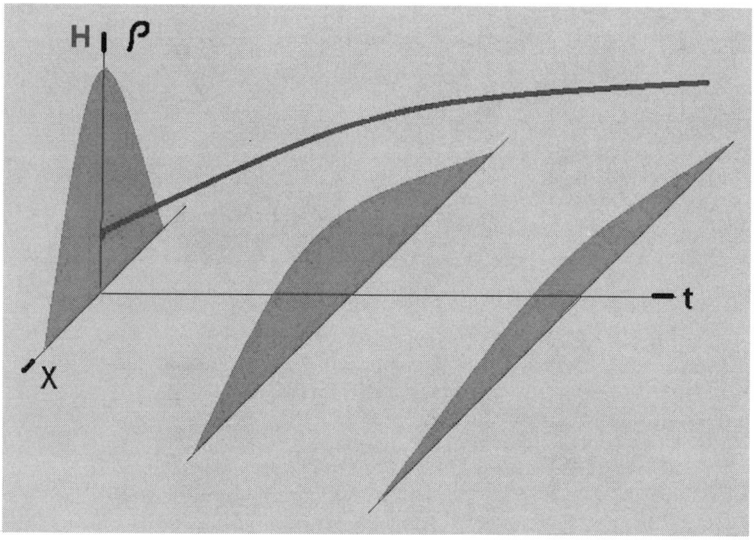

Fig. 2. Dynamics with diffusion.

Thus, the First Principles of classical physics do not capture the whole complexity of behavior of living, and in particular, intelligent systems. Nevertheless, since living systems belong to the physical world, they must obey these Principles. Hence, it seems reasonable just to extend the First Principles of Classical Physics to include living systems. The strategy for such an extension can be adopted from those in quantum mechanics. It should be recalled that classical physics can be formulated by the Hamilton-Jacobi equation (that is equivalent to Newton's second law), and the corresponding Liouville equation (that expresses the continuity of flow of the probability density), while these equations are *not coupled.* The *only* formal difference between classical and quantum worlds is in additional scalar (called quantum potential) that *couples* the Hamilton-Jacobi and Liouville equations (making them equivalent to the Schrödinger equation). Although a physical origin of this potential is not yet well understood and is still under discussion, its "purpose" is to enforce the uncertainty principle into quantum formalism [1].

In our recent publications [2,3,4,5,6], following the same strategy, we found *another* potential that extends classical physics to include living systems by *coupling* the same equations. We imposed two constraints upon this potential. Firstly, it introduced information-based (non-Newtonian) forces that originated from exchange and processing of information, thereby allowing living systems to communicate. Secondly, it affected the evolution of entropy by making it reversible. Thus, we address the problem of development of a phenomenological model that represents the fundamental property of the behavior of living systems: the ability to communicate via information flows, and the ability to increase their complexity (in terms of decrease of entropy) without interactions with the external world. (Loosely speaking, the last property can be associated with a so-called "free will").

2. The Simplest Model of Livings

The destabilizing effect of Liouville feedback is demonstrated below. We will start this section with a brief review of models introduced and discussed in [3,4,5,6,7]. For mathematical clarity, we will consider a one-dimensional motion of a unit mass under action of a force f depending upon the velocity x and time t

$$\dot{x} = f(x,t), \tag{1}$$

If initial conditions are not deterministic, and their probability density is given in the form

$$\rho_0 = \rho_0(X), \quad where \ \rho \geq 0, \ and \quad \int_{-\infty}^{\infty} \rho dX = 1 \tag{2}$$

while ρ is a single-valued function, then the evolution of this density is expressed by the corresponding Liouville equation

$$\frac{\partial \rho}{\partial t}+\frac{\partial}{\partial X}(\rho f)=0 \tag{3}$$

The solution of this equation subject to initial conditions and normalization constraints (2) determines the probability density as a function of *X* and *t*

$$\rho=\rho(X,t) \tag{4}$$

Let us now specify the force f as a feedback from the Liouville equation

$$f(x,t)=\varphi[\rho(x,t)] \tag{5}$$

and analyze the motion after substituting the force (7) in to Eq. (3)

$$\dot{x}=\varphi[\rho(x,t)] \tag{6}$$

This is a fundamental step in our approach. Although theory of ODE does not impose any restrictions upon the force as a function of space coordinates, the Newtonian physics does: equations of motion are never coupled with the corresponding Liouville equation. Moreover, as shown[2], such a coupling leads to non-Newtonian properties of the underlying model. Indeed, substituting the force f from Eq. (5) into Eq. (3), one arrives at the *nonlinear* and, in general, *non-reversible* equation for evolution of the probability density

$$\frac{\partial \rho}{\partial t}+\frac{\partial}{\partial X}\{\rho\varphi[\rho(X,t)]\}=0 \tag{7}$$

Now we will demonstrate the destabilizing effect of the feedback (6). For that purpose, it should be noted that the derivative $\partial\rho/\partial x$ must change its sign, at least once, within the interval $-\infty < x < \infty$, in order to satisfy the normalization constraint (2). But since

$$Sign\frac{\partial \dot{x}}{\partial x} = Sign\frac{d\varphi}{d\rho} Sign\frac{\partial \rho}{\partial x} \tag{8}$$

there will be regions of x where the motion is unstable, and this instability generates randomness with the probability distribution guided by the Liouville equation (7).

In order to illustrate mathematical aspects of the concepts of Liouville feedback, as well as associated with it instability, randomness, superposition, and entanglement, let us start with the following system

$$\dot{x} = -q^2 \frac{\partial}{\partial X} \ln \rho, \tag{9}$$

along with the corresponding Liouville equation (in this particular case, the Liouville equation takes the form of the Fokker-Planck equation, and therefore, it is *non-reversible)*

$$\frac{\partial \rho}{\partial t} = q^2 \frac{\partial^2 \rho}{\partial X^2} \tag{10}$$

Here q^2 is the constant diffusion coefficient.
The solution of Eq. (12) subject to the sharp initial condition is

$$\rho = \frac{1}{2q\sqrt{\pi t}} \exp(-\frac{X^2}{4q^2 t}) \tag{11}$$

Substituting this solution into Eq. (9) at X=x, one arrives at the differential equation with respect to x *(t)*

$$\dot{x} = \frac{x}{2t} \tag{12}$$

and therefore,

$$x = C\sqrt{t} \tag{13}$$

where C is an arbitrary constant. Since $x=0$ at $t=0$ for any value of C, the solution (13) is consistent with the sharp initial condition for the solution (11) of the corresponding Liouvile equation (10). The solution (13) describes the simplest irreversible motion: it is characterized by the "beginning of time" where all the trajectories intersect (that results from the violation of Lipcsitz condition at $t=0$, Figure 3), while the backward motion obtained by replacement of t with $(-t)$ leads to imaginary values of velocities. One can notice that the probability density (11) possesses the same properties.

For a fixed C, the solution (13) is *unstable* since

$$\frac{d\dot{x}}{dx} = \frac{1}{2t} > 0 \tag{14}$$

and therefore, an initial error always grows generating *randomness*. Initially, at $t=0$, this growth is of infinite rate since the Lipchitz condition at this point is violated

$$\frac{d\dot{x}}{dx} \to \infty \qquad at \qquad t \to 0 \tag{15}$$

This type of instability has been introduced and analyzed [8,9].

Considering first Eq. (13) at fixed C as a sample of the underlying stochastic process (13), and then varying C, one arrives at the whole ensemble characterizing that process, (see Figure 3). One can verify that, as follows from Eq. (11), the expectation and the variance of this process are, respectively

$$Mx = 0, \qquad Dx = 2q^2 t \tag{16}$$

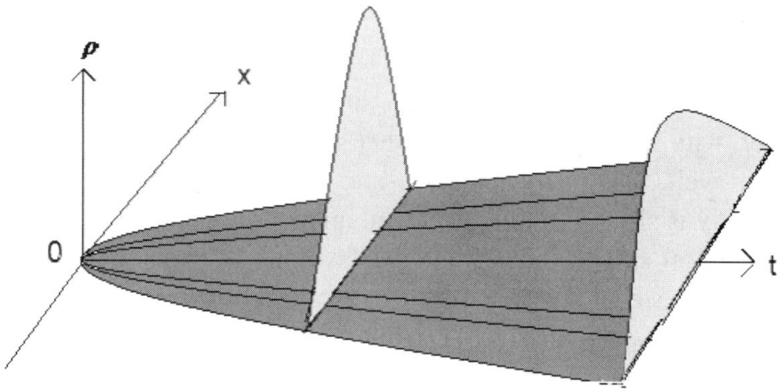

Fig. 3. Stochastic process and probability density.

The same results follow from the ensemble (13) at $-\infty \leq C \leq \infty$. Indeed, the first equality in (16) results from symmetry of the ensemble with respect to *x=0*; the second one follows from the fact that

$$Dx \propto x^2 \propto t \tag{17}$$

It is interesting to notice that the stochastic process (13) is an alternative to the following Langevin equation,

$$\dot{x} = \Gamma(t), \qquad M\Gamma = 0, \qquad D\Gamma = q \tag{18}$$

that corresponds to the *same* Fokker-Planck equation (10).

Here $\Gamma(t)$ is the Langevin (random) force with zero mean and constant variance q.

As follows from Eq. (13), all the particular solutions intersect at the same point *x=0* at *t=0,* and that leads to non-uniqueness of the solution due to violation of the Lipcshitz condition. Therefore, the same initial condition *x=0* at *t=0* yields infinite number of different solutions forming a family (13); each solution of this family appears with a certain probability guided by the corresponding Fokker-Planck equation. For instance, in our case, the "winner" solution is

$$x \equiv 0 \tag{19}$$

since it passes through the maxima of the probability density (11). However, with lower probabilities, other solutions of the family (13) can appear as well. Obviously, this is a non-classical effect. Qualitatively, this property is similar to those of quantum mechanics: the system keeps all the solutions simultaneously and displays each of them "by a chance", while that chance is controlled by the evolution of probability density (10). It should be emphasized that the choice of displaying a certain solution, i.e. *decision*, is made by the system only once, at $t=0$, i.e. when it departs from the deterministic to a random state; since than, it stays with this solution as long as the Liouville feedback is present, $(q \neq 0)$. However, as soon as this feedback disappears, $(q = 0)$, the system becomes fully deterministic, while the deterministic solution is a continuation of the corresponding "chosen" random solution, (see Figure 4).

The transition from the random to deterministic state and back, i.e. from a decision making mode to a passive state, can be automated if Eq. (9) is modified using the following logical device

$$\dot{x} = -yq^2 \frac{\partial}{\partial X} \ln \rho, \tag{20}$$

where

$$\dot{y} = -a\sqrt{y(1-y)}, \qquad a = const > 0 \tag{21}$$

Eq. (21) has two equilibrium points: y = 0 and y = 1. At both these points, the Lipschitz condition does not hold

$$\frac{\partial \dot{y}}{\partial y} = a \frac{1-2y}{2\sqrt{y(1-y)}} \to -\infty \qquad at \qquad y \to 0$$
$$and \to \infty \qquad at \qquad y \to 1, \tag{22}$$

Hence, y = 1 is a *terminal* attractor, and y = 0 is a *terminal* repeller [7,8]. Regardless of these "abnormality", the closed form solution is easily obtained by separation of variables

$$t = \frac{2}{a}\arctan\sqrt{\frac{1-y}{y}} + C,$$
$$C = 0 \quad if \qquad y = 0 \quad at \qquad t = 0 \tag{23}$$

However, this "abnormality" becomes crucial for providing a finite time T of transition from the repeller y = 0 to the attractor y = 1

$$T = \frac{\pi}{a} \tag{24}$$

It should be recalled that in classical theory of ODE, when the Lipschitz condition is preserved, an attractor is always approached asymptotically, and the transition period is theoretically unbounded. Hence, this limitation is removed due to a special form of the governing equation (21).

Qualitatively, the result (24) holds even if *a* = *a (t)*. Then the solution to Eq. (22) is

$$\int_0^t a(t)dt = 2\arctan\sqrt{\frac{1-x}{x}}, \qquad and \qquad \int_0^T a(t)dt = \pi,$$
$$i.e. \qquad T < \infty \tag{25}$$

Similar results can be obtained for *a* < 0: then the attractor becomes a repeller and wise versa. When the function *a (t)* changes its sign sequentially, the solution switches from 0 to 1 and back, respectively, while the transition time is always finite. Selecting, for instance,

$$a = \cos\omega t \tag{26}$$

one arrives at periodical switches (with the frequency ω) from a decision states to passive states and back. As shown in Figure 4, after each switch to the decision state, the system may select different solutions from the same family (13), i.e. a different decision, so that the entire solution will include jumps from one branch of the family (13) to another. However, if the probability density (11) is sufficiently sharp, different decisions statistically will be close.

3. Modified Model

The main limitation of the model discussed above is that the corresponding stochastic process always diverges approaching the thermodynamical limit. That can be illuminated by the auxiliary equation following from Eq. (10) [3,9]

$$\dot{D} = 2q^2, \qquad D = 2q^2 t, \tag{27}$$

that demonstrates an unlimited growth of the variance D due to diffusion. The first logical step to reverse this process is to introduce negative diffusion by changing the sign in Eqs. (9) and (10)

$$\dot{x} = q^2 \frac{\partial}{\partial X} \ln \rho, \tag{9a}$$

$$\frac{\partial \rho}{\partial t} = -q^2 \frac{\partial^2 \rho}{\partial X^2} \tag{10a}$$

However, the solution to Eq. (10a) is ill-posed; [2] the best illustration to that is the corresponding version of Eqs. (27)

$$\dot{D} = -2q^2, \qquad D = D_0 - 2q^2 t \tag{27a}$$

Indeed, regardless of the initial value of the variance D_0, eventually the variance D becomes negative, and that does not make sense since, by definition, the variance cannot be negative

$$D = \int_{-\infty}^{\infty} \rho \zeta^2 d\zeta \geq 0 \tag{28}$$

However, in nonlinear versions of Eqs. (9), (10), the ill-posedness can be removed. Indeed, consider the following modification of these equations

$$\dot{x} = q^2 D \frac{\partial}{\partial X} \ln \rho, \tag{29}$$

$$\frac{\partial \rho}{\partial t} = -q^2 D \frac{\partial^2 \rho}{\partial X^2} \tag{30}$$

Then, instead of Eq. (27a), one obtains

$$\dot{D} = -2q^2 D, \quad D = D_0 e^{-2q^2 t} > 0 \tag{31}$$

Thus, the variance asymptotically approaches zero remaining positive, and therefore, the density ρ as a solution to Eq. (30), approaches the sharp value displaying a negative diffusion while its evolution becomes *reversible*.

Eqs. (29) and (30) can be augmented by additional terms providing a non-zero mean $\bar{x}$ of the underlying stochastic process

$$\dot{x} = q^2 D \frac{\partial}{\partial X} \ln \rho + f(\bar{x}), \tag{32}$$

$$\frac{\partial \rho}{\partial t} = -q^2 D \frac{\partial^2 \rho}{\partial X^2} - f(\bar{x}) \frac{\partial \rho}{\partial X} \tag{33}$$

As follows from Eq. (32), the mean value approaches the value of a root of the equation

$$f(\bar{x}) = 0 \qquad if \qquad \frac{\partial f}{\partial \bar{x}} < 0 \tag{34}$$

4. General Model

The model we propose for common sense simulation is based upon n-dimensional version of Eqs. (32), (33), and (20), (21).

$$\dot{x}_i = (y-1)f_i + y(\beta_i + \alpha_i \frac{\partial}{\partial X_i} \ln \rho), \quad i = 1,2,...n, \tag{35}$$

$$\frac{\partial \rho}{\partial t} = -\sum_{i=1}^{n} \{[(y-1)f_i + y\beta_i] \frac{\partial \rho}{\partial X_i} - y\alpha_i \frac{\partial^2 \rho}{\partial X_i^{\,2}}\} \tag{36}$$

Here

$$f_i = f_i(x_1,...x_n), \qquad \alpha_i = \alpha_i(\rho), \quad \beta_i = \beta_i(\rho) \tag{37}$$

while y is defined by Eq. (21).

The solution to Eq. (36) has the same fundamental properties as the solution to its simplified one-dimensional version (10). Indeed, starting with the initial conditions

$$x_i = 0, \ \rho = \delta(X_i \to 0) \qquad at \qquad t = 0 \tag{38}$$

and looking first for the solution within an infinitesimal initial time interval ε, one reduces Eq. (36) to the form

$$\frac{\partial \rho}{\partial t} = -\sum_{i=1}^{n} \alpha_i \frac{\partial^2 \rho}{\partial X_i^{\,2}} \tag{39}$$

since

$$\frac{\partial \rho}{\partial X_i} = 0, \qquad at \qquad t = 0 \tag{40}$$

and therefore, all the drift terms can be ignored within the interval ε.

The solution to Eq. (39) is a multi-dimensional version of Eq. (11). Substitution of this solution to Eqs. (35) yields

$$\dot{x}_i = [f_i + y\beta_i + \frac{x_i}{2t}] \rightarrow [\frac{x_i}{2t}] \quad at \qquad t \rightarrow 0 \tag{41}$$

since the functions f_i are considered to be bounded. The solution to Eqs. (41) is of the form (13)

$$x_i = C_i \sqrt{t} \qquad at \qquad t \rightarrow 0 \tag{42}$$

and this can be illustrated by the same Figure 3 with the only difference that here the solution is valid only within an infinitesimal initial time interval ε. But the most important phenomenon- phase transition- occurs exactly in this interval: due to violation of the Lipcshitz condition, the solution splits into ensemble of different samples, while the probability of the solution to represent certain sample is controlled by the density following from Eq. (39). Thus, the solution to Eqs. (35) and (36) within the infinitesimal initial interval captures instability-based randomization effect.

5. Model Interpretation

From the *mathematical* viewpoint, the most significant departure from classical models is negative diffusion in dynamics of probability density, see Eqs. (30) and (36). As shown [2], the well-known ill-posedness of initial value problems for linear parabolic equations can be avoided by adding nonlinearity that prevents the variances from becoming negative.

From the physical viewpoint, the model represents a fundamental departure from both Newtonian and statistical mechanics. In particular, negative diffusion cannot occur in closed systems without help of the Maxwell sorting demon that is strictly forbidden in statistical mechanics. The only conclusion to be made is that the model is non-Newtonian, although it is fully consistent with the theory of differential equations and stochastic processes. Strictly speaking, it is a matter of definition weather the model represents an isolated or an open system since the additional energy applied via the information potential is generated by the system "itself" out of components of the probability density.

From the *biological* viewpoint, the model introduces a biological particle that, in addition to Newtonian properties, possesses the ability to process information. The probability density can be associated with the *self-image* of the biological particle as a member of the class to which this particle belongs, while its ability to convert the density into the information force - with the *self-awareness,* (both these concepts are adopted from psychology). Continuing this line of associations, the equation of motion (such as Eqs. (35)) can be identified with motor dynamics, while the evolution of density (see Eqs. (36)) - with mental dynamics. Actually, the mental dynamics plays the role of the Maxwell sorting demon: it rearranges the probability distribution by creating the information potential and converting it into a force that is applied to the particle. One should notice that mental dynamics describes evolution of the whole class of state variables (differed from each other only by initial conditions), and that can be associated with the ability of generalization and abstraction that is a privilege of living systems. It should be emphasized that the biological origin of mental dynamics as image interaction is associated with the mirror neuron. The discovery of mirror neurons in the frontal lobes of macaques and their implications for human brain evolution is one of the most important findings of neuroscience in the last decade. Mirror neurons are active when the monkeys perform certain tasks, but they also fire when the monkeys *watch* someone else perform the same specific task. There is evidence that a similar observation/action matching system exists in humans. In the case of humans, this phenomenon represents the concept of imitation learning i.e. the ability to learn by imitation, and this faculty is at the basis of human culture. Hence, a mirror neuron representing an agent A can be activated by an expected (or observed) action of an agent B which may not be in a direct contact with the agent A at all, and that can be associated with existing of self-and-nonself images as well as the self-awareness [3,5].

6. Decision Making Process

Decision making is the cognitive process of selecting a course of action from among multiple alternatives. We will distinguish two types of decision making processes. The first type is associated with the concept of a rational agent; the models of this type are largely quantitative and are based on the assumptions of rationality and near perfect knowledge. They are composed of agent's believe on the basis of evidence followed by construction and maximization of utility function. The main limitation of these models is in their exponential complexity: on the level of believe nets, the exponential complexity is caused by the fact that encoding a joint probability as a function of n propositional variables requires a table with 2^n entries; [10] the same rate complexity occurs in rule-based decision trees. The second type of a decision making process is based upon psychological models; these models concentrate on psychological and cognitive aspects such as motivation, need reduction, and common sense. They are qualitative rather than quantitative and build on sociological factors like cultural influences, personal experience, etc.

The model proposed in this paper is better structured for the second type of decision making process that exploits interaction between motor and mental dynamics in which the mental dynamic plays the role of a knowledge base replacing unavailable external information. In order to demonstrate the proposed approach to decision making, we will augmented the model (35) (36) by two additional devices. The first one, a sensor device, measures the difference between the actual and expected values of the state variables

$$\Delta = \sum_{i=1}^{n}(x_i - \bar{x}_i)^2, \qquad \bar{x}_i = \int_{-\infty}^{\infty}\int_{-\infty}^{\infty}...\int_{-\infty}^{\infty} X_i \rho dX_1 dX_2 ...dX_n, \quad (43)$$

The second one, the logical device, is based upon Eq. (21) with

$$a = A - \Delta \quad (44)$$

where *A>0* is a sensitivity threshold specified for the decision making process.

We will start the analysis of the model performance from its pre-decision mode when

$\Delta = 0,$ and therefore, *y=0* and

$$\dot{x}_i = f_i(x_1,...x_n) \qquad i = 1,2,...n, \tag{45}$$

$$\frac{\partial \rho}{\partial t} = -\sum_{i=1}^{n} f_i \frac{\partial \rho}{\partial X_i} \tag{46}$$

At this mode, the motor dynamics (45), and the mental dynamics (46) are not coupled. The performance starts with mental dynamics that runs ahead of actual time, calculates expected values of the state variables as well as the difference Δ at the end of a selected lead time *T,* (see Eq. (43), and sends this information to the sensor device (44). As soon as Δ becomes large, it changes the sign of the control parameter *a* (see Eqs. (21) and (44)), the value of *y* changes from zero to one, and the system (45), (46) switches to the decision mode taking the form (35),(36). After this switch, a deterministic trajectory defined by the solution to Eq. (45) splits into a family of the solution to the coupled system (35), (36) (see Eq. (42), and Fig. 4). But, as discussed above, the system selects only one trajectory of this family at random with the probability defined by the information forces $\alpha_i(\rho)$ and $\beta_i(\rho)$. Thus, the topology of the proposed model demonstrates *how* a decision is made, while the structure of the information forces defines *what* decision is made. This structure will be discussed in the following section.

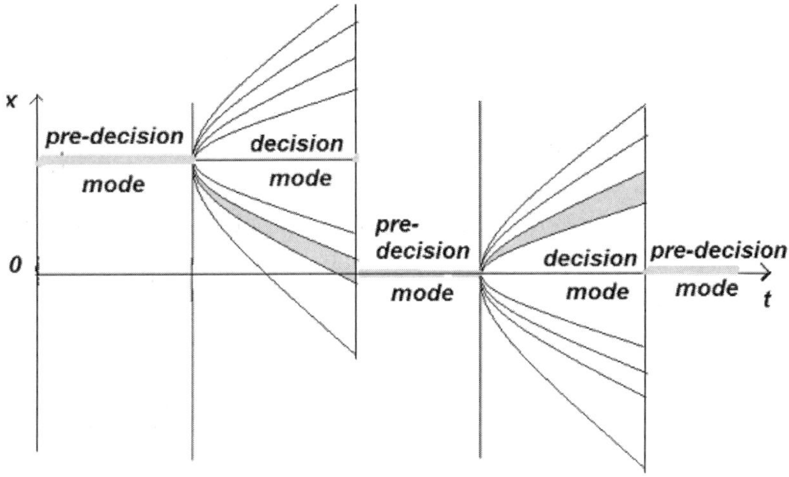

Fig. 4. Switching between pre-decision and decision modes.

7. Decision via Choice of Attractors

Let us return to the system (35), (36) in the decision mode

$$\dot{x}_i = \beta_i + \alpha_i \frac{\partial}{\partial X_i} \ln \rho, \; i = 1,2,...n, \tag{47}$$

$$\frac{\partial \rho}{\partial t} = -\sum_{i=1}^{n} (\beta_i \frac{\partial \rho}{\partial X_i} + \alpha_i \frac{\partial^2 \rho}{\partial X_i^{\,2}}) \tag{48}$$

and assume that α_i and β_i in Eq. (37) depend only upon two first invariants of the probability density ρ, namely, upon the means $\bar{x}_i$ and principle variances D_{ii}

$$\alpha_i = f_i(\bar{x}_1,...\bar{x}_n, D_{11},...D_{nn}), \quad \beta_i = \alpha_i(\bar{x}_1,...\bar{x}_1, D_{11},...D_{nn}) \tag{49}$$

As follows from Eqs. (35) and (36), the means $\bar{x}_i$ and the variances D_i must satisfy the following ODE [3,9]

$$\dot{\bar{x}}_i = \beta_i(\bar{x}_1,...\bar{x}_n, D_{11},...D_{nn}), \tag{50}$$

$$\dot{D}_{ii} = \alpha_i(\bar{x}_1,...\bar{x}_1, D_{11},...D_{nn}) \tag{51}$$

Thus, although the state variables x_i are statistically independent, i.e. $D_{ij} \equiv 0 \; if \quad i \neq j$, the time evolution of their statistical invariants is coupled via Eqs. (50) and (51).

It should be noticed that Eqs. (50) and (51) can be considered as a ODE-based approximation to the mental dynamics (48). Indeed, any probability density can be expended in a Gram-Charlier series with the first term being represented by normal distribution that is uniquely defined by the vector $\bar{x}_i$ and the tensor D_{ii}.

Next we will introduce the following structure into the functions (49)

$$\alpha_i = \sqrt{(\bar{x}_i - \gamma_{i1})(\bar{x}_i - \gamma_{i2})...(\bar{x}_i - \gamma_{in})}(\Sigma w_{ij}\Delta_j), \tag{52}$$
$$\Delta_j = (x_i - \bar{x}_i)^2,$$

$$\beta_i = \sqrt{(D_{ii} - \eta_{i1})(D_{ii} - \eta_{i2})...(D_{ii} - \eta_{in})}(\Sigma v_{ij}\Delta_j), \tag{53}$$
$$i = 1,2,...n,$$

where $\gamma_{ij}, \eta_{ij}, w_{ij}, v_{ij}$ are constants and rewrite the system (50), (51) as follows

$$\dot{\bar{x}}_i = \sqrt{[\prod_{j=1}^{n}(\bar{x}_i - \gamma_{ij})]}(\Sigma w_{ij}\Delta_j),\; i = 1,2,...n, \tag{54}$$

$$\dot{D}_{ii} = \sqrt{\prod_{j=1}^{n}(D_{ii} - \eta_{ij})}(\Sigma v_{ij}\Delta_j),\; i = 1,2,...n, \tag{55}$$

We will start our analysis with Eqs. (54). First of all, in this particular setting, these equations do not depend upon Eqs. (55). Moreover, each equation in the system (54) is coupled to other equations of the same systems only via the signs of the sums $\Sigma w_{ij}\Delta_j$ that include contribution from all the equations. Indeed, if $\Sigma w_{ij}\Delta_j > 0$, the ith equation has the terminal attractors [7]

$$\bar{x}_i^{\,0} = \gamma_{i2}, \gamma_{i4}, ... \gamma_{im}, \qquad i = 1,2,...n \tag{56}$$

and terminal repellers

$$\bar{x}_i^{\,*} = \gamma_{i1}, \gamma_{i3}, ... \gamma_{ik}, \qquad i = 1,2,...n \tag{57}$$

where $m = n$, $k = n\text{-}1$ if n is even, and $m = n\text{-}1$, $k = n$ if n is odd.

The $n^2/2$ attractors (56) can be pre-stored using n^2 weights w_{ij}. These attractors are the limit values of the means of the stochastic process that occurs as a result of switch from the pre-decision to the decision making mode, and these attractors are approached in finite time, (see eq. (24)).

Turning to Eqs. (55) and noticing that they have the same structure as Eqs. (54), one concludes that the principle variances D_{ii} of the same stochastic process approach their limiting values

$$D^0{}_{ii} = \eta_{i2}, \eta_{i4}, ... \eta_{im}, \quad i = 1,2,...n, \tag{58}$$

that can be pre-stored using weights v_{ij}.

Thus, when the system switches to the decision making mode, it chooses a sample of the corresponding stochastic process, and this sample represents the decision. The choice of the sample is controlled by the normal probability density with means and principle variances approaching their limit values (56) and (58). The choice of a particular attractor out of the set of (56) and (58) depends upon the initial values of means and variances at the end of the pre-decision period: if these values fall into the basin of attraction of a certain attractor, they will eventually approach that attractor, Figure 5.

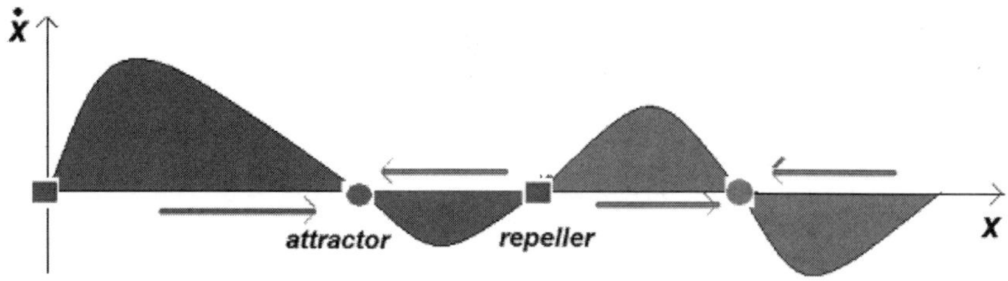

Fig. 5. Basins of attraction separated by repellers.

8. Decision via Phase Transition

In the previous section, we consider the case when pre- and post-decision states are not totally disconnected: they belong to the same basin of attraction. However, in general, decision may require fundamental change of the dynamical structure, and that can be achieved via phase transition. The distinguishing characteristic of a phase transition is an abrupt sudden change in one or more physical properties of a dynamical system. In physics, a phase transition is the transformation of a thermodynamic system from one phase to another (evaporation, boiling, melting, freezing, sublimation, etc). In engineering, such transitions occur because of change of system configuration (new load, new open valve, new switch or a logic device in control systems with variable structure, etc.).

In living systems, dynamics includes the concept of "discrete events", i.e. special critical states that give rise to branching solutions, or transitions from deterministic to random states and vice versa (for instant, life-non-life transitions). First mathematical indication of a phase transition is discontinuities in some of the state variables as functions of time. In classical approach, these discontinuities are incorporated through a set of inequalities imposing unilateral constraints upon the state variables and the corresponding parameters of the governing differential equations. The most severe changes in dynamics are caused by vanishing coefficients at the highest space or time derivatives. The best example to that is the transition from the Navier-Stokes to the Euler equations in fluid dynamics where due to vanishing viscosity, the tangential velocity on the rigid boundary jumps from zero (the non-slip condition) to a final value (the slip condition). In this section, we will modify the proposed model by introducing phase transitions that are accompanied by finite jumps of the state variables.

Let us modify Eq. (35), in the following way

$$(1-y_i)\dot{x}_i = f_i + (\beta_i + \alpha_i \frac{\partial}{\partial X_i}\ln\rho), \quad i = 1,2,...n, \tag{59}$$

where y_i is defined as following

$$\dot{y}_i = -a_i\sqrt{y_i(1-y_i)}, \quad a_i = A_i - \Delta_i, \quad \Delta_i = (x_i - \bar{x}_i)^2 \tag{60}$$

and assume that Δ_n becomes large; that changes the sign of the control parameter a_n, the value of y_n changes from zero to one, and the system (45), (46) switches to the decision mode in the form

$$(1-y_i)\dot{x}_i = f_i + (\beta_i + \alpha_i \frac{\partial}{\partial X_i}\ln\rho), \quad i = 1,2,...n-1, \tag{61}$$

$$0 = f_n + (\beta_n + \alpha_{.n} \frac{\partial}{\partial X_n}\ln\rho), \tag{62}$$

$$\frac{\partial\rho}{\partial t} = -\sum_{i=1}^{n}[(f_i + \beta_i)\frac{\partial\rho}{\partial X_i} - \alpha_i \frac{\partial^2\rho}{\partial X_i^{\,2}}], \quad i = 1,2,...n-1 \tag{63}$$

Since Eq. (62) is *algebraic*, one of the state variables, for instance, x_n can be expressed via the rest of the variables. Hence, formally the decision making mode is described by Eqs, (47)-(55), but the dimensionality of the system drops from n to *n-1*; obviously, the functions $f_i(x_1,...x_{n-1}), \alpha_i(D_{11},...D_{n-1,n-1}), \beta_i(\bar{x}_i,...\bar{x}_{n-1})$, become different from their n-dimensional origins, and therefore, the location of the attractors (56) and (58) become different as well , and this difference may be significant. The number of decision making modes associated with phase transitions is equal to the number of possible sub-spaces, and it grows exponentially with a linear growth of the original dimensionality of the model.

9. Discussion and Conclusion

Decision theory is a body of knowledge and related analytical techniques of different degrees of formality designed to help a decision maker choose among a set of alternatives in light of their possible consequences. Decision theory can apply to conditions of certainty, risk, or uncertainty. In this paper, we concentrate on the last case when the probability distributions over different alternatives are unknown. Decision theory recognizes that the ranking produced by using a criterion has to be consistent with the decision maker's objectives and preferences. However, it is not concerned with defining objectives, designing the alternatives or assessing the consequences; it usually considers them as given from outside, or previously determined. Given a set of alternatives, a set of consequences, and a correspondence between those sets, decision theory is suppose to offers conceptually simple procedures for choice. For the case of uncertainty, decision theory offers two main approaches. The first exploits criteria of choice developed in a broader context by game theory, as for example the where we choose the alternative such that the worst possible consequence of the chosen alternative is better than (or equal to) the best possible consequence of any other alternative.

The second approach is to reduce the uncertainty case to the case of risk by using subjective probabilities, based on expert assessments or on analysis of previous decisions made in similar circumstances. In particular, when the external world does not provide sufficient information, a human turns for "advise" to his experience, and that is associated with a common sense. The simplest representation of human experience is by if-then rules. However, in real world situations, the number of rules grows exponentially with the dimensionality of external factors, and implementation of this strategy is hardly realistic. One of the ways to fight such a combinatorial explosion is to represent rules in a more abstract and more generalized form by removing insignificant details and making the rules more inclusive. This procedure is usually accompanied by coercion, i.e. by reducing uncertainties about the world via forcing it into a known state regardless of the initial state. Indeed, many natural and social phenomena exhibit some degree of regularity only on a higher level of abstraction, i.e. in terms of some invariants. Within the mathematical formalism of the model of mental dynamics, this means that only variables of the highest level of abstraction are capable to classify changes in the external world and send them to the corresponding attractor. Based upon that, the model reacts to these changes , and that can be associated with common-sense-based decision. Collaboration, competition and games are examples of those types of human activities that require common sense for prediction of intentions of a collaborator or a competitor in order to make a right decision when objective information for a rational decision in incomplete.

Exploiting these concepts, the paper presents a methodology for efficient synthesis of dynamical model simulating a common-sense decision making process. The approach is based upon the extension of the physics' First Principles that includes behavior of living systems. The new architecture consists of motor dynamics simulating actual behavior of the object, and mental dynamics representing evolution of the corresponding knowledge-base and incorporating it in the form of information flows into the motor dynamics. The autonomy of the decision making process is achieved by a feedback from mental to motor dynamics. This feedback replaces unavailable external information by an internal knowledgebase stored in the mental model in the form of probability distributions. Representation of the dynamical model in a parameterized form reduces the task of common-sense-based decision making to a solution of the following hetero-associated memory problem: store a set of m stochastic processes given by their probability distributions in such a way that when presented with an unexpected change in the form of an input out of the set of M inputs, the coupled motor-mental dynamics converges to the corresponding pre-assigned stochastic process out of the given m processes, and a sample of this process represents the decision.

Acknowledgment

The research described in this paper was performed at Jet Propulsion Laboratory California Institute of Technology under contract with National Aeronautics and Space Administration.

References

1. M. Zak, *Foundations of Physics Letters*, **15**, No. 3. 229-243 (2002).
2. M. Zak, *Chaos, Solitons & Fractals*, 1019-1033 (2005).
3. M. Zak, *Chaos, Solitons & Fractals*, **28**, 5. 616-626 (2006).
4. M. Zak, *Chaos, Solitons & Fractals*, 1154-1167 (2007).
5. M. Zak, *Chaos, Solitons & Fractals*, **34**, 344-359 (2007).
6. M. Zak, *Chaos, Solitons & Fractals*, **19**, 645-666 (2004).
7. M. Zak, *Physics Letters A*, **133**, No. 1-2, 18-22 (1989).
8. M. Zak, *Int. J. of Theoretical Physics*, No. 32, 159-190 (1993).
9. H. Risken, *The Fokker-Planck equation*, Springer, N.Y (1989).
10. J. Pearl, *Probabilistic reasoning in intelligent systems*, Morgan Kaufmann Publ. San Mateo, CA (1986).

Part IV

Synchronization

CHAPTER 20

SYNCHRONY IN DYADIC PSYCHOTHERAPY SESSIONS

Fabian Ramseyer[1], Wolfgang Tschacher[2]

Department of Psychotherapy, University Psychiatric Services
Laupenstrasse 49, 3010 Bern, Switzerland
E-mail: [1]ramseyer@spk.unibe.ch; [2]tschacher@spk.unibe.ch

Synchrony is a multi-faceted concept used in diverse domains such as physics, biology, and the social sciences. This chapter reviews some of the evidence of nonverbal synchrony in human communication, with a main focus on the role of synchrony in the psychotherapeutic setting. Nonverbal synchrony describes coordinated behavior of patient and therapist. Its association with empathy, rapport and the therapeutic relationship has been pointed out repeatedly, yet close evaluation of empirical studies suggests that the evidence remains inconclusive. Particularly in naturalistic studies, research with quantitative measures of synchrony is still lacking. We introduce a new empirical approach for the study of synchrony in psychotherapies under field conditions: Motion Energy Analysis (MEA). This is a video-based algorithm that quantifies the amount of movement in freely definable regions of interest. Our statistical analysis detects synchrony on a global level, irrespective of the specific body parts moving. Synchrony thus defined can be considered as a general measure of movement coordination between interacting individuals. Data from a sequence of N = 21 therapy sessions taken from one psychotherapy dyad shows a high positive relationship between synchrony and the therapeutic bond. Nonverbal synchrony can thus be considered a promising concept for research on the therapeutic alliance. Further areas of application are discussed.

1. Introduction

When people celebrate a special occasion, they often organize a gathering where guests interact in a pre-structured manner. It is during interactive behavior that phenomena of synchrony can be observed. The occasion that marks the launch of this book is an anniversary and the form of the gathering is a collection of papers assembled in a book. Now, if we imagine a more traditional birthday party, we may easily detect various examples of synchronized behavior. These behaviors usually occur in different modalities and under a variety of conditions. This paper focuses on nonverbal behavior, so a very prominent display of nonverbal synchrony would be a dancing crowd at such a birthday party. Other manifestations could be observed from the beginning of the festivities e.g. the imitation of postures during small-talk at the *apéro*[1], the simultaneous sipping of champagne[2], similarities of hand gestures between speakers[3] or the convergence in prosodic peculiarities when guests with different accents begin a dialogue[4]. Further examples of synchrony could be detected towards the very end of the party such as contagious yawning[5,6], which would signal that the end of the evening is drawing near. A primary question that pops up when considering such examples is whether we may recognize a specific function or purpose in these synchronized actions or not. In this chapter, we will attempt to clarify the connection between synchronized movement and 'rapport' (i.e. relationship formation). To support the hypothesis that such a connection exists, we will focus on the specific setting of psychotherapy and will present as an empirical example of synchrony the patient-therapist dyad.

Our questions thus indirectly address possible functions of imitative behavior, the synchronization to a shared rhythm, be it at a birthday party, in everyday social interaction or in psychotherapy sessions.

"Monkey see, monkey do …"[7] and "Imitation is the sincerest form of flattery"[8] are proverbs that point to the importance of imitative behavior for human beings. As a starting point, we can pick an arbitrary social situation (such as the birthday party described above), because there is a very high likelihood that we will be able to observe phenomena of synchrony on various levels and in different sensory modalities.

In neuroscience, the so-called mirror neurons – first discovered in macaque primates[9,10] – have inspired numerous studies dealing with imitation in primates as well as humans. This has finally resulted in a body of evidence in humans that has associated neurobiological structures with concepts such as empathy, theory of mind and autism[11].

Coordinated nonverbal behavior made its appearance in psychotherapy research in the late 1960s. The seminal work of Condon & Ogston[12,13] was the starting point for empirical research dealing with synchronized (imitative) behavior during human interaction. Several psychotherapy schools have subsequently pointed out the significance of nonverbal attunement (e.g. NLP, Hypnotherapy) and there seems to be a consensus among practitioners that nonverbal coordination plays an important role in the development of the therapeutic relationship. We have set out to empirically clarify whether signs of synchronized nonverbal behavior can be detected in videotaped sessions of dyadic psychotherapies and how they relate to the perceived quality of the relationship between patient and therapist.

We view embodiment and embodied cognition as the larger framework of the investigation of synchrony and like phenomena[14]. Cognitive functions as well as interpersonal attributions are not sufficiently understood as computation or information processing and, therefore, the computer metaphor of the mind has been found to be ill-advised. Modern psychology has proceeded towards acknowledging the intrinsic embedding of the mind in its body and in its environment: Lewin's *Lebensraum*[15] concept. Contemporary systems theory[16] with its emphasis on self-organization provides the formal foundation of a dynamical system approach to the mind[17], which may shed a new light on attributes of cognition such as intentionality[18] as well as attributes of interpersonal functioning, which is the topic of this chapter.

2. Synchrony in Nonverbal Behavior

The significance of synchrony for human interaction is exemplified in the following citation: “Behavioral synchronization is a form of coordinative interaction which is thought to be present in almost all aspects of our social lives, helping us to negotiate our daily face to face interaction.”[19]. Considering human interaction globally, it is evident that some form of coordination or mutual influence plays a crucial role. Cappella[20] makes the following summary: “Coordination is arguably the essential characteristic of every interpersonal interaction. ... Interpersonal communication requires the coordination of behavior.” This usually subtle form of coordination which tends to operate outside conscious awareness[21,22] has been given many different names that all share a similar higher-order meaning. Terms such as interactional synchrony[12], nonverbal mirroring[23], shared rhythmicity[24], motor mimicry[3] or chameleon effect[21] embrace the underlying dimension of coordination between two or more individuals in the domain of nonverbal action. Because of the diversity in terminology, we prefer to briefly clarify various uses and meanings of coordinative behavior. According to Bernieri & Rosenthal[25], we may classify most of the manifestations of synchrony in the human domain using the term *interpersonal coordination.* This expression is loosely defined as "... the degree to which the behaviors in an interaction are nonrandom, patterned, or synchronized in both timing and form."[25]. Since there are many different labels or descriptions for such behavior, a basic categorization of interpersonal coordination is needed. Such an attempt can be based upon the focus of the observer – the classical distinction between quantitative versus qualitative measurements. Studies that emphasize temporal aspects like simultaneous movement, rhythm, or meshing of nonverbal behaviors, mainly regard quantitative characteristics. Because of this stress on kinetic qualities, we classify this type of synchrony as *movement synchrony.* It deals with " ... the precise timing and coordination of movements between individuals ... while the nature or similarity of movements is irrelevant."[26]. An illustration of movement synchrony would be a therapist who matches his speed of movement to that displayed by the patient. Irrespective of the specific parts of the body that are involved, global, quantitative elements such as speed, duration, or complexity of movement are synchronized between the two interacting individuals. Alongside these quantitative elements, we may say that *movement synchrony* always contains a *dynamic* element.

If, in contrast, the focus lies on static or mostly qualitative features of an interaction such as postures, mannerisms, or facial displays, we classify these behaviors under the general term of *behavior matching.* While matching is a very global term seldom used in studies concerned with synchrony, it can be viewed as the core meaning of terms like postural mirroring[27], mimicry[28], congruence[1], or the chameleon effect[21].

Unfortunately, the proposed distinction is not easy to make, because these two categories are not disjunctive. In real interactions it is common to observe a mixture of both: If patient and therapist share the same posture (i.e. static synchrony, behavior matching) and then change their bodily configuration with a coupled timing (dynamic synchrony, movement synchrony), we see synchrony on the quantitative as well as on the qualitative level. Such an instance is depicted in Figure 1, where on the level of dynamics we can see nonverbal synchrony because of the simultaneous movement of patient and therapist and, on the other hand, we see the patient imitating the postural configuration of the therapist's arm position (frame number 1 vs. frame number 8). This simultaneity makes it evident that either the concept is insufficiently defined or there is no such thing as a single-faceted occurrence of synchrony.

The majority of empirical studies concerning synchrony in adulthood have dealt with two similar phenomena: empathy and emotional contagion or rapport in general. Most research efforts have been invested on the link between emotional closeness and imitation. This connection has a historical tradition and can be found in the scientific writings of Charles Darwin[29], who used the term "sympathy" to refer to imitation, and Gordon W. Allport[30] who stated that " ... our understanding of other people is derived from our capacity to imitate, usually in imperceptible ways, the behavior of the person we are trying to understand ... ." "Empathy becomes simply 'kinesthetic inference'." The connotation of synchrony and empathy remains very strong these days, maybe due to the fact that a better understanding of this connection proposes possible benefits during interpersonal encounters or could even be used as a form of manipulation strategy. Condon – who coined the term "interactional synchrony" – stated that "synchrony and other forms of behavioral sharing express degrees of closeness or distance between interactants."[31].

Taken together, the spectrum of findings covers many domains of human life. This multi-faceted propensity of the phenomenon of synchrony can be illustrated with an example right from the beginning of an individual's life: Mother-infant studies on imitative behavior[32] have shown that neonates and older infants imitate basic facial expressions. The same holds true for the opposite direction of influence, as was shown in mothers who opened their mouths in response to the open mouth of their infant whom they were about to feed[33]. This finding implies that the phenomenon has the quality of bilateral influence.

There is accumulating evidence for a strong link between psychopathological phenomena, including autism and schizophrenia and deficiencies in empathy and theory of mind (ToM). "I believe that infants are given a jump-start in developing a theory of mind through their primordial capacity for nonverbal imitation"[34]. It is assumed that a deficit surrounding the mirror-neuron-system impedes a child's development[35]. Interestingly, even simple mechanisms such as the contagious nature of yawning show marked differences in healthy children versus children with autism disorder[36].

In view of this, Ramachandran[37] stated " ... that mirror neurons might do for psychology what DNA did for biology: they will create a unifying framework and help explain a host of mental abilities that have hitherto remained mysterious and inaccessible to experiments."

To summarize the role of synchrony in nonverbal behavior, we claim that embodiment, mimicry and mirror neurons are tightly interrelated. Together they constitute our social, emotional, and interpersonal space[38].

3. Synchrony in Psychotherapy

The analysis of nonverbal behavior has a long tradition in psychotherapy research, dating back to the beginnings of psychology itself. Diverse authors such as Sigmund Freud and Gordon W. Allport noted the connection between nonverbal behavior and the quality of dyadic relationships. "A path leads from identification by way of imitation to empathy, that is, to the comprehension of the mechanism by means of which we are enabled to take up an attitude at all towards another mental life."[39].

If we were to ask practising psychotherapists at random, there is a high probability that many would share the opinion that nonverbal behavior plays a key role in the development and maintenance of a favorable therapeutic relationship[40,41]. In contrast to this shared undisputed practical knowledge, empirical research of *synchrony in psychotherapy* never had a real blooming phase. Although the initial work of Scheflen[1,42,43] suggested promising ways to conceptualize empathy, rapport and the quality of the therapeutic bond, *naturalistic* studies have been very scarce. A critical review of published contributions leaves the impression that "synchrony in psychotherapy" never transcended the stage of descriptive or even solely anecdotal evidence[44]. Those empirical studies with experimental variations of diverse nonverbal features connected to the phenomenon of synchrony have either used student subjects for their therapy-analogue interaction sequences, or have taken very short clips (in the range of seconds or few minutes) of "real" psychotherapies. We conclude that in spite of the large number of studies following these traditions, a solid, baseline- and chance-controlled investigation of synchrony in "real" psychotherapy is still lacking.

Moving to the field of social psychology, the state of research is more advanced as there is clear evidence for the beneficial effect of synchrony on rapport. During the last 10 years, a large body of results that was based on good experimental designs has accumulated[21].

This state of affairs encouraged us to initiate a project in order to circumvent the methodological and conceptual problems listed above. Using new technology and a rich database consisting of thousands of videotapes from an ambulatory psychotherapy research centre[45], we have set out to clarify the relevance of nonverbal synchrony in everyday psychotherapy settings with real clients.

3.1 Motion Energy Analysis (MEA)

Summing up the methodology employed in studies of synchrony so far, a clear dominance of observer ratings is apparent. Although a human observer has several advantages insofar that 'synchrony' as a gestalt phenomenon is relatively easy to detect[46-48], we think that contemporary multimedia equipment makes computerized quantification of movement readily accessible and provides a valuable extension of the methodological arsenal.

One such simple way of measuring motion in a recorded film sequence is based upon an image-differencing algorithm[49,50]. Karl Grammer and his research group at the University of Vienna have successfully implemented this method in a number of empirical studies, e.g. of courtship communication[26,51], physical attractiveness[52] and lovesickness[53].

Such a computer-based system removes several problems commonly encountered when assessing nonverbal behavior by means of observer ratings. The most obvious advantage concerns the amount of time needed to assess synchrony: behavioral observation is very time-consuming due to the fact that a rating system has to be developed, observers have to be enrolled in intensive training and all the chosen sequences of an interaction require manual coding (see e.g. Condon & Ogston's[12] frame-by-frame analysis at 48 frames per second). In addition to such considerations, setting the appropriate time and category resolution can be an even bigger obstacle. The determination of categories may lead to an "atomization of behavioral units"[26], ultimately culminating in one behavioral category for every specific behavior. The same holds true for time resolution, but here a different problem arises: time-lagged or non-stationary phenomena can be detected only with the appropriate resolution. Changes may pass unnoticed if the time span exceeds the attentional capacity of a human observer. Another critical issue lies in the domain of measurement theory, notably in the objectivity of measurements: in order to obtain sufficient objectivity, it is indispensable to train observers in a lengthy procedure. Furthermore, the level of resolution or abstraction of rating instruments is limited by the maximal complexity an observer's mind can handle.

Image differencing in the context of motion energy analysis (MEA) is a tool that avoids most of the mentioned shortcomings inherent with human observers. The method is based on the fact that each individual picture (frame) of a black-and-white movie sequence has a fixed number of pixels that represent a distribution of gray-scale values ranging from 0 (black) to 255 (white). With a fixed camera shot and nothing moving in view of the camera, each pixel retains its gray-scale value from one frame to the next. When, however, a person or an object moves, temporary changes of the gray-scale distribution emerge and can be quantified. The amount of movement from one frame to the next equals the amount of gray-scale change from one frame to the next (thus the name 'image differencing'). Image differencing, therefore, is a simple method to quantify movement in a video stream. In spite of its simplicity there are some caveats that have to be addressed: First, the camera shot has to remain perfectly steady throughout the sequence; second, lighting conditions should be kept stable (otherwise erroneous amounts of change would be quantified); third, the method solely quantifies movement: it is blind to its direction or location.

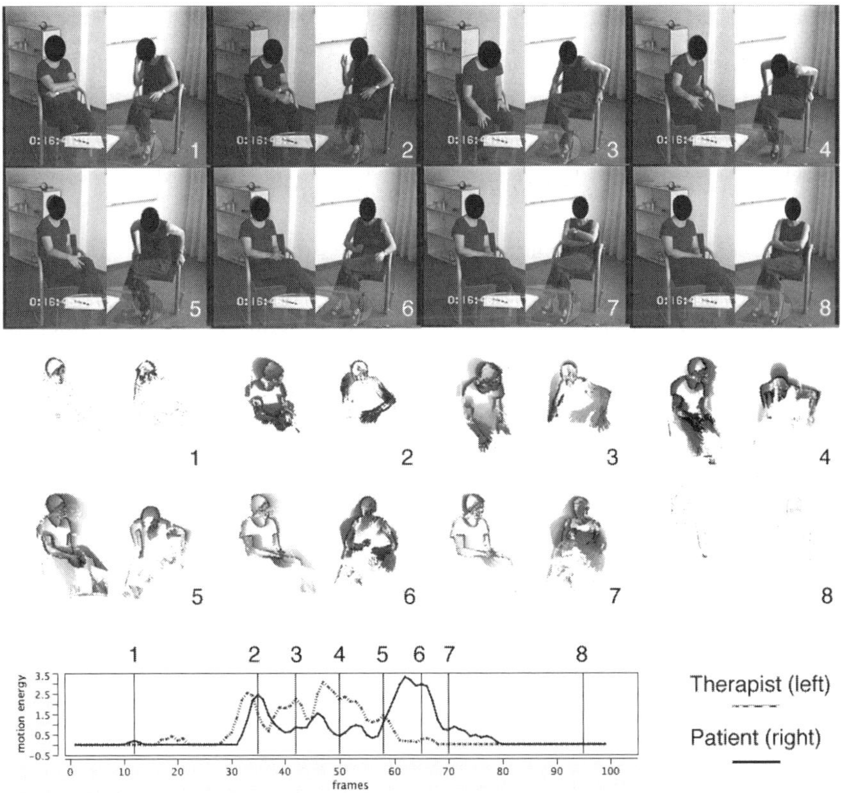

Fig. 1: A sequence (duration, 10 seconds) taken from a "natural" dyadic psychotherapy. The upper 8 pictures portray selected frames of the original movie, the lower pictures are the corresponding motion energy images, where white depicts no movement at all and black pixels show regions where changes have occurred from one frame to the next. Gray shaded areas show movement from previous frames (movement trail). Bottom graph shows amount of movement energy for the 2 interacting individuals. Numbers in the graph correspond to numbers of the pictures.

To monitor movement of two persons in a psychotherapy session, we defined two regions of interest (ROI) where changes in gray-scale values will be detected and separately recorded as numerical streams of data. We therefore produced two continuous time series measuring the amount of movement in the regions defined beforehand. Figure 1 shows 8 frames of a nonverbally synchronized sequence within one therapy session.

Taken together, by image differencing, we have achieved a modern, fully automatic implementation of the manual frame-by-frame coding techniques used by Condon & Ogston[12].

3.2 Cross-correlation of Nonverbal Movement Behavior

Prior to statistical computations, the two streams of raw data were edited with several procedures and filters in order to eliminate video-noise (resulting from poor tape quality). We then standardized the values in relation to the size of a ROI (thus, the data are independent of the size of each individual's ROI), smoothed the data with a moving average and finally calculated a threshold that separates genuine movement (movement bursts) from random noise (random fluctuations in the video tape). The primary statistical analysis we then implemented is based on cross-correlations. We programmed a time-lagged cross-correlation algorithm[54] that quantifies the association of the two streams in a range of +/- 5 seconds. This measure is computed windows-wise, i.e. each minute of a psychotherapy session with duration of 50 minutes is analyzed separately to take into account time-dependent changes in the associations between patient and therapist (thus allowing for the non-stationarity of the phenomenon). The resulting correlation coefficients are charted in a color-coded cell-plot (see Figure 2) that shows the direction of association for one member of the dyad. Values to the left show instances where the patient followed the therapist in her movement (= 'pacing') and correspondingly to the right where the patient was followed (imitated) by the therapist (= 'leading').

The matrix of correlation values may be considered an estimate of the total amount of structure in the sequence analyzed (i.e. the quantity of coordination between the two partners). This measure of structure is simply the absolute mean of all correlations of the graph. Additionally, we calculated the amount of structure for pacing and leading behaviors separately (only the left or right sides of the plot).

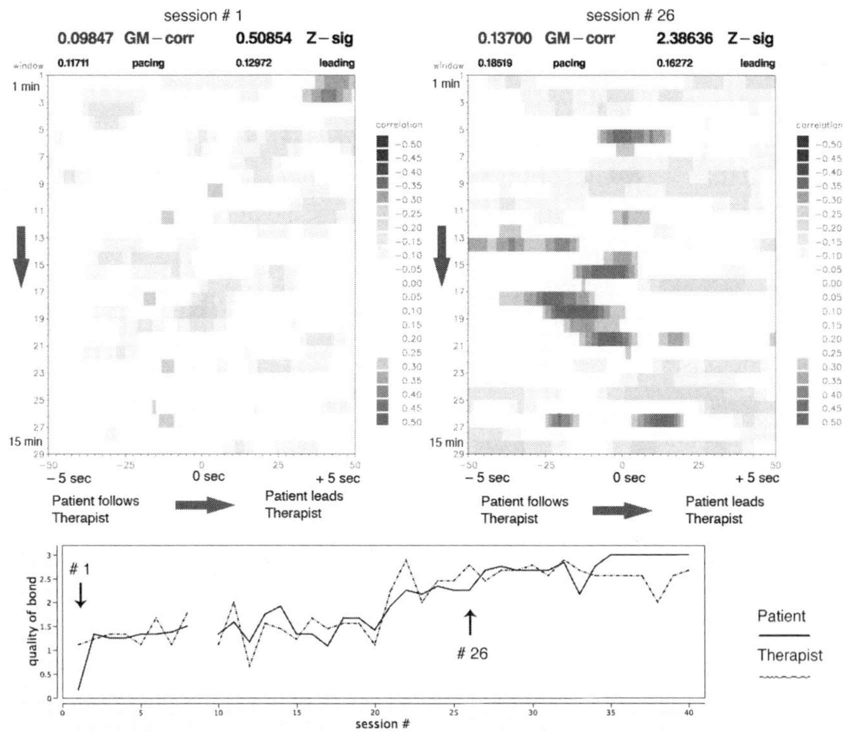

Fig. 2: Color-coded cross-correlations of two psychotherapy sessions (sessions 01 & 26 of 40) taken from the same therapy dyad. Different amounts of movement synchrony are visible: more regions with dark values indicate a higher degree of synchrony. The direction of entrainment is shown with values being either to the left (patient follows the movements of the therapist) or right side of the center. Instantaneous synchrony shows up in the center region (i.e., 0 seconds delay, middle of the graph). GM-corr, grand-mean-correlation; Z-sig, standardized individual significance. Bottom graph shows evolution of the factor "quality of therapeutic bond" for both patient and therapist. Arrows indicate chosen sessions.

A final, quite important step in our analysis is to rule out that the detected movement synchrony may have occurred solely by chance. We therefore corrected for random contingencies between the two movement streams. There has been a short debate about the genuineness of Condon

& Ogston's[12] findings on interactional synchrony in the 1970s[55,56]. We fully support this critical consideration of synchrony findings and thus implemented a novel statistical mechanism that would prevent falsely positive detection of synchrony in psychotherapy sessions. To accomplish this, for each therapy session, we generated N = 100 surrogate datasets by shuffling the genuine data. In order to not destroy the original structure of movement bursts, we shuffled each time series windows-wise. In this way, the original structure inside one window remains intact, but due to the shuffling of the windows' position, it is paired with another window from a random position in the therapy. In such a way, the motion energy values of the therapist's behavior from e.g. the 5th minute are paired with the movements of the patient from e.g. the 9th minute of this same session. The genuine cross-correlation coefficients were ultimately contrasted with the absolute mean of the shuffled coefficients and considered significant if they exceeded 2 standard deviations from the randomized means.

3.3 Evolution of Synchrony in Dyadic Psychotherapy

The psychotherapy research centre at the University of Bern offers a unique opportunity to study synchrony under field conditions. The centre is an ambulatory psychotherapy clinic open to the general public that serves two purposes: an educational purpose, because psychotherapists in training and under close supervision work there as part of their psychotherapy training, and a community service purpose, because of the services delivered to the community. The therapeutic principles are independent of classical psychotherapy schools and based on Grawe's consistency theory[45,57]. Predominantly, diagnoses are from the neurotic-affective spectrum. Each therapy session held in this setting is recorded by two video cameras that are joined into a split-screen image of full-body views of patient and therapist. Owing to the habitual recording of therapy sessions, we were able to select therapies that were conducted completely independent of our scientific study. The selected therapy took place during the year 1999/2000 and neither patient nor therapist had any knowledge of the concept of synchrony whatsoever. Our post hoc test of the relationship between synchronized movement and session outcome is therefore free of any possible influences on the normal behavior of the dyad.

Another special feature of the psychotherapy research centre of the Institute of Psychology at the University of Bern consists of the regular administration of post-session evaluations both of patient and therapist. Evaluation forms are completed immediately after the termination of every single therapy session. They consist of 22 items (patient) and 27 items (therapist), respectively. Individual items (e.g. "My therapist and I understand each other well"; "Today, I felt comfortable with the patient") are rated on a 7-point Likert scale. In this analysis, we focused on the factors "Quality of therapeutic bond Patient" rated by the patient (8 items) and "Quality of the therapeutic bond Therapist" rated by the therapist (8 items).

To test if global synchrony shows changes during dyadic psychotherapy and whether the total amount of "structure" is correlated with micro-outcome, we have chosen a single therapy course from a large sample of recorded psychotherapies. This therapy had a duration of N=40 sessions. It was characterized by a gradual increase of both patients' and therapists' factors assessing the quality of the therapeutic bond. We limited this pilot study to one female same-sex dyad. The patient was 37 years old and suffered from recurrent major depression (DSM-IV: 296.3).

3.4 Analysis

Our main objective was to collect a sample of therapy sessions that would give us access to nonverbal synchrony in its purest form. We wished to rule out extraneous influences or affordances within the therapy situation to a maximal extent. This meant that out of 40 sessions of this particular dyad, we selected 21 sessions that met our rigorous criteria. These minimal conditions for inclusion in the study were the following:

Technical considerations:

- Clear, full body view of both patient and therapist (optimal camera positions, no occlusions due to e.g. disfunctional split-screen adjustments)
- Stable conditions in light source
- No movement of cameras
- Acceptable quality of VHS signal (camera image not too noisy)

Thematic prerequisites:

- Seating position throughout sequence, i.e. not leaving the room
- No discussion of questionnaires
- No joint filling-out of forms
- Exclusion of arranging for next meeting (taking out personal agendas, planning next session)

These criteria were quite strict in order to capture ‘pure’ synchronized interaction that occurred during ‘normal’ verbal interaction only. We chose such rigorous measures because we had noticed, in a previous pilot study, that the amount of synchrony tended to be exaggerated by situational affordances, such as the finding of a new appointment or the discussion of the results of a questionnaire.

3.5 Results

Global movement characteristics were calculated for each 30-seconds window, resulting in a total of 630 values for the N = 21 therapy sessions. The analysis of these movement patterns showed that the patient moved significantly more than the therapist ($T_{(629)} = 2.14$; $p < .05$). The corresponding relative amount of time with movement behavior was 41.5% for the patient and 36.6% for the therapist. Complexity, maxima and speed of movement, however, did not differ significantly between patient and therapist, indicating that, except for time spent in movement, patient’s and therapist’s movement characteristics were comparable.

The amount of synchronized movement (grand mean of cross-correlations) was positively correlated with the patient's evaluation of the quality of the therapeutic bond ($r = .60$; $p < .01$; $R^2 = 35.5$; $F_{(19)} = 9.9$) and the therapist's evaluation of the therapeutic bond ($r = .69$; $p < .001$; $R^2 = 48$; $F_{(19)} = 16.6$). A comparison of pacing and leading indicated that the patient's predominant pattern was the imitation of the therapist, i.e. there was more pacing than leading ($T_{(20)} = 2.64$; $p < .05$).

As the therapy progressed from session 1 to 40, so did the amount of synchronized movement. These findings are illustrated in Figure 3.

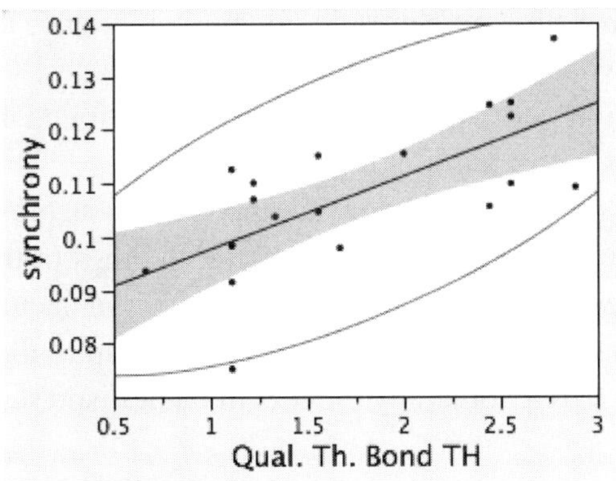

Fig. 3: Correlation between synchrony and "quality of the therapeutic bond" from the therapist's perspective.

For the evaluation of genuine synchrony versus synchrony as a chance occurrence, we pooled all 21 therapy sessions to conduct the surrogate test. The amount of synchrony was significantly higher in genuine therapy sessions compared to the surrogate samples with windows shuffled ($T_{(20)} = 2.31$; $p < .05$).

4. Discussion

Our results show that nonverbal synchrony is a highly promising concept for psychotherapy research in general. It offers a new way to look at the therapeutic bond. We could also demonstrate that interactions in this setting are embodied in the sense that simple motor parameters reflect psychological variables that are fundamental for psychotherapy.

We have to bear in mind, however, that for the field of clinical psychology and psychotherapy, the evidence does not yet justify a general acknowledgment of the synchrony concept in the way that has been claimed by some practicing psychotherapists of e.g. NLP[58,59]. Although synchrony is an intriguing concept for psychotherapy research, it would be premature to claim for it an exclusive role in the formation of empathy and in the development of the therapeutic bond. Further well-controlled empirical research will be necessary before such steps can be taken. Primarily the question of causality should be addressed in the near future, because our correlative measures do not allow inference about the causal direction of synchrony. It is thus still a hypothesis that the mechanism of "imitation breeds liking" that has been found in social psychology[21,60] may also be at work in psychotherapy.

It is our hope that novel methodological approaches such as motion energy analysis may help reevaluate the interesting findings of the 1960s and the encouraging results in other fields of research. An accumulation of solidly based empirical evidence should, therefore, inspire the revival of nonverbal research of psychotherapy process in naturalistic settings.

References

1. A. E. Scheflen, *Psychiatry,* 27 (1964).
2. A. G. Billings, C. A. Gomberg, B. H. Nash and S. Weiner, *Journal of Studies on Alcohol,* 39 (1978).
3. J. B. Bavelas, A. Black, N. Chovil, C. R. Lemery and J. Mullett, *Human Communication Research,* Vol 14 (1988).
4. R. Neumann and F. Strack, *J Pers Soc Psychol,* 79 (2000).
5. R. R. Provine, Bulletin of the Psychonomic Society, 27 (1989).
6. R. R. Provine, Bulletin of the Psychonomic Society, 27 (1989).
7. E. Partridge, A Dictionary of Catch Phrases: American and British, from the Sixteenth Century to the Present Day, Ed. P. Beale, Scarborough House Publishers, Lanham, Maryland (1992).
8. G. Y. Titelman, Random House Dictionary of Popular Proverbs and Sayings (Random House, New York, 1996).
9. V. Gallese, L. Fadiga, L. Fogassi and G. Rizzolatti, *Brain,* 119 (1996).

10. G. Rizzolatti, L. Fadiga, V. Gallese and L. Fogassi, *Brain Res Cogn Brain Res,* 3 (1996).
11. L. M. Oberman and V. S. Ramachandran, *Psychol Bull,* 133 (2007).
12. W. S. Condon and W. D. Ogston, *Journal of Nervous and Mental Diseases,* 143 (1966).
13. W. S. Condon and W. D. Ogston, *Journal of Psychiatric Research,* 5 (1967).
14. P. M. Niedenthal, *Science,* 316 (2007).
15. K. Lewin, *Principles of Topological Psychology.* McGraw-Hill, New York (1936).
16. H. Haken, Information and self-organization: A macroscopic approach to complex systems Springer, Berlin (2000).
17. W. Tschacher, and J. P. Dauwalder, The dynamical systems approach to cognition: concepts and empirical paradigms based on self-organization, embodiment, and coordination dynamics. World Scientific, New Jersey, p. 330 (2003).
18. W. Tschacher and H. Haken, *New Ideas in Psychology,* 25 (2007).
19. A. Kendon, R. M. Harris, and M. R. Key, *The organization of behavior in face to face interaction* Mouton and Co., The Hague, p. 3 (1975).
20. J. N. Cappella, in *The sourcebook of nonverbal measures*, Ed. V. Manusov, Lawrence Erlbaum Associates, Mahwah, NJ, p. 383. (2005).
21. T. L. Chartrand and J. A. Bargh, *J Pers Soc Psychol,* 76 (1999).
22. J. L. Lakin and T. L. Chartrand, *Psychol Sci,* 14 (2003).
23. E. J. Charny, *Psychosom Med,* 28 (1966).
24. A. Kendon, *Acta Psychologica,* 32 (1970).
25. F. J. Bernieri and R. Rosenthal, in *Fundamentals of nonverbal behavior. Studies in emotion & social interaction*, Eds. R. S. Feldman and B. Rime, Cambridge University Press, New York, p. 403 (1991).
26. K. Grammer, K. B. Kruck and M. S. Magnusson, *Journal of Nonverbal Behavior*, 22 (1998).
27. M. LaFrance, Personality and Social Psychology Bulletin, 11 (1985).
28. J. L. Lakin, V. E. Jefferis, C. M. Cheng and T. L. Chartrand, *Journal of Nonverbal Behavior,* 27 (2003).
29. C. Darwin, *The expression of the emotions in man and animals*, Eds. P. Ekman, P. Prodger, Oxford University Press, Oxford (1998).
30. G. W. Allport, *Personality: a psychological interpretation.* pp. 531, 532, Holt, Rinehart, & Winston, New York (1937).
31. W. S. Condon, in *The relationship of verbal and nonverbal communication*, Ed. M. R. Key, Mouton Publishers, The Hague, Paris, New York, p. 49-75 (1980).
32. A. N. Meltzoff, and W. Prinz, *The imitative mind: Development, evolution, and brain bases*, Cambridge University Press, New York (2002).
33. R. O'Toole and R. Dubin, *J Pers Soc Psychol,* 10 (1968).
34. A. N. Meltzoff, Journal of Communication Disorders, 32 (1999).

35. J. H. G. Williams, A. Whiten, T. Suddendorf and D. I. Perrett, *Neuroscience and Biobehavioral Reviews,* 25 (2001).
36. A. Senju, M. Maeda, Y. Kikuchi, T. Hasegawa, *et al., Biol Lett,* (2007).
37. V. S. Ramachandran, Mirror Neurons and Imitation Learning As the Driving Force Behind "The Great Leap Forward" in *Human Evolution,* 69 (2000).
38. P. M. Niedenthal, L. W. Barsalou, P. Winkielman, S. Krauth-Gruber and R. Ric, *Personality and Social Psychology Review,* 9 (2005).
39. S. Freud, *Group psychology and the analysis of the ego,* W.W. Norton & Company, New York, p. 53 (1959).
40. P. Philippot, R. S. Feldman, and E. J. Coats, *Nonverbal behavior in clinical settings*, Oxford University Press, London (2003).
41. P. Philippot, R. S. Feldman, and E. J. Coats, *The social context of nonverbal behavior*, Cambridge University Press, New York (1999).
42. A. E. Scheflen, Psychiatry: Journal for the Study of Interpersonal Processes, 28 (1965).
43. A. E. Scheflen, *Psychosom Med,* 28 (1966).
44. F. Ramseyer and W. Tschacher, *Constructivism in the Human Sciences,* 11 (2006).
45. K. Grawe, *Psychological Therapy*, Hogrefe, Seattle, Toronto (2004).
46. F. J. Bernieri, S. Reznick and R. Rosenthal, *J Pers Soc Psychol,* 54 (1988).
47. F. J. Bernieri, J. M. Davis, R. Rosenthal and C. R. Knee, *Personality and Social Psychology Bulletin,* Vol 20 (1994).
48. F. J. Bernieri, , J. S. Gillis, J. M. Davis and J. E. Grahe, *J Pers Soc Psychol,* 71 (1996).
49. M. Sonka, V. Hlavac, and R. Boyle, *Image processing, analysis, and machine vision*, Chapman & Hall Computing, New York (1993).
50. A. F. Bobick and J. W. Davis, IEEE Transactions on Pattern Analysis and Machine Intelligence, 23 (2001).
51. K. Grammer, R. Honda, A. Schmitt and A. Jütte, *J Pers Soc Psychol,* 77 (1999).
52. K. Grammer, V. Keki, B. Striebel, M. Atzmüller and B. Fink, in *Evolutionary aesthetics*, Eds. E. Voland and K. Grammer, Springer, Heidelberg, Berlin, New York, pp. 295-324 (2003).
53. M. Bechinie, BHS - The "broken heart syndrome" - ethological aspects of lovesickness, University of Vienna, Vienna (1998).
54. S. M. Boker, M. Xu, J. L. Rotondo and K. King, *Psychological Methods,* 7 (2002).
55. J. J. McDowall, *J Pers Soc Psychol,* 36 (1978).
56. J. B. Gatewood and R. Rosenwein, *Journal of Nonverbal Behavior* 6 (1981).
57. K. Grawe, *Neuropsychotherapie*, Hogrefe, Göttingen, p. 509 (2004).
58. J. R. Plazo, Maximum influence advantage. 2 day persuasion workshop, BookSurge Publishing, USA (2006).

59. S. Lankton, *Practical Magic*, Crown House Publishing Ltd, Carmarthen, Wales (1980).
60. C. M. Cheng and T. L. Chartrand, *J Pers Soc Psychol,* 85 (2003).

CHAPTER 21

TEMPORAL PERSPECTIVE FROM AUDITORY PERCEPTION

Gerold Baier[1], Thomas Hermann[2]

[1]*Manchester Interdisciplinary Biocentre, University of Manchester, Manchester UK*
E-mail: Gerold.Baier@manchester.ac.uk

[2]*Neuroinformatics Group, Bielefeld University, Bielefeld, Germany*
E-mail: thermann@techfak.uni-bielefeld.de

Dynamically complex diseases with distributed and multi-scale interacting physiological rhythms require a more refined temporal perspective of the scientific observer than is currently provided by visual displays of physiological data. We argue that sonification, the auditory inspection of experimental data, provides a unique approach to the representation of the temporal aspects of the data as it addresses the human sense of listening. The ear's capacity to detect temporal patterns of sameness and differences, of coincidence and coordination – widely exploited in listening to music and spoken language – creates a new temporal perspective in the scientific observer. We briefly describe some examples of sonifications of biomedical data and discuss their value in recovering the temporality of complex physiological processes. Auditory Gestalt formation can be exploited for the classification and differentiation of diseases. Finally, we stress the complementarity of auditory and visual representations and argue for combined audio-visual displays in order to adequately deal with complex phenomena, as in the case of dynamical diseases.

1. The Challenge of Complex Dynamical Diseases

In 1977, Glass and Mackey introduced the concept of *dynamical diseases*[1]. In this and later publications a dynamical disease was characterized by the operation of dynamical instabilities in physiology that impose new self-organized temporal patterns on the body's healthy dynamics[2]. While the concept is plausible and can be substantiated with quantitative and qualitative mathematical models of known physiologic processes, an important new problem arises from the introduction of this concept. In mathematical analysis it is convenient to focus on few variables and on specific time scales. However, the very concept of instability requires that all feedback mechanisms operating at a given time have to be accounted for, as they might influence the point and type of instability substantially. This leads to the call for more and, in particular, multivariate measurements, i.e. parallel registration of multiple variables in their temporal dependence, which is needed to construct extended and detailed mathematical models. Doing this, however, leads to data of formidable complexity, as it turns out that, even under healthy conditions, many variables tend to show rhythmic organizations. The rhythms can be of very different temporal scale, ranging from milliseconds for neural firing to minutes for intracellular messenger signaling to hours for episodic hormone release to years in growth and aging. In addition, they tend to have complex relationships among themselves (somewhere between independent and fully synchronized) as, for example, in the relationship between breathing and the heartbeat. If the concept of dynamical disease is then viewed from a Systems Biology point of view[3], the claim is that, in fact, an integrated understanding of the disease requires that many or all parallel rhythms of the body must be studied in order to gain understanding of the dynamic changes that accompany disease. The isolated focus on a single rhythm is not sufficient.

This claim presents a new challenge to the observation that precedes diagnoses as it requires a multisource/multilevel monitoring of human physiology. While, in principle, this can be (and often is) done by displaying data as parallel time series on one screen or on various screens at the same time, it becomes clear that in practice this approach is limited by the inadequacy of the sense of seeing to properly perceive and analyse multiple rhythmic events. However, if one thinks of the relative ease with which a trained listener is able to follow the individual voices of complex polyphonic music, this suggests that the human sense of listening might be able to deal with this kind of data in a better way – if only the data were presented in the appropriate auditory manner. This would introduce a much-needed temporal perspective to the perception of physiologic processes (which is lacking in numbers that represent concentrations or levels as if they were fixed entities and in pictures or photographs) and it will enable the observer to form a data-driven hypothesis about the dynamics of the processes on the level of interaction of processes (rather than that of isolated processes).

Such kind of reasoning leads us to propose the introduction of the auditory representation of data as a means of observing and analyzing multirhythmic processes, for example, those associated with dynamical diseases.

2. Listening to Complex Events

Listening is an activity that is hardly exploited when dealing with the perception of natural phenomena as a scientist. It might (at first glance) even appear weird if one were asked to listen e.g. to an experiment, rather than to look at it. However, that is by no means due to any predetermined superiority of the human sense of seeing. It is by convention that visual observation of an experiment and the simultaneous or subsequent visual representation of recordings are preferred. Clearly, it is counterintuitive to imagine listening to e.g. human anatomy. In the first place, there is nothing to be heard, and in the second place, one can argue that a 3D structure is best captured by the sense of seeing. Things become less obvious when one tries to deal with processes that are of a complex temporal nature. Human physiology, for one, provides a large number of examples where one can argue that both senses are at a comparable disadvantage to start with, as in the case of electric activity of nerve cells. There is nothing to be heard, yet there is nothing to be seen either. In fact, neurophysiologists who record the electric activity of neurons, groups of neurons, and brains, traditionally have their recordings displayed on a screen *and* at the same time connected to a loudspeaker to be able to listen to them. And, while classical printed journals do not offer any means of conveying an audio recording to the reader of an article, more recently, lecture demos and web sites on the Internet have been taking advantage of the possibilities to include audio in combination with visual representations[4]. The main reason is that, in this case, temporal structure is key to the understanding of the results.

However, this combined audio-visual approach is the exception rather than the rule. In particular, none of the classical natural sciences like biology, biochemistry, chemistry, physics, and others, has a habit of using auditory displays routinely in the laboratory and neither do they prepare audio demos of their results for communication among peers. Yet, as stated above, there is no a priori reason to restrict scientific observation to visual displays. In fact, it is fundamental to the very endeavor of science that the modeling and formation of theories based on empirical observation is essentially *independent* of the mode of perception, i.e. the sensory channel by which one becomes aware of a phenomenon. The law that governs the acceleration of a ball on an inclined plane will be the same whether one watches it rolling down or listens (as Galileo presumably did in his experiments that led to its first formulation[5]) to sound produced by the ball touching strings attached at certain points along the track.

So, while it is a widespread habit to use visual displays in science, it is fair at least to investigate whether there are alternative means of achieving the same or perhaps even better results. Visual displays are transformations of the data into a geometric representation that is convenient to perceive. This representation may preserve geometrical aspects of the data (as, for instance, in imaging) but can also be entirely unrelated to the original geometry as e.g. in abstract phase space representations. It should then be equally possible to create auditory displays of data that convey the important information readily for the sense of hearing and compare its performance with visual displays.

In fact, medicine as we know it today, a discipline of outstanding importance, historically was transformed into a truly scientific discipline with the aid of the human sense of hearing. In the 18th century, the foundations for the development of a modern discipline based on empirical observations were laid by the studies in pathological anatomy, exemplified by the work of investigators like Giovanni Battista Morgagni (1682-1771) and, later, Xavier Bichat (1771-1802). But while their work provided a new and differentiated coordinate system for the localization of pathologies within the body, their observations were all made on dead corpses. Autopsy was not possible in the sick body. In order to be able to combine the knowledge from pathologic anatomy with observations of the living body (which could not be opened for visual inspection), two means of auditory analysis had to be introduced: percussion and mediate auscultation.

3. Listening in 19th Century Medicine

Leopold Auenbrugger (1722-1809) published a treatise called "Inventum Novum" in 1761 in which he advocated the method of treating the human body as a resonator. Like any physical resonator, it does not produce sounds unless perturbed. Therefore it has to be perturbed physically, e.g. by drumming against it with the fingers or the hand; or by shaking it in order to make internal liquids move. Then listen to the response. In this procedure, the body's internal, self-produced noises are ignored and the focus is instead on the response to the external perturbation. The perturbation needs to be short and abrupt for percussion in order to allow the perception of the elicited transient sounds, which are possibly of short duration. Similar to the fate of the work of Morgagni, which was more widely accepted only some decades after his death, the work of Auenbrugger was essentially ignored at first (notwithstanding a translation into French by Rozière de la Chassagne in 1770) until its central idea was picked up by Nicolas Corvisart around 1800 in Paris. Corvisart again translated Auenbrugger's essay and advertised the method as a means of arriving at an improved diagnosis of diseases. Applying percussion to his patients Corvisart actually became famous for being able the predict the outcome of autopsies for diseases of the lung.

Effectively, percussion tests for mechanical properties at a given moment and can be assumed to provide information about the body's state. While in principle the temporal change of sound characteristics could be monitored by repeated perturbations, in practice, percussion was not thought to inform about a body's temporal characteristics.

One of Corvisart's students was Rene Theophile Hyacinth Laennec (1781-1826). Laennec was familiar with the concept and practice of percussion by way of his teacher and he obviously also conceived of the human body as a physical entity capable of producing sounds. But Laennec, in addition, put emphasis on the fact that, apart from the sounds elicited by external perturbation, there are the bodies spontaneously produced internal sounds. He therefore extended Auenbrugger's idea that acoustic properties provide information about the body's internal state to these internal sounds. An important example is the sound of respiration. As this sound can be perceived from the outside by pressing the ear against the chest of a patient, this means that the lungs serve as a resonator for the body's own muscular activity. If the lungs are damaged or filled with water, this does not only change the acoustic properties of external knocking against the chest but also the acoustic properties of the breathing as modified by the lungs.

Laennec then faced two main difficulties when trying to use the sense of hearing for this new way of approaching physiology. First, the quality of the sounds perceived when pressing the ear against the clothes or the skin of a patient was not good. And, second, the direct bodily contact was uncomfortable for both physician and patient, and in the case of women also inappropriate. To overcome these problems, Laennec introduced the instrument he initially called "the tube" and which soon came to be known as the stethoscope.

After its introduction, the use of the stethoscope spread quickly in Europe and Northern America and throughout the 19th century the auditory assessment of disease was dominant. Due to the lack of methods of visual display of the interior of a patient, it was listening that helped to "render the interior of the human's body a dynamic field of interaction"[6].

Laennec set the ground with his book "On Mediate Auscultation or: Treatise on the Diagnosis of the Diseases of the Lungs and Heart", published in 1819. He stressed two important aspects. First, the observer has to refine his listening skills. In this respect, he is aided by the use of the stethoscope. The body's sounds as perceived with the stethoscope are clearer: they can be differentiated better from external sounds, and it is easier to locate the source of a sound. Second, in order to be useful as a means of diagnosis, auditory clues have to be communicated. Therefore, a new language is required that allows acoustic properties to be characterized so that they can be used as indicators of pathologies, both in the individual diagnosis and in the scientific discourse with peers. However, both the perception of acoustic properties and the language to communicate their finding was directed at invariant properties, or states.

4. The Fate of Mediate Auscultation

Throughout the 19th century, auscultation made listening a big success in medicine while at the same time neglecting the key aspect of temporal perspective. The key to its success was localization and identification, because a hypothesis about a specific pathology at a specific site was what could be confirmed by autopsy. Anatomical pathology provided the new theory of disease and it consisted of visible alterations of tissue at specific locations. Thus, while being valuable as a means of obtaining clues about pathology from the living body, in the course of the 19th century, listening became subordinate to the visually inspired model of disease. As Foucault puts it, the ear became a temporary substitute organ for the eye as long as it was not possible to look inside the body[7]. This explains the moderate emphasis made on the temporal structure of the acoustic features as compared to its localization and identification purposes. Autopsy, the ultimate reference point for the confirmation or rejection of diagnosis, stripped the disease of its dynamic features. In all of Laennec's discussions, it is obvious that his reference point when talking about disease is provided by the findings obtained or expected from autopsy. Like Corvisart, he depended on dissection of the dead body and the visually perceptible confirmation of altered or damaged tissue for his method to be accepted. Therefore, a temporal perspective on human physiology was neither sought after nor intended.

An instructive example of this is the treatment of the human voice by Laennec, which was inspired by another of his teachers, Matthieu-François-Régis Buisson (1776-1804). Buisson introduced the distinction of the functional voice, like the voice in speech and the voice in song, from the “voice as such” or the voice “strictly speaking”. The latter refers to the acoustic properties of the human voice and Buisson explicitly stated that these properties allowed conclusions about the “biology of its owner”[8]. Laennec exploited the sonic features of the voice in his investigations on diseases of the lung. He ordered the patients to speak while he himself listened to different parts of the chest and derived a number of acoustic characteristics that would represent certain pathologic states. In his treatment of the voice in terms of its sound qualities, it becomes evident how the temporal qualities were almost entirely neglected. In his “Treatise”, Laennec does not mention what the patients said nor does he comment on any temporal variation of sound qualities. Similarly, rhythmic patterns do not play a significant role in his descriptions of “mucous rattle” and “metallic tinkling” when he characterizes the auditory quality of breathing. A condition like the Cheynes-Stokes breathing, which is characterized by strong periodic modulations of the breathing volume (and therefore cited as an example of a dynamical disease by Glass and Mackey[1]), is not part of this auditory effort.

This is the reason why we started our discussion by mentioning “dynamical diseases”. This concept specifically highlights temporal features and uses alterations of dynamics as a symptom of disease. It is a concept that calls for a renewal and refinement of temporal perspective when dealing with human physiology and its pathologies.

5. Sonification

While, in current scientific medicine, the auditory mode of perception hardly has any importance and the scientific discourse is dominated by visual displays, this does not, in our opinion, prove a belief in the primacy of the sense of seeing. First, nowadays there is a general tendency to put less weight on the qualitative results of the doctor’s physical examination (and by implication her or his sensory perceptions) in order to put more weight on the quantitative data obtained from laboratory analysis. This is equally true for all sensory modes of perception. Second, throughout the 20th century, it was easy to distribute results from e.g. X-ray examination, microscopy etc. in visual form, i.e.

using pictures, graphics or sketches. Even when auditory examination was made, as in the case of cardiac arrhythmias, the choice of representation was the electrocardiogram rather than a tape with the recording, simply for convenience. And third, up until the 1990s, hardly any technique was available for the physician and clinician to get a proper auditory representation of physiologic events that were not audible. Whereas imaging techniques were developed to render visual displays of basically any technique that records spatial data (computer tomography, PET, NMR and others), for a long time no such possibility was available in the auditory domain. Taken together with a general emphasis on visual displays, both in scientific practice and scientific teaching, this contributed to the dominance of visual representation. However, with the wide accessibility and manageability of digital media, it is now possible to revive the auditory approach to physiology.

The new technique that makes this possible is called sonification. Sonification is the synthesis of sounds from data for a scientific purpose. While analog methods were employed historically, nowadays "synthesis" normally is meant to imply digital rendering of sounds. "Data" then can be digitally stored numbers from any source, particularly from processes that by themselves do not produce any pressure waves in the surrounding medium. This is, of course, crucial for physiologic and biochemical processes that do not involve mechanical movement, like metabolite concentrations or electric nerve cell activity. In that sense, sonification is on a level of mediation with imaging techniques which generate visible images of processes that are not visible. This is achieved by attributing pseudo-colors to quantitative data. Parameter mapping, the predominant sonification technique, in turn, attributes acoustic properties to quantitative data. The "scientific purpose" in the definition of sonification is important to distinguish it from sound production for e.g. an aesthetic purpose, as in music. This is important because, in contrast to visual representation, auditory representation of data is not commonly trained in scientific teaching and education and only vague notions exist about how analytic listening can be used to extract scientifically useful information about natural phenomena from synthesized sound.

The conference proceedings of the first "Workshop on Auditory Displays" summarizes the state of the art in sonification up to 1992[9].

Since that year, regularly held conferences on Auditory Displays have provided a forum for the presentation of advances in the field and the resulting publications are the main portal to the literature of the field[10]. Thus, tools are readily available for the application of sonification to biomedical data in general and physiologic data in particular[11].

6. The Sonification of Physiologic Data

There exists a large number of sonification techniques but we do not want to review them here. Rather, we would like to focus on one special technique that was devised to highlight the temporal features of the data in connection with pathologic alterations of the underlying physiologic processes. Event-based sonification (EBS) was introduced to deal with abnormal rhythmic activity in human electroencephalogram recorded from patients with epilepsy[12]. While first it was applied to output from data-driven model dynamics[12], in a second step, it was shown that event-based sonification can be done directly on the data[13]. With data-motivated adjustments, it is possible to use EBS for the display of pathologic rhythms, as distinct from normal background activity[14]. It is the latter technique in its clinical application that concerns us in the present context.

As a simple example, let us discuss Sound 3 of Ref. 15. To understand and interpret the sound, we first need to give a brief description of the underlying data and the sonification technique used.

The data used for the sonification were obtained by electroencephalography. The electroencephalogram (EEG) is the parallel measurement of electric potentials from different sites on the human scalp. Potentials are typically recorded from 19 or more electrodes with a sampling rate between 100 and 500 Hz. The potential amplitudes are in the range of microvolt and the dominant frequencies are between 0.1 and 30 Hz. The recording of the EEG is crucial in the case of patients with epilepsy because i) epileptic seizures, in most cases, are accompanied by abnormal rhythms in the EEG; and ii) because the EEG in phases between seizures often has characteristic rhythmic abnormalities that can be associated with the disease. Both the seizures and the activity between seizures have to be understood as being transient, i.e. being defined by its temporal dynamics rather than its "state". While this has often been neglected in the past (epilepsy being often described by a binary model which distinguishes the "epileptic" from the "non-epileptic" state), recently there have been increased efforts to characterize the evolution and the corresponding transient features of epileptic rhythms as measured on the scalp in more detail[16,17]. Thus, awareness is increasing that epileptic activity presents multiple time-dependent temporal rhythms with changes often occurring on different time scales. For such complex cases, it has been argued that EBS might be a technique to appropriately represent the underlying spatio-temporal dynamics[18].

Event-based sonification requires the definition of *events*. In the case of rhythms, these can be local maxima or minima of a time series or some other feature. For our sample sound, maxima in two time series recorded in parallel were identified. The next step is to use the events to trigger sound events. In real time sonification, sound events are triggered according to the moment when they occur. Thus, temporal relationships between maxima within a time series and between time series are strictly preserved. Two more features were included for the sound synthesis: the amplitude of a rhythmic unit (an oscillation) was measured as the distance between the potential at a maximum and the potential at the preceding minimum. This amplitude was then used to linearly control the level of the sound event. And the temporal difference between a maximum in one channel and the previous maximum in the other channel was evaluated and used to control the number of harmonics (audible as brightness) of a sound event. Thus, a dull sound indicates that the maxima in the two channels are widely separated, whereas a sharp sound indicates small separation or coincidence of maxima.

The data segment chosen for the synthesis of the sound consists of about 7 seconds of background activity followed by about 12 seconds of seizure activity (a so-called absence seizure), picked by both channels. After the seizure, the activity returns to the background activity. Background activity is characterized by its ongoing, irregularly occurring sound events. Both the timing and the volume distribution resemble a stochastic process. No clear rhythmic component is perceptible. The seizure activity, in contrast, is recognizable by way of its increased intensities of the beats and its pronounced regularity. A punctuated rhythm consisting of a repeating sequence of one fast and one slow inter-beat interval is salient. The transition from and to background activity marks the beginning and the end of the attack. As is typical for absence seizures, the changes occur in both recorded channels. Thus, by virtue of the sonification, the epileptic seizure becomes an audibly perceptible unit or gestalt.

In this special case, sonification (especially if implemented in real time[19]) allows the following observations: epileptic activity can be perceived without effort and is clearly salient from its dominant periodicity and rhythmic structure. These features also allow its differentiation from equally loud but noisy muscle artifacts. During the course of a single seizure, the maintenance of the level and a slow decrease in main frequency is perceptible. Listening to different seizures, one is able to estimate the length of the seizure and is thus capable of e.g. detecting whether the average duration changes in the course of days (a typical time scale for clinical monitoring of patients). Comparison of the two channels allows the degree of synchronization between the channels in a stereo setting to be perceived and thus also aids the detection of changes in the relationship between two sites during seizure activity. The sound synthesis can easily be extended to 19 or more EEG channels. However, it is then advisable to use more elaborate multi-speaker output (as in virtual auditory environments) to be able to discern more complex rhythmic relationships between multiple channels[20]. For example, if the sonification of the present example is extended to the 19 electrodes of the original recording, it becomes obvious that there are "waves" of activity from frontal to occipital regions on the scalp. In other types of epileptic activity, the so-called focal seizures, it is also possible to recognize hemispheric preferences and subgroups of coordinated electrodes among the whole set[20]. We have furthermore produced sonifications of abnormal rhythms associated with the ongoing activity of an epileptic focus when no clinical seizure occured[14,15]. The results indicate that sonification is indeed capable of providing the required information from the data.

7. Some Implications

What are the implications of an auditory representation of such (and related) physiologic recordings for the scientific observer?

First, temporal processes are produced or reproduced along a temporal coordinate and perceived as such. In auditory reproduction of data, the time scale can be chosen arbitrarily; it can be stretched or compressed but sequences of events in the data maintain their original ordering. For regular or entirely random sequences of events, this may not be of major concern. For transient behavior, where one specific sequence of events is followed by another, this will lead to enhanced attention. Auditory perception of data may lead to a shift in focus from the permanent to the transitory.

Second, the auditory sense strongly relies on grouping principles or gestalt construction in order to make sense of an ongoing sound stream. Analogous to the concept of an object in the visual domain, it was proposed that a concept of *auditory object* should be used to describe a temporarily limited, identifiable, and recognizable pattern within a complex soundstream[21]. This means that the human auditory sense is capable of extracting higher-level or composite information from complex data streams and thus can be trained to perceive e.g. pathologies in physiologic data in the form of such auditory objects. At this point, it is important to note that, while it is essential to agree on standards for sonification in order to communicate results, this does not exclude the experimental modification of sonification strategies in order to optimize such gestalt perception. It has been emphasized that interactivity might play a critical role in the successful use of sonification for specific tasks[22].

Third, the sonification of multivariate data, as in the case of the human EEG, demonstrates that a particularly strong feature of auditory perception is the detection and monitoring of complex correlation patterns. The simultaneous occurrence of events is only an extreme case of a continuum of correlations between events. The analysis of complex auto-correlations (correlations between successive events in a single sound source) is performed almost without effort in speech perception and the analysis of cross-correlations (correlations between events in distinct sound sources) is continuously at work in the perception of music with multiple voices, be it homophonic or polyphonic. In particular, the latter feature may prove valuable when applied to multivariate physiologic data where there is complex interplay between events and where there are interactions on different time scales. Again, simple regular or completely random patterns will only form the extreme cases, the complex patterns being the rule in living organisms.

8. A Multimodal "Gaze" at the Human Body

The application of the stethoscope to perceive sounds from within the body was termed "mediate" auscultation by Laennec. Similar to the use of microscopy in anatomy, the term "mediate" refers to the use of a mechanical instrument needed in order to detect what the immediate application of the ear and the eye cannot reveal. With the advent of imaging techniques and sonification, a second level of mediation is introduced between the diseased body and the scientific observer. This time, the mediation is not mechanical but digital. While it might appear that this second mediation removes the observer even further from his or her object, in fact it is intended to bring her or him closer. The reason for this is that both imaging and sonification methods are currently being used in connection with refined non-invasive measurement techniques online, i.e. while the recorded physiological processes are happening. The time from the actual happening of a change to its auditory or visual perception is being decreased as much as possible. For example, online sonification of the human EEG is easily realized with delays that are in the order of one tenth of a second[19]. As the temporal resolution of the sonification is even higher, it is now possible to listen to temporal changes and patterns of any variable that can be monitored continuously almost as soon as they occur. If this information is combined with a visual display that tells *where* the change is going on, the observer is able to form a gaze that is as integrative and comprehensive as possible. The medical "gaze" (incidentally, a word that implies lack of temporality) can now be endowed with a multimodal, "plurisensorial structure"[23] in a much broader sense than at the beginning of the 18th century. Specifically, it may now explicitly include the complex temporal patterns under normal and abnormal conditions. While listening, we now can hear much more than before and thereby better appreciate the full spatio-temporal complexity of both healthy and pathologic physiology.

References

1. M.C. Mackey, L. Glass, *Science*, **197**, 287 (1977).
2. See e.g. L. Glass, M.C. Mackey, *Ann. NY Acad. Sci.* **316**, 214 (1979); M.C. Mackey, J.G. Milton, *Ann. NY Acad. Sci.* **504**, 16 (1987).
3. c.f. L. Alberghina, H.V. Westerhoff (eds.), *Systems Biology: Definitions and Perspectives* (Springer, Berlin, 2005).
4. See e.g. the webpage http://williamcalvin.com/1970s/NeuralSounds.htm
5. S. Drake, Galileo at Work, Chicago University Press, Chicago 1978, p. 78. Cited after: F. Dombois, *Wann Hören? Vom Forschen mit den Ohren. In: Chemie – Kultur – Geschichte* (Verlag Chemie, Weinheim 2002).
6. J. Sterne, *The Audible Past* (Duke University Press, Durham & London, 2003), p. 127.
7. M. Foucault, *The Birth of the Clinic* (Vintage Books, New York, 1973), p. 165.
8. After J. Sterne in Ref. 6, p. 122.
9. G. Kramer, *Auditory Displays – Sonification, Audification and Auditory Interfaces,* (Addison Wesley, Reading 1994).
10. See http://dev.icad.org/
11. See e.g. the tutorial on "Sonification of Biomedical data" presented at ICAD 2006, London: http://www.sonification.de/projects/eegson-icad2006/index.shtml
12. G. Baier, T. Hermann, in: S. Barress (ed.) *Proceedings of 10th International Conference on Auditory Displays*, Sydney, Australia, June 6-9 (2004).
13. G. Baier, T. Hermann, S. Sahle, U. Stephani, in: T Stockmann (ed.), *Proceedings of 12th International Conference on Auditory Display ICAD2006*, London, UK, June 20-23 (2006), p. 148.
14. G. Baier, T. Hermann, U. Stephani, *Clin. Neurophysiol.* **118**, 1377 (2007).
15. Online version of Ref. 13, doi:10.1016/j.clinph.2007.01.025, also available at: http://www.sonification.de/projects/eegson/CLINPH2007/index.html
16. S. Schiff, T. Sauer, R. Kumar, S.L. Weinstein, *NeuroImage* **28**, 1043 (2005).
17. K. Schindler, H. Leung, C.E. Elger, K. Lehnertz, *Brain*, **130**, 65 (2007).
18. G. Baier, T. Hermann, M. Müller, in: E. Banissi et al. (eds.), *Information Visualization*, IEEE Computer Society, Los Alamos, CA (2005), pp. 5-10.
19. T. Hinterberger, G. Baier, *IEEE Multimedia* **12**, 70 (2005).
20. G. Baier, T. Hermann, U. Stephani, in: W. Martens (ed.) *Proceedings of the 13th International Conference on Auditory Display*, Montreal, Canada, June 26-29 (2007), pp. 491-496.
21. T.D. Griffiths, J.D. Warren, *Nature Review Neuroscience* **5**, 887 (2004).

22. T. Hermann, A. Hunt, *IEEE Multimedia* **12**, 20 (2005).
23. Ref. 7, p. 164.

CHAPTER 22

PERCEPTION OF SIMULTANEOUS AUDITIVE CONTENTS

Christian Tschinkel

Department of Musicology, University of Graz
Mozartgasse 3, A-8010 Graz, Austria
E-mail: tschinkel@hof-productions.com

Based on a model of pluralistic music, we may approach an aesthetic concept of music, which employs dichotic listening situations. The concept of dichotic listening stems from neuropsychological test conditions in lateralization experiments on brain hemispheres, in which each ear is exposed to a different auditory content. In the framework of such sound experiments, the question which primarily arises concerns a new kind of hearing, which is also conceivable without earphones as a spatial composition, and which may superficially be linked to its degree of complexity. From a psychological perspective, the degree of complexity is correlated with the degree of attention given, with the listener's musical or listening experience and the level of his appreciation. Therefore, we may possibly also expect a measurable increase in physical activity. Furthermore, a dialectic interpretation of such "dualistic" music presents itself.

1. Introduction: Musical Space

Musical concepts have always also been concerned with the space in which the music is to be played. From the early days of unison music, step by step, the need to raise music to a higher level evolved, which led, as a result of a "true" polyphony, to an extension of space. Therefore, as early as the Renaissance, music was closely connected with architecture, as the positioning of the individual musicians was composed along with the musical work. In order to achieve this, the equality of all parts is a basic precondition. Harmonic thinking, which peaked in Viennese classical music is hierarchical and temporarily pushed aside the form of polyphony, which was based on major motifs (melody) and minor ones (accompaniment).[1] However, acoustic and auditive perception of space remain to this day an important composition parameter. Today, music is, again, closer to the discourse of art and science: The information-theoretical aspect, in particular, characterizes sound as the carrier medium of physical information about space, apart from visual exploration of the latter. Space modifies sound in accordance with its characteristics and vice versa.[2] Contemporary sound installations plumb body-environment interaction, for instance, in the shape of *walk-in music*.[3] Alvin Lucier's 1970 work *I am sitting in a room* is a famous example of spatial performance art.

Electro-acoustic methods allow us to simulate spaces and to virtually nest highly differing spatial properties. By means of the arbitrary setting of digital codes, impossible spaces can be created and analyzed in the framework of psychoacoustics. Audification and sonification are the new buzzwords by means of which the listener tries to advance into both microscopic and macroscopic spaces.

2. Listening to the (Electro-acoustic) World

Differences in volume and run time, as well as certain filtering effects of sound, form the physical basis for our everyday binaural hearing with two healthy ears.[4] However, during the course of the 20th century, electro-acoustic techniques have created a completely new type of listening rendered by means of speakers: As a result of its different kind of emission characteristics, the speaker is, on principle, not regarded as really being the source of the sound, but rather as representing the idea of a sound projection. As a result, most varying combinations have arisen which influence our listening on a psychological level. In particular, stereophony as a special kind of two-channel rendering has established itself, which opens up a virtual space before the listener by means of an auditory mapping of the binaural visual field with the help of a pair of speakers. If earphones are used, this space stretches to a maximum of 180°. A stereophonic recording is always created from stereomicrophony, in which two microphones, which replace our ears, point into two different directions. Just as in binaural listening, different portions of the same sound events reach the two recording channels at the same time. A sound source which is moving towards the left thus loses energy in the right channel, while the sound level indicator of the left channel rises. However, this effect can also be achieved artificially, i.e., by means of sound techniques, by means of a non-moving sound source or a mono-recording (a single microphone). Afterwards, the signal is continuously faded out (the volume is turned down) in one channel, while it is simultaneously faded in (the volume is turned up) in the other. Our brain thus receives nearly the same information as it would in binaural listening.

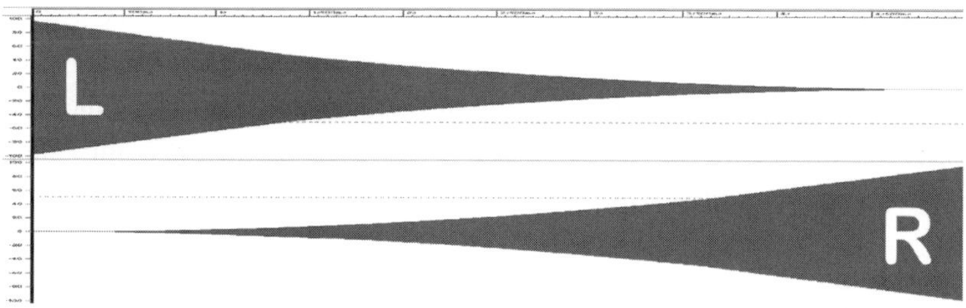

Fig. 1 Schematic representation of a stereophonic sound motion from left to right.

True two-channel arrangements, however, are achieved by means of two monaural "sources". Two mutually independent sound events are emitted from both speakers. No part of either signal is contained in the other. However, free listening via the speakers of a stereo sound system would, again, correspond with a binaural situation, because the signals mix with each other on their way to the ear. Finally, such tonal artifacts also occur in everyday life: The bell on the clock tower strikes while somewhere else a car horn sounds. Musically speaking, this would correspond to the rendering of a duet, real three-channel arrangement to that of a trio, and so forth. An entire speaker orchestra named

Acousmonium[5] celebrates, within the framework of the *musique acousmatique*, a multiphone multi-channel arrangement, which closes the historical circle of the previous section.

A totally different and actually quite utopian situation occurs if one listens to a two-channel recording with closed headphones. This situation does not resemble any "natural" hearing/listening, as completely differing contents may be played to each ear. This special listening situation is called dichotic hearing. Neuropsychology, in particular, discovered (through Kimura in 1961) that dichotic hearing can be used to carry out tests concerning hemispheric laterality. In this context, asymmetric processing in the brain is studied. Laterality also manifests itself in, for example, our left- or right-handedness.

3. Dichotic Listening and Auditive Information Processing

As a result of the fact that the two auditory pathways cross in the brain, the left ear leads the signal to the right hemisphere and the right ear transmits the stimulus to the left hemisphere (contralateral neural connection). A dichotic listening test is derived from experimental lab conditions, in which a different sound event is played simultaneously to each ear of a test subject by means of earphones. Originally, the idea was to play two fairy tales simultaneously to the test subjects. A listener would automatically focus his attention onto one audio channel only. Therefore, when the preferred tale was retold, it was possible to draw conclusions about laterality. In recent years, short successions of syllables[6] uttered by an individual speaker, followed by two simultaneous number sequences or sequences of tunes as test items, have become established in the study of language lateralization. Comprehension of these sequences is then assessed quantitatively. In this procedure, test results are not always conclusive, a fact which, in the case of musical tunes played, also reveals the complexity of music as stimulus material. A very rough summary concerning auditive information processing may be formulated as follows:

Monaural stimulation shows that musicians can process a tune better with the right ear and non-musicians with the left ear. These results correspond to those derived from the differentiation of language and music. According to this, music, just like language, is "analyzed" by musicians in the left hemisphere (right ear), but is perceived by laymen in a contoured, holistic way (left ear).[7] Concerning musical processing *per se*, dichotic listening tests do not show functional differences between the two hemispheres. It is true that musical laymen display high ear laterality, but process music either on the right or on the left. With increasing musical experience, there is an increase also in the degree of parallel participation of both hemispheres (which corresponds to low or no ear laterality). In addition to this bihemispheric processing, music is also processed subcortically. A connection with the limbic system[8] explains the potentially strong emotional involvement when we listen to music, although one should not necessarily conclude that this is where emotions are located or centred[9], as here, too, the interplay of many partial areas manifests itself.

4. Pluralistic Music

"The impressionist model of polytonality is the intersounding of spatially separated music at the fairground."[10] What Theodor W. Adorno described in a relatively tranquil image is nowadays probably a familiar experience to almost anyone who has walked among the rows of shelves of a media market: From many hi-fi systems, many different types of music are blared out simultaneously, which merge into *a single* utopian mixture of sounds, which would leave any futurist turning green with envy. That world of sound they ideologically proclaimed at the beginning of the 20th century but was not realizable technologically then, is today clattering down on us with high intensity.[11] Harmonic sound relations seem to dissolve as a result of this complex interplay and the ordering systematics of simple pop structures takes on a chaotic form. In spite of this, by now this sound mélange is perceived by many – particularly by young "recipients" – not as disruptive, but rather as stimulating. De-coupled from the traditional bourgeois concert situation, this scenario of mass-cultural music consumption is undertaken happily and willingly.

This phenomenon is rooted in 20th-century collage thinking[12] which, according to Enjott Schneider, started with the substitution of the static world picture by a fluid one, which is, again, strongly connected with the idiosyncrasies of today's *multi-tasking*. In combination with modern production methods, technologically fixed sounds[13] lead, via the ideological path of *musique concrète*, to both a "concrete" form of polyrhythmics and polytonality, which no longer need to be conceived in an abstract manner as a mute sheet of notes. Today's creators of music model the sound itself in a recording studio, like a sculptor. Polyphonous composing is thus no longer the "objectivization of the We"[14], but generates the sound of polyphony in a solo effort. Increasing compositional complexity in connection with the acousmatic spatial design of polyphone multi-channel arrangements inevitably directs music towards the pluralistic, so that it is therefore always also a kind of spatial composition. Pluralistic music does not stand for the polyphone sound of the individual parts of a piece of music, but for the simultaneous sound of many pieces in the same place. There are several examples of this form in musical history. Four of them are introduced in the following:

- Wolfgang Amadeus Mozart: *Don Giovanni* (1787)
  In one passage of this opera, the sound of three musical pieces can be heard simultaneously. However, all of them still adhere to a common metre.

- Bernd Alois Zimmermann: *Requiem for a Young Poet* (1969)
  This is a monumental work for orchestra, jazz combo, three choirs, soprano and baritone solo, electronic sounds, organ and speakers.

  "The simultaneity of various levels of experience has reached (here) a culmination of musical entanglement. (...) To Zimmermann, it is the most extreme attainable consequence of an idea which he himself described as the *spherical nature of time*."[15]

- Gian Piero Reverberi (*Rondo Veneziano*): *Controdanze* (1988)
  "*Controdanze* is a special composition with three dances: If you listen to the mono-left only, you can hear the Minuetto (3/4), if you listen to the mono-right only, you hear the Polka (2/4), and on

stereo, you hear the Gavotta (4/4). This technique is called musical holography."[16]

Up to now, academic literature on music is really not familiar with the concept of musical holography, a fact which suggests that it has only been employed in this musical work. However, this term may be regarded as an extension of stereography (a spatial representation in a visual domain). If one listens to the track with earphones, a dichotic listening situation, as described earlier, emerges.

- Neurosis: *Times of Grace* (1999) & Tribes of Neurot: *Grace* (1999)
  "*Grace* is sonically designed to be played simultaneously with *Times of Grace*. (...) The ideal setup is two independent stereo hi-fi systems placed on opposite sides of the listening space. (...) You must attempt to have them start playing as close to the same moment as possible (...)."[17]

  The rock band *Neurosis* and its alter-ego *Tribes of Neurot* create the textural soundscape of *Grace*, from electro-acoustic pieces of their album *Times of Grace*. If one plays both CDs simultaneously, according to the group, the result is a "multidimensional sound experience (which is) performed by the two distinct personalities of this group."[18]

The specific feature of the last two examples is the fact that the directions on the CD cover encourage the listener to try out experimental listening with one or two ordinary stereo systems. In the case of *Rondo Veneziano*, fading out one of the two stereo channels by means of the balance control leads to only one of the mono-channels being audible. The entire stereo track therefore contains three different pieces of music: Two mono pieces and one "dualistic" piece (with headphones, actually a "dichotic" one).

5. Defining "Dichotic Music"

If we consider Reverbi's 1988 description of musical holography, it seems hard to maintain that I am supposed to be the "inventor" of "dichotic music". However, against the background of my work on experimental music, I first formulated in 2004 aesthetic principles for a new kind of music on the basis of psychological testing and a lab situation and have already expressed these in several compositions and production techniques.

The simple starting point of dichotic music is the fact that it has become technically possible to feed sound into two ears simultaneously, independently of one another. However, at first sight, the concept is limited to a dualism which should be easy to perceive even without headphones. The listener should register immediately that he is exposed to two simultaneously played pieces of music, which is why a polarization of the audio material is required. "Seeking out the opposite" as well as "overcoming contradictions" inevitably leads to the concept of dialectics, which has always been the driving force in the discussion and analysis of art. Thesis and anti-thesis combine into a synthesis. Comparable to the rhetorical figure of the *oxymoron* (example: "bitter-sweet"), the dramaturgy reaches fulfillment in the psychological interplay from the left and the right, which is oriented both on the conventional image of the *montage* and on that of the multi-perspective *collage*, without actually completely belonging to either. As a result, the individual parts lose their value in favour of the end product, which is not at all the intention of "dichotic music". Likewise, the possibility of a shared tonality and a shared tempo must also be preserved, which is not possible in the collage method. The basis of this hybrid musical design is the intention of achieving a balanced relation between two states: On the one hand, the dichotic presentation should not drift into the abstract two-channel arrangement of classical electro-acoustic music and thus obstruct the differentiation and, with this, the identification of the two individual pieces. On the other hand, two pieces should not be arbitrarily combined (as in the media market situation described above), as a synthesis may only emerge as a compositional design at best only by chance. The two musical pieces should be clearly separate/separable but still interconnected. This paradoxical task seems to be solvable only by constructing similarity relations, in which a mutual reflexivity between them is considered especially necessary. A "reflex-aphoric" and "holographic tension"[19] thus emerges, which may also be assigned to most other areas of art. The notions of iteration and self-similarity, which the Austrian composer Bernhard Lang has dealt with in detail[20], lead this musical concept into the vicinity of fractal research.

Thus far, there are no empirically established results concerning the reception of such a musical structural model and its cognitive representation. This is because, apart from sound sketches, only two experimental pieces by the author exist, and these are not enough for empirical purposes. For the time being, the aesthetic concept of this type of music forms not much more than a utopian scenario: music-theoretical debate, which takes account of psychoacoustics appropriate to electro-acoustic music, remains speculative. The new concept may establish itself as a musical form as did the fugue. However, it may do so as a composition principle rather than a separate musical form. To extract the headphone situation – which is actually seemingly autistic and asocial[21] – into open space in the shape of ordinary two-channeled arrangements or a live performance, such as two simultaneous staged pieces, does not, admittedly, amount to a convincing innovation. However, in a certain sense, it is a generalization of the total concept.

6. Speculation and Outlook

Another question which remains unanswered at present is that of exactly where dichotic musical thinking begins. The left- and right-handedness of two-part inventions by Bach is a conceivable starting point. With the playing of the second part, the increase in complexity in this music is comparable to that in a dichotic presentation, in particular as it is actually possible to reproduce the two-part counterpoint also dichotically. Here, the whole is also more than the sum of its parts and two-part music could, aside from the principles formulated above, be understood as a rudimentary form of "dichotics".

As our listening perspective is based on a sensitive internal differentiation regarding neuronal connections in order to steer our perception, it is not out of the question that aestheticized dichotic listening would have to be newly trained and cultivated. This should be considered in both the phylogenetic (i) and the ontogenetic (ii) sense, as is best described by means of two simple examples:

(i) If today we watch an old movie of, say, the 1920s, we notice that a resolution of 12 images per second no longer provides a continuous film experience while, according to reports of that time, various movies worked on the audience of the time like a kind of drug. This means that irreversible habituation effects towards technological developments are involved, which apparently follow a "continuous adaptation to the more complex" and converge in today's multi-tasking.

(ii) Concerning the individual development of listening, I would like to give, what for me, has become a much-loved example, because it is a personal one. It is taken from my experience of my parents' 45 rpm single record which seemed to contain the same piece (Richard Clayderman's *Ballade pour Adeline*) on both sides: Approximately at the age of four, I often compared the A and B sides without being able to make out any kind of difference concerning the music. As other single records contained different songs, I rated this as a pure waste of space. Only a few years later did I realize that one side contained a version for solo piano and the other a version for orchestral accompaniment. This example shows that, as children, many of our perceptional performances are still very limited and we thus establish a completely different world. My personal experience described above has, in the meantime, been reflected in the results of many studies on the topics of "the origins of music" or "infantile perception of music".[22] With the perception of dichotically presented music, similar behaviour to these two examples could occur.

"Dichotic music" is a complex stimulus which, as a result of introducing multiple perspectives, provides a space with a greater degree of freedom for the arbitrary guidance of attention. Comparison with the cocktail party effect comes to mind, which, incidentally, only works binaurally. Dichotic perception of music is primarily determined by the hemispheric dominance of the listener and, in addition, possibly by the degree of trained guidance of attention. Selective processes constitute themselves from gestalt-theoretical principles such as similarity, proximity, continuity and linearity. Transferred to musical forms, they manifest themselves predominantly in musical grouping and phase structure, as well as possibly in the contour of melodies.[23] With the dichotic concept, these forms are used in a special manner. A future investigation into the perception of "dichotic music" is envisaged by using the tool of auditive scene analysis.[24]

Finally, a few comments on the "healing effect" of commercial headphone music should be made. Dubious sound-carriers propagate hemispheric synchronization for stress reduction with stereophonic relaxation music (however, so far, this has not been shown to have any scientific relevance).

- Kamal: *The Quiet Earth – Dusk* (1993)

> "If headphones are used, the two brain hemispheres are synchronized, i.e., their level of activity is equalized, and frequencies are induced. (...) The synchronization of the two hemispheres reduces the polarity between the two areas in order that they do not unconsciously hinder each other any longer. (...) These recordings (...) support a meditative state during listening."[26]

In contrast to such an esoteric interpretation of stereophone headphone music, possible pathological effects of "dichotic music" have already been discussed which could emerge as a devastating but unlikely counter argument. It may possibly have the potential to trigger epilepsy or schizoid tendencies. In this case, an artistic output would literally lead to *chaos*.

So-called *driving effects*[27] cannot be excluded. These are known in the visual (induced by light flashes) and auditory domain (induced by dynamicized rhythm) and could play a similar role in connection with dichotic presentation. Furthermore, an increase in physical activity, similar to that observable in pop music and described in pop theories, would therefore be conceivable.

References

1. A. Webern, *Der Weg zur Neuen Musik*, Vienna 1980, p. 17-20, (1932).
2. W. Jauk, *Der musikalisierte Alltag der digital culture*, [post-doctoral thesis], Graz, p. 497, (2005).
3. R. Minard, *Sound Installation Art*, in *Beiträge zur Elektronischen Musik*, Sonderband zur Ringvorlesung "Die Klangwelt am Rand der Datenautobahn", Vol. 6, edited by R. Höldrich, Graz, p. 23, (1996).
4. H. Bruhn and D. Michel, *Hören im Raum*, in *Musikpsychologie. Ein Handbuch*, edited by H. Bruhn et al., Reinbek bei Hamburg, p.652, (1993).
5. F. Bayle, *Musique acousmatique. Propositions, positions*, Paris, p. 44, (1993).
6. K. Hugdahl and A. Asbjørnsen, *Dichotic Listening with CV-Syllables. Manual*, Department of Biological and Medical Psychology, University of Bergen, p. 1, (no publication year).
7. C. Fassbender, *Funktionsteilung im Cortex*, in *Musikpsychologie. Ein Handbuch*, edited by H. Bruhn et al., Reinbek bei Hamburg, p. 622-627, (1993).
8. G. Roth, *Das Gehirn und seine Wirklichkeit. Kognitive Neurobiologie und ihre philosophischen Konsequenzen*, Frankfurt am Main, p. 194, (1997).
9. C. Fassbender and G. Frommer, *Ohr-Asymmetrie in der Wahrnehmung dichotisch dargebotener Melodien als Funktion von musikalischer Begabung und musikalischer Erfahrung*, in *Musikpsychologie. Empirische Forschungen – Ästhetische Experimente*, Jahrbuch der deutschen Gesellschaft für Musikpsychologie, Vol. 6, edited by K.-E. Behne, G. Kleinen and H. de la Motte-Haber, p. 87-104, (1989).
10. T. W. Adorno, *Philosophie der Neuen Musik*, Frankfurt am Main, p. 141, (1958).
11. D. Kaufmann, *Die Präsenz futuristischer Ideen in der elektroakustischen Musik*, in *Der musikalische Futurismus. Ästhetisches Konzept und Auswirkungen auf die Moderne*, edited by Otto Kolleritsch, Graz, p. 51, (1976).
12. E. Schneider, *Von der "niederen" Popularkultur zur Mutter aller Künste. Film, Multimedia, Collage als Ausdruck der postmodernen Ästhetik*, in *Was ist Pop? Zehn Versuche*, edited by Walter Grasskamp et al., Frankfurt am Main, p. 146, (2004).
13. J.-N. von der Weid, *Die Musik des 20. Jahrhunderts. Von Claude Debussy bis Wolfgang Rihm*, Frankfurt am Main, et al., p. 316, (2001).

14. W. Jauk, *Aspekte der Popularisierung der digitalen Künste. Pop-Music und Computer-Musik*, in *Zur Wahrnehmung zeitgenössischer Musik*, edited by E. Schimana and J. Gründler, Graz, p. 87, (2002).
15. CD B. A. Zimmermann, *Requiem für einen jungen Dichter*, Wergo Schallplatten GmbH, Mainz, CD-Booklet, p. 2, 13, (1989).
16. CD Rondo Veneziano, *Concerto*, Baby Records, CD-Inlay, (1988).
17. CD Tribes Of Neurot, *Grace*, Neurot Recordings, CD-Booklet, (1999).
18. CD Neurosis, *Times Of Grace*, Relapse Records Inc., CD-Booklet, (1999).
19. J. Briggs and F. D. Peat, *Die Entdeckung des Chaos. Eine Reise durch die Chaos-Theorie*, Munich, Vienna, p. 92-116, 127-172, 301-310, (1990).
20. B. Lang, *Diminuendo. Über selbstähnliche Verkleinerungen*, in *Beiträge zur Elektronischen Musik*, Sonderband zur Ringvorlesung "Die Klangwelt am Rand der Datenautobahn", Vol. 7, edited by R. Höldrich, Graz, p. 17-20, 41, (1996).
21. R. Schönhammer, *Walkman*, in *Musikpsychologie. Ein Handbuch*, edited by H. Bruhn et al., Reinbek bei Hamburg, p. 181-187, (1993).
22. S. Trehub, G. Schellenberg and D. Hill, *The origins of music perception and cognition: A developmental perspective*, in *Perception and Cognition of Music*, edited by I. Deliège and J. Sloboda, East Sussex, Psychology Press Ltd., p. 103-128, (1997).
23. D. Temperley, *The cognition of basic musical structures*, Cambridge et al., p. 55-65, (2001).
24. A. S. Bregman, *Auditory Scene Analysis. The Perceptual Organization of Sound*, Cambridge et al., The MIT Press, (1990).
25. A. D. Patel and I. Peretz, *Is music autonomous from language? A neuropsychological appraisal*, in *Perception and Cognition of Music*, edited by I. Deliège and J. Sloboda, East Sussex, Psychology Press Ltd., p. 192, (1990).
26. CD Kamal, *The quiet earth – Dusk –*, Nightingale Records, CD-Booklet, (1993).
27. G. Harrer, *Das „Musikerlebnis" im Griff des naturwissenschaftlichen Experiments*, in *Grundlagen der Musiktherapie und Musikpsychologie*, edited by G. Harrer, Stuttgart, p. 40, (1975).

CHAPTER 23

COMPUTER SIMULATIONS AS HIDDEN TIME-ECOLOGIES

Gus Koehler

Dept. of Regional Planning & Public Policy
University of Southern CA, Sacramento Center and
Time Structures, Inc., 1545 University Ave., Sacrament, CA 95825
E-mail: rhythm3@earthlink.net

Computer simulations are used to investigate epidemics, social phenomena, decision-making, supply networks, and physical and chemical processes among other things. This chapter examines how time is incorporated into such simulations. This method is criticized by showing why it is inadequate, leading to potentially erroneous conclusions. The concepts of "time-ecology," and "heterochrony", and J.T. Fraser's theory of the nested hierarchy of time are used to propose solutions.

1. Introduction: Simulations in Silica

NetLogo is a convenient way to give us a taste of how time is incorporated into simulations and how it is interpreted as a simulation proceeds. NetLogo is among the most user-friendly and widely distributed of simulation software.[1]

"[It]... is a programmable modeling environment for simulating natural and social phenomena. NetLogo claims to be particularly well suited for modeling complex systems developing over time. Modelers can give instructions to hundreds or thousands of independent "agents" all operating concurrently. This makes it possible to explore the connection between the micro-level behavior of individuals and the macro-level patterns that emerge from the interaction of many individuals."[1]

NetLogo is used to simulate without temporal distinctions emergent phenomena involved in Aids epidemic, altruism, ethnocentrism, chemistry and physics processes, earth sciences, segregation, art, mathematical problems, and ant activity.

In a digital computer, the simulation instructions include the following: "…in the form of COMMAND (ADDRESS) where the address is an exact (either absolute or relative) memory location, a process that translates informally into "DO THIS with what you find HERE and go THERE with the result." Everything depends not only on precise instructions, but on HERE, THERE, and WHEN being exactly defined."[2] The NetLogo world is constructed by programming interactions between three interrelated types of agents, all controlled by a computer "timer" – the when – with a resolution in milliseconds or so, depending on the underlying Java Virtual Machine. The first, an "agent", appears to move (actually, different pixels light up) according to programmed rules over the second, called "patches" or a landscape. There can be "breeds" of agents such as foxes, chickens, and grass that interact in set ways. The third is an "observer" which "hovers" over the unfolding agent-landscape and uses the clock to demarcate agent and landscape activities. "[All three] are beings that can follow instructions. Each…can carry out its own activity, all simultaneously."[2] Agents execute commands asynchronously but are programmed to "wait" for the others to complete their commands before going on. Groups of agents (our foxes and chickens, for example) are started randomly every time the program is initiated so as not to favor one agent or segment of the landscape over another. Landscapes may have a preferential point like a pile of grass for a chicken agent, but its constituent patches are set.

NetLogo explicitly uses three temporalities to simulate five very different realms in its world: mathematics, physical, biological, psychological, and the social. The first NetLogo temporality is the computer's "timer", which regulates all three agents. The second is that of programmer-observer, who controls the number of variables and generates and interprets results. The third is asynchronous and random agent programmed control strategies across the entire NetLogo "world". These often very simple programs and rules can lead to the emergence of very complex patterns.

2. Time the Familiar Simulation Stranger

The theologian and philosopher St. Augustine nicely summarized our dilemma. "What is time? If no one asks me, I know but if I wanted to explain it to the one who asks me, I plainly do not know."[3] Going farther, we find the dilemma is much deeper. For example, Augustine, Avicenna, and Bergson argued that time is a psychological phenomenon that is a feature of memories and expectations, not something objective and "out there". On the other hand, people with the same information generally agree, at least at a superficial level, on the time order of events – even if they don't agree on their duration, tempo, rhythm and other characteristics – suggesting some sort of a unique and, to some degree, single observer-independent succession. Unfortunately, this view is inconsistent with the psychological and neuropsychological literature.[4] Apparently, consciousness requires an ongoing brain-based sense of time, melting past, present and future together, creating a flow.[5] One would suspect that such blending and flows are not the same for everyone. Lloyd's work supports Husserl's view that individuals' formulate unique conscious perceptions that are in constant temporal flux. Port and Van Gelder assert that cognition is a time-based dynamical system and not discrete-event based even if it involves multiple agents on a large landscape like NetLogo, which is the current orthodoxy used by most agent-simulation researchers.[6,7]

We begin to see the dilemma for time and simulation construction in three critical areas: the temporal state of the agent, landscape event states, and the direction of the flow of time for the agent, for the landscape, and their interaction. First, measurements of the duration of an event in NetLogo are defined by the intervals of the measuring instrument (here the computer's clock). For Newton, "absolute, true and mathematical time, of itself, and from its own nature, flows equitably without relation to anything external[a]."[8] Time is "background dependent" as is the case in NetLogo. Absolute or mathematical time – Newton's pure time – is the sole concept of time used to show and develop all simulations. But, as Newton has pointed out, time should not be confused

[a] From what vantage point do we stand that permits us to know this "flowing"?

with its sensible measuring method. So what is a simulation's, its agent's and landscape's temporality if it is not the "measuring" clock?

In contrast to background-dependent time, relational time emerges from changes occurring between two agents or agents and their landscape, or both with an observer. Time is instantiated in a *relationship, in its mode of causality,* and is found only there. As such, time is demarcated by events, by flows of relationships with varying durations. For our purposes, time is taken to be background-dependent, as a complex flow of interacting events that, following an emergent rule tied to their scale of interaction, continuously structures a form, be it an "agent" or "landscape" or "observer".

Then there is the difficulty of tensed and tenseless time. From this perspective, time does not flow as an independent background. Things and events exist in time only in a unique and separate state of the past, of the present or of the future. In a tenseless universe, the past, present, and future all exist in the same "time block". Time is not relational, things are ONLY in the state of past, present or future or everything in time is tensed in relation to a single reference. An agent and its landscape are tenseless, lacking a past-present-future as programmed electronic objects. Such physics-based systems have no necessary temporal orientation.

Tensed time in local relationships has no directionality. Local past-present-future are not analytically basic; temporal priority emerges from local contextual causality which establishes what comes first and what comes second.[8,9] The future itself may be problematic in that there is no necessity that there will be a future at all. For example, in certain possible worlds, some propositions simply cannot occur because they are about individuals absent from those worlds (thus no future). "The very fact that individuals come into being [are emergent as in a simulation] makes it impossible for us to formulate crucial statements about such individuals in a satisfactory way before they have been brought into being. ... [Such propositions] are simply not stateable".[10]

Each of these contrasting concepts of time – time-background dependent, background-independent, tensed and tenseless time – produce serious difficulties with describing where in time an event – particularly a simulated one – is located. This is significant for how we conceptualize agents, their interactions, their virtual landscapes, the observer's role, and what the progress from one computer screen to the next as a pattern of relational temporal events may mean.

Virtual space is an electronically constructed space-time visualized according to various electronic requirements constrained by a particular human visual, auditory, and tactile system. Cultural icons and other semiotics – including a "living" agent – reveal multimedia patterns of information: "a consensual hallucination experienced...as lines of light ranged in the nonspace of the mind, clusters and constellations of data."[11] Virtual reality is an artifact of an electronic Newtonian space with its own temporal characteristics affecting the perceptual psychology of the viewer in certain predictable ways. This display is "seen" by the "virtually trained eye" as a world of its own.[12,13] Here information, occupying virtual space, is perceived to be free to create diverse spatial-time patterns. It appears to be free-floating, but is in fact constrained virtual reality's own temporal characteristics.

I will hold, with Michon, that time is local and relational in that an agent and its networks do not live in background space-time: the flow of their patterned relational structures – an event stream – *as* space-time.[14] Time is strictly local and relational, suggesting that tensing, circularity, etc., are also strictly local as the agent-landscape-observer. "[Agents], flames, waves, vortexes – none are supervening on underlying constituents. They are more like knots or twists in the flow. Nothing persists other than the organization [pattern] of the knot itself. They are topological entities [a metapattern of continuous flows], not substantive entities".[15]

3. Time-Ecology: The Agent on its Landscape

Generally, we are trying to simulate a complex adaptive system's dynamics with emergent properties of psychological, governmental, social or economic relationships that influence the development and growth of each by triggering emergent processes and structures as they continuously co-evolve out of differing pasts into differing futures in geographic and virtual space presents.[16] We will call the temporal structure of such a natural complex system a time-ecology whose operation will be central to our examination of simulation temporality.

A natural time-ecology is the multiple emergent spatial-time pattern(s) formed by five temporal nested heterochronic event flows (temporal nesting will be developed below).[17,18] Time is the local relationship(s) of process rates that regulate flows in the space of a time ecology. A single local point may have one or many temporalities (a desert has one; a man walking on a desert has two, and so forth). Each of the five temporalities unfold according to their own causalities with different rates of complexification.[19] Landscapes and agents are themselves a kind of turbulence, an ordered field (attractor basin of some kind) with flows, eddies in space defined by morphodynamic rules. Nothing has an independent existence outside of these configurations; all is oscillation as orderly heterochronically combining event streams. From an ideal observer's global and relational perspective, they flow this way and that in time.

Simon points out "...the frequencies of interaction among elements in any particular subsystem of a system are an order of magnitude or two greater than frequencies of interaction between the subsystems".[20] This suggests that a time-ecology can be decomposed into component systems with boundaries defined according to their number and rate of interaction. This is not to say that fewer interactions or lower frequencies are unimportant; sensitivity to initial conditions is dependent on such relationships and is one causal factor leading to emergence. It is just *this* collection – perhaps even a limited number – of heterochronic flows between temporally based causalities which maintains a time-ecology.

In a natural time-ecology, an agent is defined by its varying time-independent relational state.[17,18] A "state" is an instantaneous statement of heterochronic frequencies of internal processes of the form being generated. For example, sexual maturity will be such an internal measure for an animal. Inputs and outputs (information, energy, resources, etc) are subsets of states which are coupled to other parts of the time-ecology that are close or far away via sensors and effectors with their own processing and extension velocities, much like the agents they connect to.

Heterochronically controlled processes represent an autonomously executed, naturally selected mapping that changes an agent's state but is not *directed* from "outside". Coupling and entrainment to the environment and to other agents may be weak or strong or intermediate or take other forms.[21] The range of each process' frequencies are not unlimited but are confined by their unique velocity cones. The whole is a pattern built up of diverse streams of resources, information, and energy flow into and shaping the agent or its landscape. The complexity of any one layer can be measured by an internal increase in statistical complexity, as the amount of information and minimal sufficient statistics to describe that agent as a state.[19,22]

Ecological layers as defined by Simon above are interdependent in such a way as to constrain the range of morphodynamic rules of emergent structures from the top down or induce emergent structures from the bottom up. Present, past, and future have no necessary directionality in such hierarchies if all five temporal levels are involved (these levels are discussed below).

Change is continuous in a time-ecology, involving linked but varying causalities across linked temporal dimensions (developmental, phenomenological, social, clock are examples of temporal dimensions). Change varies according to its time dimension and velocity cone propagation characteristics.[23] For example, backward causation is possible in nootemporality. It may be that the nested temporally based causalities of different time-ecology layers may account for new emergent properties.

Finally, in terms of morphodynamic organizational structuration time, an agent, being a time-ecology and situated in one as well, will seek to achieve a constant rate of entropy production even though not all subsystems are necessarily optimal. This is the basis for varying speeds and tempos of heterochronic structuration across a layered ecological time and space system.

Novel time-ecology emergents can be seen, as Tooley says "[As s]omething new coming into being with each instance of some level or pattern of lower level constituents. ...[A]n ontological emergent would be a temporal emergent the first time an instance appeared."[9] Emergence presupposes levels: "...[T]he nature of levels remains that of ontological inclusiveness: higher levels include lower levels as constituents... Any causal consequence of a higher level of organization will affect lower, constituent patterned levels. Causal emergence implies downward and upward causation. This causal theory of temporal order and direction in relationships produces a contingent absolute space being space-time points that stand in causal relations to one another. This makes it possible for an event that is not causally related to any other [local] events to have, nevertheless, spatial-temporal location. ...[I]f the direction of time is given by the direction of causation, and space-time points themselves stand in causal relations, then time is, as one naturally thinks of it as being, an all-pervasive feature of the world [and of a time-ecology]."

4. Time Ecology Levels and Temporal Causal Complexity

J.T. Fraser's hierarchical theory of time recognizes five stable, hierarchically nested temporal levels in nature with five different forms of causation.[24] Each integrative level subsumes the functions and structures of the level or levels beneath it. Each adds additional degrees of freedom to its predecessor. Each has its own characteristic causality, which is also nested. Starting from the most complex, the five levels and their forms of causation are:[24,25]

- *Nootemporality*: the temporal reality of human mind or noetic intentionality. Its level-specific causation is "noetic intentionally in the service of symbolic causes" and is reversible through feedback loops that redefine earlier essential contexts.
- *Biotemporality*: the temporal reality of living organisms. Its characteristic distinction is the biological ability to distinguish past and future relative to its own organization. Its level-specific causality is "intentionality in the service of biological needs." "In sum: organic intentionality and its corollary, freedom of choice, came about with biogenesis. It cannot be identified in the inanimate world...."[25] For biotemporality, there is "…no von Neumann address matrix, just a molecular soup, and the instructions say simply "DO THIS with the next copy of THAT which comes along." The results are far more robust. There is no unforgiving central address

authority, and no unforgiving central clock. This ability to take general, organized advantage of local, haphazard processes is exactly the ability that (so far) has distinguished information processing in living organisms from information processing by digital computers".[2]

- *Eootemporality*: the universe of large scale matter. Its time is the physicist's *t*. Causality is deterministic, making possible predictions and retrodictions.
- *Prototemporality*: the time of elementary objects. It is undirected and discontinuous. Its level-specific causality is probability where predictions and retrodictions can only be stochastic.
- *Atemporality*: or the mode of time characteristic of a black hole. There is no mode of causation because it is a level of absolute chaos or pure becoming. There is no "now" and "then", no "here" and "now" – all are discontinuities.

Fraser's well-accepted work raises very serious questions about social simulations. For starters, we can see that physics *t* is not appropriate for either nootemporality nor biotemporality, but is nested in both. This suggests that each agent positioned on its landscape, and the landscape itself for that matter, could exhibit quite different nested forms of time and causality. These difficulties raise questions about how these temporal levels are causally tied together as local background-independent, time-influencing event processes unique to one level while constraining others perhaps at other levels.

Fraser's nested theory of time shows how time-ecology relationships and processes are governed by each time-ecology level and give rise to additional levels, including the evolution and emergence of new forms of time and causation. It suggests that simulations that approach their subject from a process standpoint and utilize the appropriate levels and linkages of nested time and causality may discover entirely new worlds.

Generally, we are interested in simulating the heterochronic flows in agents and landscape of three types of time: nootemporal time, agent time as biotemporal, and eootemporal or clock time. Agent time as noological and biological recognizes that everything, be it social (nations, communities, roads, IT, air transport, fashion, styles of consumption, etc.) or social constructs, goes through developmental cycles with a beginning, middle, and end. These processes can be accelerated only up to a point without fundamentally restructuring the agent itself. For example, agent time may include aging, human migration and related factors which if accelerated lead to different behaviors and have implications for other local agents and landscape. Human agents not only develop and grow, but also have characteristic temporal noetic intentionality accompanying each stage. For example, older people are not in the same temporal reality as younger people, which creates practical synchronization problems.[26] Disruption of agent time leading to the agent's fundamental restructuring (bifurcation point, particularly if it is at the edge of chaos) could be a response if a noetically rooted chaotic or catastrophic future horizon like a natural disaster (eootemporality temporality) is seen to be approaching.

Nootemporal time is discrete and punctuated. Nootemporally, one thinks about the past, present or future but not all at the same time or in any order relative to the arbitrary point of taking action in the present. There is a "space" between these temporal reorientations: consciousness involves time-space chunking.[27,28] This suggests that the direction of causality in nootemporality is highly complex indeed. For example, if we mark a present event and then mark the following causally connected event, such a mark can emerge noetically in the past or in the future. In the past because we are constantly reinventing it.[29] The past is embedded in the present as noted by Lloyd's findings or extends into a created, approaching future which then changes the present. (The dynamical characteristics of these relationships will be briefly touched on in the discussion of the temporal signature below.)

For current simulations, it would be difficult to place biotemporal or nootemporal elements in an ordered temporal context by level and as extended in geographic space unless their particular modes of temporally based causality are properly linked. This weakness makes it difficult to specify how independent and dependent temporal variables are heterochronically embedded in the process.

The problem is made worse by the adoption of Newtonian or eootemporal time as the sole time for extended simulation landscapes. This reduction to a single abstract time dimension seems to ignore not only other temporal levels and causality with their accompanying measurement methods but also temporal qualitative or subjective values (this is a "good time" or "these are bad times") that may be tied to a particular patch on a landscape but not to another. We are not denying that clock time permits multiple events at different scales across an extended geographic and historical space to be tied together, only that an inadequate specification of temporal independent levels and variables provides an inadequate specification of key casual factors. This is particularly so for agent-based biotemporal variables associated with development and growth, and noetic intentionality. The notion that a particular time-ecology's local present is the result of the heterochronic interaction of various, quite different temporalities and causalities in varying space settings, including biological process as well as the semiotic characteristics carried forward by cultural artifacts of whatever size in multiple temporal realms, appears to have been overlooked in current simulations.

5. The Temporal Signature: Local Dynamic Nootemporality

Nootemporality for the individual person (often an agent in a simulation) results from a particular and unique temporal signature. The temporal signature has two parts: temporal perspective, and temporal progression. Generally, a person's sense of identity depends on "continuity in temporal perspective, especially future time perspective. Overall, the character of the "how" – the structuration of the flow and its movement into the future – is an essential part of one's future image even though it has previously been unrecognized as such. ...[P]erspectives on the future are more permanent than other elements of time... . If the continuity of the future perspective is disrupted, one becomes estranged from one's self, leaving an uneasy feeling of strangeness and unfamiliarity."[30,31] The ability to plan future activity is disrupted.

Temporal progression is the dimension of process and motion which describes, among other things, movement or flow of the present into the future. The question here is "how" the present is moved into the future. Two components describe the speed and nature of this "how" in motion: Tempo and rhythm. Tempo refers to the pace of activity. With respect to future events, tempo includes both the rate at which the future approaches and the speed of onset of specific events. (With respect to past events, it is how fast they fade away.) Tempo also bears upon relativism or independence of flow, i.e. the role of external forces, and/or other dimensions such as space on the "flow of life." The dimension of tempo then, covers "fastness" and whether that "fastness" has independence for the actor (background-dependent or independent, for example).

Rhythm refers to the regular recurrence of certain features of time. It asks whether time moves in "pulsating" or periodic cycles, whether its motion is smooth, or whether it is irregular and unpredictable. Is it truly unpredictable or only apparently unpredictable? A distinction can be drawn between the existence of rhythm and the nature of change within that rhythm. It may be that, within a rhythmic system, change occurs gradually or in sharp disjunctive motions, bringing unexpected novelty. Is the rhythm or the cadence within the rhythm smooth, choppy, disjunctive, etc?

Both the temporal perspective and temporal progression "move" time in different ways. From a temporal background-independent viewpoint, they establish dynamical relationships with the past and future that inform what and where the present is. Each agent's local rules are contextualized by the temporal signature.

6. Agent Local Time as Patterns of Networks of Temporal Level Appropriate Velocity Cones

A "velocity cone" refers to the velocity that any event's causal process propagates at; it is the limit of the velocity of all causal propagation.[29] At maximum speed, an event can only propagate a specific distance either into the future or from the past at a temporally limited rate with the distance narrowing the closer it comes to the immediate present. All events that can be reached by a particular eootemporal, biological, or nootemporal signal traveling into the future, or coming from the past, can be depicted on their velocity cone.[32,33]

Spatial-temporal events reveal the ordered motion of a landscape and its agent's multiple "velocity-cones". This spatial-temporal limitation defines and delimits an agent and the influence that various events propagating at different speed can causally have on its structuration rate(s), depending in turn on linkages to velocity cones of resources, energy, and information.[34] Here the focus is on the continuous heterochronic flow of a "wave front" as it is received at different rates by agents and networks across a landscape.

Our discussion of an agent's temporal levels suggests that there are two fundamental velocity cones: that of biotemporality or allometric velocities, and that of nootemporal or noetic velocities. Propagation of ordered relational motion is distinct in these temporalities and is not instantaneous as in a computer simulation. Further, individual heterochronic flows vary in their timing: some are continuous and some punctuated, some are cyclical, some proceed out of the past, some approach from the future as projections and emergent properties, etc. Structuration of a local agent is maintained by synchronization, entrainment, etc., of such varying heterochronic rates.[21]

Large scale spatio-temproal phenomena are significant for simulations. Pipes and various surfaces such as political jurisdictions extend beyond a single node's velocity cones.[35] Large portions of a landscape can be influenced by overlapping surfaces that change local time. Simultaneity can occur at multiple local points beyond the extent of any one agent's velocity cone via pipes like the internet or other electronic communication traveling at the speed of light, but the signal can only be processed as information at the rate it can be transduced and interpreted by biological and noetic systems. Agent life-worlds are the dynamic extensions and entrainments of their temporal signatures upon such locally extended surfaces.

For a local agent, the past is embedded in the present in different ways as memories, artifacts and procedures (letter writing and e-mail) differ from one local agent to another across the landscape. We know from the agent's time signature that the past recedes at different rates for different agents as does the approach of the future (clear, complicated, complex). These same conditions also apply to local time on an adaptive landscape. Thus, a social world is a dynamical world because "what facts there are depends upon what time it is".[9]

Thus at any one point, time-space causality may be the result of multiple local heterochronic agent-landscape-pipe past-present-linkages, each with their own dynamical pacing, and temporal perspective, and temporal progression, all embedded in biotemporality. The result is complex action where local relationships indicate temporal priority between two events by identifying the direction of causation via the direction of local time. Keeping in mind Fraser's nested hierarchies of time and causality, and the essential characteristics of nootemporality (no necessary and common ordering in temporal perception of perceived progression), as well as the concept of temporal signature, causality becomes very complex indeed, particularly when these conditions are spread out across space. Again, we are looking at this as a multilayered heterochronic flow of events that structurate agents and landscapes.

7. Space-Time of Landscapes

"Input and output of the various Agents are coupled to elements of state, but the environment does not distinguish which elements of state are so coupled. That distinction depends on the agents that exist at any moment and the capabilities of their sensors and effectors. The lack of a distinguished input and output means that the environment, unlike an agent, is unbounded. The fact that an environment, like an agent, has state and process means that putting a boundary and associated input and output values around an environment (and its associated agents) yields a higher-level agent."[33] A number of temporal issues emerge: is the environment in fact unbounded and without temporality, or is it locally temporally textured? The momentary local activity of an agent suggests a local temporal connection with the environment, a kind of boundary. What can be said of this boundary making and closing process?

Current simulations seem to consider space as a sort of "geographic extension" displayed in virtual space. Actually, virtual space and its navigation has its own political and temporal implications. Such spatial temporal assumptions are not consistent with social or public policy-making structures. For example, temporal/spatial geographic analysis, using Geographical Information Systems and other computer based tools, depict dynamic relationships with extending or contracting temporal social patterns. This represents a kind of geographic space/time plasticity in, of, and between social entities at all scales, be they cities or their social-culture groups.[35,36] Multiple parallel nested temporal processes, including nootemporally differing pasts, presents and futures, interpenetrate, feeding forward and backward at different points across the multiple layers of this geographic time-ecology.

Geo-temporal plasticity and seemingly backward causality are important limitations that do not appear to be part of the spastic-temporal characteristics of current simulation landscapes. For example, a linear concept of the flow of history, depicted in the way the simulation moves local patches of the landscape into the "future", would tend to minimize such nootemporal plasticity. It gives the appearance of two dimensional temporal uniformity across the landscape rather than the necessary patchy geophysical-temporal complexity.

Temporally, the local relationship of agent-landscape events that are accessible to the agent vary in their duration, scheduling (time and sequence in the network), and by their own unique temporal characteristics.[35] Fraser's temporal and causal nesting defines such accessibility in that this nesting is local and selective, suggesting a complex, layered landscape with varying fundamental temporal characteristics that causally interface with each other in complex ways.

8. Simulation Causality Issues and Suggestions for Improving NetLogo

Fraser states that "[it] is not possible to make predictions about the conduct of a human group, acting in the service of concrete or symbolic causes, without allowing for noetic and biological intentionalities, as well as the deterministic, probabilistic and chaotic contributions".[24] This warning suggests a number of complex challenges to casual explanations based on two-dimensional simulations like NetLogo:

- Temporally complex agents and landscapes may interact across particular temporal levels both locally and via pipes to other local portions of the time-ecology. This produces a 3 D to 5 D "spaghetti-like" connectivity.

- Causal relationships between each layer of the nested causalities for an agent or its landscape or across a time-ecology are difficult to identify and specify.
- Fraser suggests that the Feigenbaum Map traces the various forms of causality leading from formal unpredictability, to self-similarity or emergence of patterns, ending with the intentional noetic creation of additional patterns.[24] Such an approach could lay the foundation for determining the role that time plays in emergence.[37]
- Visualization and analysis of these complex temporal-spatial flows of events in multiple dimensions are particularly problematic.

To summarize, local accessibility and varying nested hierarchies of time, the temporal signature, heterochrony, and the limitations imposed by velocity cones together suggest that a simulation's virtual time-ecology landscape, at least as far as social simulations like NetLogo, could be enriched by following properties:

- Be topographically highly complex in the temporal nesting and levels, each with a kind of boundedness and connectedness leading to intermingled causality;
- Be highly variable in its heterochronic rates of temporal/spatial characteristics as limited by velocity cones;
- Visualization techniques showing that events on such a surface will vary locally from moment to moment in highly complex and dynamic ways, involving emergent highly local patterns driven by agent that are simultaneously existent in various temporalities, at varying tempos and rhythms in varying pasts-presents-futures, and regulated by nested causalities.
- Evolutionary selection of temporal boundaries, heterochronic flows and rates, and layers or variations in their boundaries, including their velocity cones could produce emergent higher-level agents or lower level ones that are substantially different due to new morphodynamic rules derived from such nested heterochronies.

References

1. U. Wilensky, NetLogo. http://ccl.northwestern.edu/netlogo/. Center for Connected Learning and Computer-Based Modeling, Northwestern University, Evanston, IL. NetLogo3.1.2 User Manual (1999).
2. G. Dyson, http://www.edge.org/3rd_culture/dyson05/dyson05_index.html (2005).
3. J. T. Fraser, *Of Time, Passion and Knowledge*. (Princeton University Press, Princeton 1975).
4. D. Antonia, Scientific American, 66, September (2002).
5. D. Lloyd, *Journal of Cognitive Neuroscience*, 14, 818-831 (2002).
6. R. Port and T. Van Gelder, *Mind as Motion: Explorations in the Dynamics of Cognition*. (MIT Press, Cambridge 1995).
7. H. Parunak and H. Van Dyke, Annals of Operations Research, Special Issue on Artificial Intelligence and Management Science (2003).
8. C. Callender, C. and R. Edney, R. *Introducing Time*. (Aerial Press, Santa Cruz, California 2001).
9. M. Tooley, *Time, Tense and Causation*. (Claredon, Press Oxford 1997).
10. P. Ohrstrom, *The Uncertainty of the Future*. Paper prepared for the International Society for the Study of Time Conference, Gargonza, Italy (2001).
11. W. Gibson, Neuromancer, in P. Nayar, *Virtual Worlds*. Sage, New Delhi (2004).
12. K. Hayles, *How We Became Posthuman*. (University of Chicago Press, Chicago 1999).
13. K. Hayles, *My Mother was a Computer*. (University of Chicago Press, Chicago 2005).
14. M. John, *Models of Local Time*. Paper presented at the XIth International Conference of the European Society for Cognitive Psychology, Ghent, Belgium (1999).
15. M. Bickhard and D. Campbell, *Emergence*. (http://www.lehigh.edu/~mhb0/emergence.html 2/18/99).
16. J. Holland, *Hidden Order*, (Addison-Wesley, Redwood City 1995).
17. G. Koehler, Gus, *A Framework for Visualizing the Chronocomplexity of Politically Regulated Time-Ecologies*. Paper prepared for the International Society for the Study of Time Conference, Gargonza, Italy (2001).
18. G. Koehler, Time, Complex Systems, and Public Policy: A Theoretical Foundation for Adaptive Policy Making, *Nonlinear Dynamics, Psychology and the Life Sciences*, Vol. 7, 1 (2003).
19. J. T. Fraser, *Time, Conflict, and Human Values*. Appendix B. (University of Illinois Press, Chicago 1999).
20. H. Simon, Foreword in W. Callebaut and D. Rasskin-Gutman, Modularity: *Understanding the Development and Evolution of Natural Complex Systems*. (Boston: MIT Press 2005).

21. A. Winfree, *The Geometry of Biological Time.* (Springer, New York 1980).
22. C. Shalizi, K. Shalizi, and R. Haslinger, *Quantifying Self-Organization with Optimal Predictors.* arXiv:nlin.AO/0409024 v1 10 Sep 2004.
23. G. Koehler, *Sorting out the Temporal Confusion of Computer Simulations.* Paper presented at the International Society for the Study of Time, Conference, Cambridge, UK (2004).
24. J.T. Fraser, From Chaos to Conflict, in J.T. Fraser, M. Soulsby, and A. Argyros: *Time, Order and Chaos.* (Madison, Conn.: International University Press 1998).
25. J. T. Fraser, *Human Freedom,* KronoScope, Vol. 2, No. 2, pp. 223-247, (2000).
26. P. Hancock, *The Time of Your Life: One Thousand Moons.* KronoScope, Vol. 2, no. 2, p. 135, (2002).
27. W. Lawrence, *Dynamical Cognitive Science,* (MIT Press, Boston, Chapter 5, 2002).
28. D. Lloyd, *Journal of Cognitive Neuroscience,* 14, 818-831. (2002).
29. H. Reichenbach, *Space and Time.* (General Publishing Company, Canada 1958).
30. E. Victoria, *The Use of Temporal Constructs as a Model for Understanding Perceptions of Environmental Hazards* (Las Vegas, NV: dissertation, Department of Sociology, University of Nevada, Los Vegas, 1997).
31. V. Koehler-Jones, *The Temporal Signature of Small Manufacturer,* Time Structures, Sacramento (2003).
32. W. Kaufmann, *Relativity and Cosmology.* (Harper and Row, New York 1973).
33. L. Smolin, *Three Roads to Quantum Gravity.* (Basic Books, New York 2001).
34. H. Parunak, H. Van Dyke, S. Brueckner, *Entropy and Self-Organization in Multi-Agent Systems.* Agents'01. Montreal Quebec, Canada (2001).
35. V. Tyler, *Metapatterns: Across Space, Time, and Mind.* (Columbia University Press, New York, 1995).
36. D. Janelle, *Embedding Time in Accessibility Analysis.* Center for Spatially Integrated Social Science, University of California, Santa Barbara (2002).
37. G. Koehler, *EU Transport Foresight Planning: Comments on How Chronocomplexity Limits Such Strategic Planning Efforts.* Paper presented at the Society for Chaos Theory in Psychology and Life Sciences, 12 Annual Conference, Portland State University, Portland, Oregon (2003).
38. G. Koehler, *The Feigenbaum Diagram: A Metapattern for the Social Construction of Time.* Paper presented at the Sixth Annual International Conference of the Society for Chaos Theory in the Psychology and the Life Science, University of California, Berkeley (1996).

CHAPTER 24

ANTI-FLARING: HOW TO PREVENT THE MARKET FROM OVERHEATING

Artur P. Schmidt

Chalet Gessenay, CH-3792 Saanen, Switzerland
artur.schmidt@wissensnavigator.com
www.wissensnavigator.com, www.unternehmercockpit.com

Otto E. Rössler

Institute of Theoretical Chemistry, University of Tübingen
Auf der Morgenstelle 8, 72076 Tübingen, Germany

A new damping method for the rational exploitation of flaring-type economic phenomena is proposed. The steering parameter for Anti-Flaring is called the Livermore Indicator. The approach in general is described as endonomics.

1. Introduction

Considering the power of systems theory, there exists an opportunity for the central banks to avoid so-called bubbles or superflares in the economy. Economics shows a special type of instability which was called "flaring" by Georg Hartmann.[1] (For the mathematical gourmet: Flare Theory is a simplified version of Milnor Attractor Theory).

The best example of a flare attractor is to be found in a book by the Wienerwald founder Friedrich Jahn: "A life for the Wienerwald - From Waiter to Millionaire and Back."[2] He followed the strategy of using a fixed percentage of his earnings for expanding his business by spawning new Wienerwalds all over the world. Any similarities with McDonalds are purely coincidental.

2. Timing

Anti-Flaring is a new phrase and a new strategy. If the desirable flaring process has got out of hand – due to overheating beyond the point where the debt spiral that is part of it has reached a level that can't be healed anymore – then the idea of anti-flaring can be the method of last resort.[3] Ideally, the Federal Reserve would adjust the interest rate and/or the money supply just in time, much as China is having to do at the time of writing. However, there is a special risk involved here. If the right moment has been allowed to pass inadvertently, the effect can be counterproductive and even disastrous (famous examples being the Great Crash of 1929, the Nikkei Crash in 1990 and the Nasdaq Bubble a decade later). But would not choosing the right moment amount to a miracle of clairvoyance?

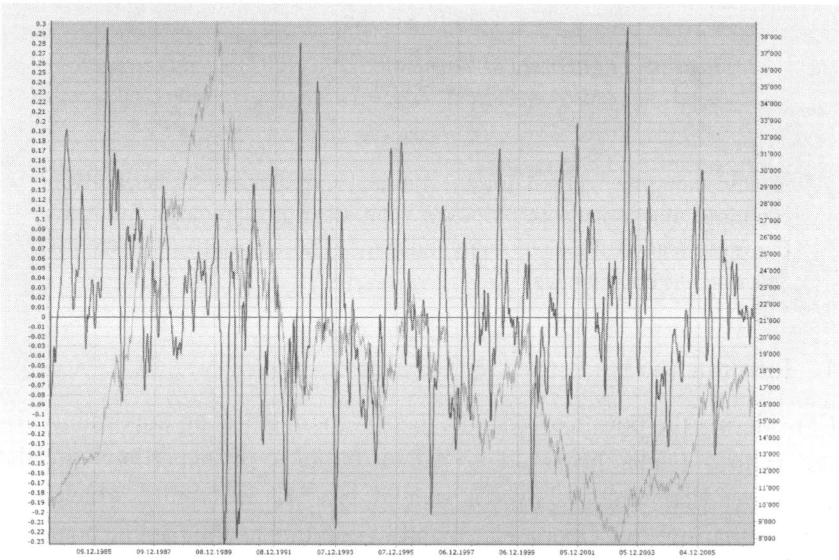

Fig. 1: Nikkei Livermore Indicator Flares 1984 - 2007; The dark line represents the Livermore Indicator, the grey line is the Nikkei Index.
Source: www.unternehmercockpit.com

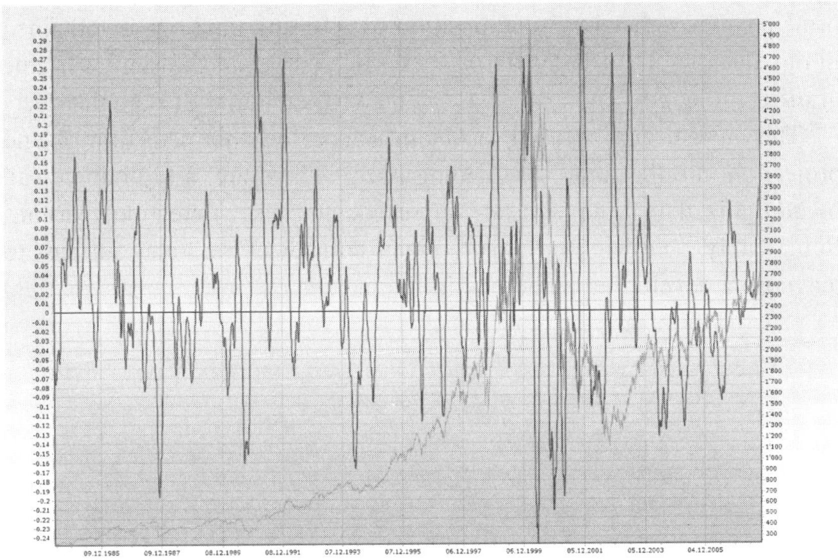

Fig. 2: Nasdaq Livermore Indicator Flares 1995 - 2007; The dark line represents the Livermore Indicator, the grey line is the Nasdaq Index.
Source: www.unternehmercockpit.com

3. Matrix Analysis

This miracle is to be described in the following - but please, do not tell anybody.

There exists a mathematical algorithm whose matrix-theoreticial fundamentals are very simple but easy to be questioned by a specialist. The inventor discovered it on the basis of long empirical and theoreticial practice, but non-cyberneticians cannot yet quite understand how it works.[4] Nevertheless, the principle and the steering by the so-called Livermore Indicator is very simple.

It is based on the principle of autocatalysis in dynamical systems.[5] The best introductory book is still Robert Rosen's 1971 international bestseller (especially in Japan) "Dynamical System Theory in Biology". The principle is based on Oscar Lange's 1969 book of Recursive Matrix Dynamics[6], which was first implemented in an economic context by Michel Godet in the eighties.[7] Autocatalysis is the one leg of the new principle. The other is a superposition of differing time curves obtained in this fashion on different flaring-type time curves of the same variable of different length. This makes use of the self-similarity of flare theory.

This insight can be used to obtain a more reliable detection of and projection into the future of ongoing flares. Therefore, a big bubble can be recognized in time with an unprecedented early-detection margin.[8] Remarkably, following-up a quite small number of variables suffices to diagnose a developing pathology in the market.

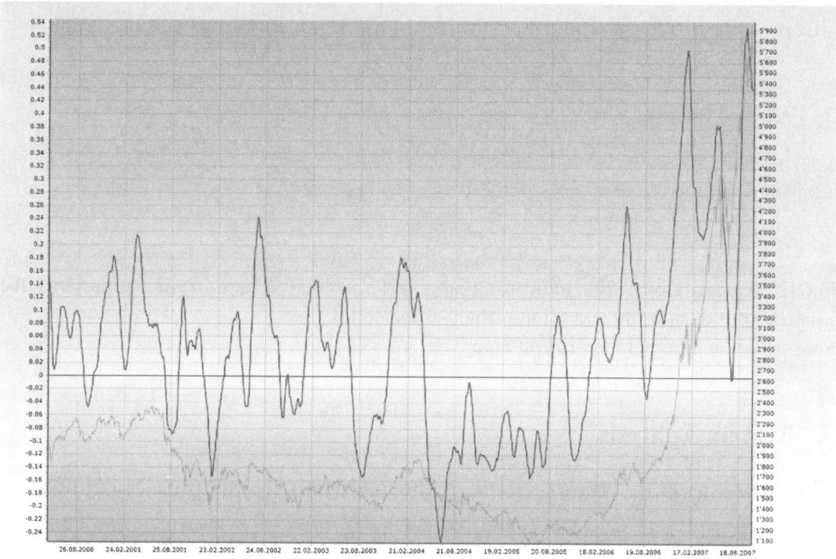

Fig. 3: Chinese A-Share Livermore Indicator Flares 2005 – 2007[9]; The dark line represents the Livermore-Indicator, the grey line is the Chinese A-Share Index.
Source: www.unternehmercockpit.com

Why should you believe this claim of ours? An empirical proof could be seen to lie in the performance of the *Unternehmercockpit* (www.unternehmercockpit.com), which was founded by the first author. But of course, empirical success, with its necessarily finite time-base is no theoretical proof.

Without the new flare theory with its self-similar height-structure on all magnitudes, as described in Hartmann's thesis, a similar confidence that this is really the structure of an economic system would not be possible to the same extent.

Flares also have a surprising implication: they are scale-free. Therefore, quite innocent short specimens can reveal an ongoing bigger process, formed my many little elements via summation. This confirms and supports the fractal economic theory of Benoit Mandelbrot, admittedly in an idiosyncratic manner.[10]

Another short observation. Adjusting the zero-level, through a kind of low-pass filter, allows us to eliminate the influence of negative/positive fluctuations – in the near-zero area.

4. Anti-Flaring

When these indicators are used, an overflow is successfully avoided and the corresponding sector of the partial or full economy in question can be steered much in accord with the early hopes of Norbert Wiener.

If this prediction were borne out, the main effect would be that speculation can effectively be taken out of the market. This is an apparently new option of Endo-Management, which, of course, comes ethymologically from end-o'-management, like in will-o'-the-wisp.[11]

In this way, the Hedge Fund speculation can be reduced to absurdity, so that such funds cannot perturb the very system they prey on any longer. Helmut Schmidt's concern about the danger of uncontrolled hedging could thereby be mitigated. The new indicator, which allows a related second upturn to be distinguished from a transient last positive wave during a downturn, exposes the hidden structure of the market. Even before professionals in the market have any clue what's going on, the newly invented Livermore Indicator (www.unternehmercockpit.com) will show the new direction of a cybernetic trend.[12]

If everybody were in possession of this new diagnostic instrument - for example, if Google were to make it available - this would amount to a very effective artificial damping of non-productive fluctuations in the financial markets. If every movement of the market is based not on speculation but on effectiveness, the new indicator will push the whole economy towards entrepreneurship and knowledge orientation, as anticipated in the book "Medium of Knowledge" by the present authors.[13]

The effect that can be achieved with this new method will be reminiscent of that which banker Alfred Herrhausen anticipated in 1989, along with his 'unreasonable' view of turning around the Third World economy by intelligent debt escape. This endo-approach also provides a philosophical background for a healthier economic future based on cybernetic risk management[14] that allows an early anti-flaring course to be steered in the complex system known as the economy.[15] We have named this new research field endonomics.[16]

5. Sum-up

Human activists in the market as described by Ludwig von Mises are pathological animals.[17] The have a choice between positive and negative feedback loops, between flaring and anti-flaring. Endo-Management, the style of governance which allows complexity to be steered, shows a way toward a healthier and less erratic development of the global market. If applied correctly, the market participants, especially the Federal Reserve, have the choice of taking the red or the blue pill, as in the famous counterculture movie 'Matrix'.

References

1. G.C. Hartmann and O.E. Rössler, Coupled flare attractors – a discrete prototype for economic modelling, *Discrete Dynamics in Nature and Society* **3**, 1, pp. 25-32 (1999).
2. F. Jahn, *Ein Leben für den Wienerwald - Vom Kellner zum Millionär und zurück.* private publishing (1993).
3. W.B. Arthur, The End of Certainty in Economics, *Einstein Meets Magritte*, D. Aerts, J. Broekaert, E. Mathijs, eds., Kluwer Academic Publishers, Holland (1999).
4. N. Wiener, *Kybernetik*, Düsseldorf (1963).
5. W.R. Ashby, *Einführung in die Kybernetik.* Frankfurt a. M. (1985).
6. O. Lange, *Ganzheit und Entwicklung in kybernetischer Sicht.* Berlin (1969).
7. M. Godet, *Scenarios and Strategic Management*, London (1987).
8. D. Farmer and S. Joshib, The price dynamics of common trading strategies, *Journal of Economic Behavior & Organization,* **49**, 2, pp. 149-171 (2002).

9. E. Ahmed/Rosser, J. Barkley/Uppal, Jamshed Y., Evidence of nonlinear speculative bubbles in pacific-rim stock markets, *The Quarterly Review of Economics and Finance* **39**, 1, pp. 21-36 (1999).
10. B. Mandelbrot, The inescapable need for fractal tools in finance, *Annals of Finance* 1, pp. 193–195 (2005).
11. A.P. Schmidt, *Endo-Management*, Bern (1998).
12. J. Foster, Why is Economics not a Complex Systems Science? *Discussion Paper*. No. 336, School of Economics, The University of Queensland (2004)
13. A.P. Schmidt and O.E. Rössler, *Medium des Wissens*. Bern (1999).
14. A.P. Schmidt, *Wohlstand_fuer_alle.com*; Muenchen (2001).
15. O.E. Rössler, *Endophysics - The World as an Interface*, World Scientific, Singapore (1998).
16. A.P. Schmidt, Return on Knowledge, *GDI-Impuls* 3/01, pp. 45-51, (2003).
17. L. von Mises, *Human Action. A Treatise on Economics*. London (1949).

CHAPTER 25

LEVERAGING THE FUTURE – EXISTENCE OF A NEW ENDO-REALITY IN ECONOMICS

Artur P. Schmidt

Chalet Gessenay, CH-3792 Saanen, Switzerland
artur.schmidt@wissensnavigator.com
www.wissensnavigator.com, www.unternehmercockpit.com

Otto E. Rössler

Institute of Theoretical Chemistry, University of Tübingen
Auf der Morgenstelle 8, 72076 Tübingen, Germany

A new attractor in economics is identified. It is ultimately based on the Godet Correlation (matrix of interdependencies), which originally was introduced intuitively in a sociological context. Here, for the first time, a purely observational, quantitative and multitime version is introduced. It amazingly reveals the existence of an organizing force vector in complex dissipative systems such as the economy and, presumably, other living and non-living systems.

1. Introduction

Economies today resemble airplanes without pilots. It seems that they are running on a strange autopilot called neoliberalism. Economists, who generally have no clue of the trends they predict, would describe this as a perfect situation, because the market should control the flight. Nevertheless, the actual debt situation and the meltdown of the middle-class show another picture: Systems can get out of control and collapse if the complex interdependencies are not understood.

A situation where no one is in the cockpit leads to a Darwinistic economy, which should be replaced by one with that well-known type of cockpit-approach called the social market economy or, as Ludwig Erhard would have put it, "prosperity for everyone"[1].

2. Looking Behind the Curtain

To bring back this focus, we searched for an analytic tool which shows us the existence of a new reality in economics through identifying the real attractors which are hidden behind the complexity curtain. If we observe the long-term cybernetic feedback between interest rates (short and long-term), the money supply and the major market indices (such as the German DAX), we see some very interesting hidden patterns. Under special market situations with a high interest-spread (between short and long term) and inflationary money supply, markets show a behaviour which will eventually lead to market crashes. Therefore, it would be very useful for market participants to have an early-warning system when systems overshoot their normal behavioural patterns.

What most economists do not see are the hidden patterns behind economic trends. These patterns, which we predict to have a common attractor, can ruin many a market participant in no time and make a small number of investors very rich during the same period. An old rule formulated by Ludwig von Mises says that if the relative increase of money supply is positive, interest rates should climb and if money supply is negative, they should fall.[2] This is based on a so-called linear causality, which is only true if hidden factors do not lead to the inverse results as anticipated. But as complexity in the economy is increasing faster on any delta t, we may consider, as an alternative, a new analytic tool for describing markets in all their variety. For this purpose, a software called I-Matrix, which is based on non-linearity, was designed to reveal the real force behind economic trends. Cybernetics is not about predicting the future: its aim is to show the hidden patterns of the Now, because the Now is nothing but the "master market" itself.

3. Leverage versus Risk

How to create the new out of the blue? An endo-observer approach is described for steering complex systems by minimizing risk while maximizing growth.[3] The biggest challenge in the procedure is to avoid creating opportunity costs as a result of wrong timing in volatile markets. Everybody knows that the market is always right, but hardly anybody seems to know that high gains are the biggest risk protection against further declines. The most popular measure of risk is the beta-factor, which measures volatility. The general thinking is that volatile stocks are riskier than non-volatile ones. But if we can find out the real forces behind a stock or a market, risk and volatility will not equate.

Economies, just like enterprises and individuals, can increase their opportunities if they use means of leveraging such as the Internet, navigation tools or access to capital. However, the biggest leverage for management and entrepreneurship is the synergy between mathematical cybernetics and economics. This marriage allows for a better understanding of complex systems and uses the power of volatility for superior gains. The real risk in the financial market is not volatility but missing the immense forces which drive economies, markets and stocks to new all-time highs. Therefore, when it comes to investing, the real risk is not volatility. For the money manager Ron Muhlenkamp, risk is "the probability of losing purchasing power over time."[4]

4. Calculating the Feedback Loops

What is needed to use the power of volatility is a systems approach that can identify the hidden feedback loops that steer the multidimensional interdependencies behind the market curtain. Unfortunately, the high-dimensional orchestration, which may be described by matrix multiplication, is virtually unknown up to now. The variety of hidden feedback loops is the real engine which drives complex ecological and economic systems. Therefore, only a calculation of these loops allows us to avoid a decline of purchasing power. Intuitive understanding of these lags behind the technical opportunities provided by the computer and the Internet. Thus, it is not quite clear how the "Livermore Indicator" really functions. So far, there is no mathematical theory of iterative multiplicative matrix operations. In the 1980s, Michel Godet described in his books the power of recursive matrix multiplications for the prediction of correlated trends in complex interactive technical and social systems.[5] One precondition is non-negativity, which means that any set of variables assumes only positive values. But the trick is obvious: Any variable oscillating around zero can be shifted up toward positive values, for example, through accumulation.

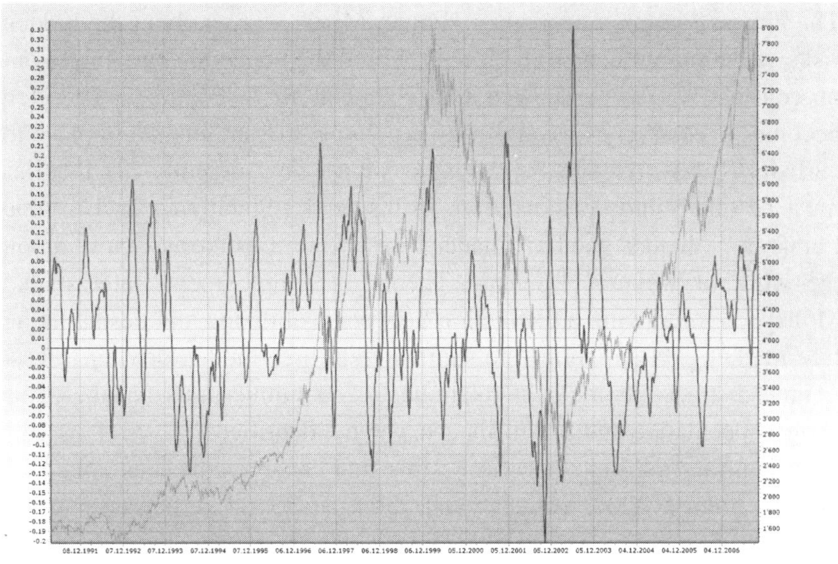

Figure 1: German Stock Index Dax from 1990 to 2007; The dark line represents the Livermore Indicator, the grey line is the Dax Index.
Source: www.unternehmercockpit.com

Then the cross-impact Godet-Matrix (interdependancy-matrix) can be iterated seven times (or more) yielding a stable vector.[6] This is an attractor describing the mutual impact of the different variables. Over different time scales, slightly different stable vectors will emerge. But they will appear like perturbations of a single vector. The vector is a living entity. It describes a kind of "force" hidden in the unknown system in question. It can, for example, explode or shrink to almost zero. It describes a magic holistic feature of the system in question. Like a hologram, adding more variables and resolution does not greatly change the precision of the vector's identity and strength. The hidden force of an index can be calculated from no more than 12 key economic indicators. Figure 1 shows the development of the German stock index Dax from the 1990s to the year 2007.

5. Avoiding Panic

The almost forgotten economist Hyman Minsky wrote about the hidden risks an economy faces if the forces behind the economic curtain get out of control. His work showed that financial market exuberance often becomes excessive, especially if no regulatory constraints are in place to curb it. Today's Hedge Fund jungle shows how urgently his theories must be taken into consideration. In his book "Stabilizing an Unstable Economy"[7], he designed a "financial instability hypothesis", building on the work of Keynes' "General Theory of Employment, Interest and Money."[8] In drawing a distinction between stabilizing and destabilizing free market debt structures, he summarized three hidden driving forces: "Three distinct income-debt relations for economic units ... labeled as hedge, speculative and Ponzi finance, can be identified."

When the first hints become visible that cheap credit is ending, as it did in August 2007, insiders sell their stocks. Bad news or panic reactions can roll over quickly into chain reactions and create the ultimative Black Day (a synonym for a mega-crash) in the markets.[9] Euphoria can transform in milli-seconds into revulsion, panicking investors into selling. The hidden forces will then create a downward momentum which is far greater and faster than the force that leveraged the prices to new all-time highs. Therefore, there is no other solution to avoid the risk of bubbles than to find the real attractors of economic developments and to act in advance to reduce the risk of overheating. If the risk is not reduced, participants in the markets have only one chance if they do not want to lose control of their assets. They need a system which shows them when the vectors in the markets change direction. Leveraging the return of single stocks can be a good means for profit maximization for investors, but leveraging the entire economy into a status of irrational exaggeration (known as the Greenspan Approach) is extremely risky for all market participants if they do not know anything about the hidden forces.

5. Conclusion

Our attractor approach to economic analysis offers a unique opportunity to change the rules of the game before the game gets out of control. It is tempting to conjecture that this type of approach has many applications not only in economics and finance, but also in other nonlinear systems such as the weather, the brain and the physiological status of an organism in health or disease. The vector can be used not only for trading or diagnostic purposes, but also for therapeutic intervention. Both in micro- and in macroeconomics, important applications come immediately to mind. The Federal Reserve is desperately seeking a diagnostic and therapeutic instrument of such power. The "heart" of the economy can be made visible for the first time, including all its palpitations. Even a heart attack of the economy can be detected before it occurs. Once such an almost over-optimal diagnostic and therapeutic instrument is available, the world economy will have an excess value available for benevolent intervention, which could be used to actively promote the achievement of economic health across the planet at an earlier point in time. Therefore, we pledge that Lampsacus[10], hometown of humankind in the internet, should be financed from the new attractor. The new force vector is like that of a Yedi Knight, so let the Force be with you!

For J.O.R.

References

1. L. Erhard, *Wohlstand für Alle*, Düsseldorf (1957).
2. L. von Mises, *Human Action. A Treatise on Economics.* London (1949).
3. A. P. Schmidt, *Endo-Management*, Bern (1998).
4. http://www.muhlenkamp.com/methods/pdf/methods_07.pdf
5. M. Godet, *Scenarios and Strategic Management*, London (1987).

6. A.P. Schmidt, *Entwicklungstrends der Verkehrsflugzeugbranche und deren Projektion bis zum Jahr 2010 unter Anwendung einer integralen Szenariotechnik*, Bern (1992).
7. H. Minsky, *Stabilizing an Unstable Economy*, New Haven (1986).
8. M. Keynes, *General Theory of Employment, Interest and Money*, New York (1936).
9. http://www.heise.de/bin/tp/issue/r4/dl-artikel2.cgi?artikelnr=20194&mode=print
10. http://www.lampsacus.com

CHAPTER 26

ENDONOMICS: LOOKING BEHIND THE ECONOMIC CURTAIN

Artur P. Schmidt

Chalet Gessenay, CH-3792 Saanen, Switzerland
artur.schmidt@wissensnavigator.com
www.wissensnavigator.com, www.unternehmercockpit.com

Otto E. Rössler

Institute of Theoretical Chemistry, University of Tübingen
Auf der Morgenstelle 8, 72076 Tübingen, Germany

An interface is presented that was not visible before we had developed the new software I-Matrix, which allows us to look behind the economic curtain. Forces that no economist can predict are found to be present. We studied economic developments from 1931 to 2007 with the Livermore Indicator solely, taking only the inflation rate into account. Relative price movements are the hidden forces responsible for the up and down turns in the economy. Even in a deflation, consumer prices can fall more slowly in relation to the economic downturn, which means there is an inflation even when the prices fall in absolute terms. We therefore propose that there is an "economic relativity theory" based on relative prices changes.

1. Introduction

Never believe statistics you did not forge yourself. Similarly never believe an economist. The problem is that economic thinking is mostly based on linear thinking. Economies, however, behave irregularly in a nonlinearly way, which does not mean irrationally but deterministic-chaotically, based on the underlying attractors.

2. Deflation Revisited

The original definition of deflation as an absolute decrease in prices is just as misleading as the classical definition of inflation as an absolute increase in prices. We therefore suggest an inflation theory based on relative price differences:

A relative inflation decrease stimulates the economy.
A relative inflation increase weakens the economy.

What economists really mean when they talk about deflation is a period in which there is a relative positive inflation with absolute prices falling. Not more and not less. If we were to define deflation in relative terms, we could understand, for example, the real benefit of the IT revolution. The deflationary trends in subparts of the economy resulted in an overall negative inflation in relative terms. The latter has stimulated the economy both through the last decade of the 20th century and the first decade of the 21st century.

3. Economic Analysis

Figure 1 illustrates that our theory makes sense throughout the last 81 years, including even the Great Depression. Considering only the development of the Dow Jones Industrial Index and the Consumer Price Index (CPI) Rate (including energy and food prices) as analysed with our software I-Matrix, the apparent correctness of the proposed theory can be demonstrated. Economic downturns were always preceded by a bottoming-out of the Livermore Inflation Indicator and economic expansion periods always started at the turning point of a flare when the relative inflation began to decrease again. If this oberservation can be generalized, the major economic theories of the last 100 years are due for reconsideration in the context of a relative-motion theory in economics.

While relative increase of money supply is positive and interest rates do not climb the inflationary pressure tightens up. But when the relative increase of money supply is negative and interest rates are not reduced, the inflationary pressure loosens. The first case means an economic downturn, while the second means an economic upturn which together form a normal economic cycle.

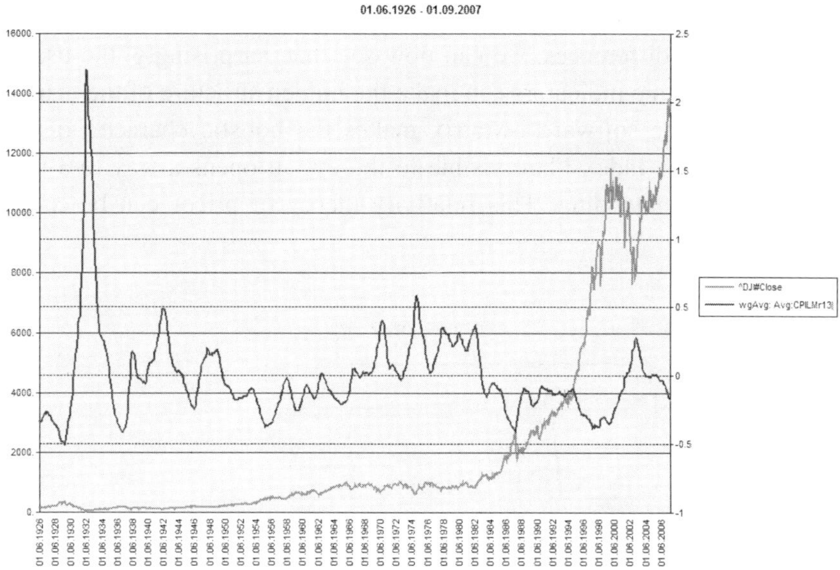

Figure 1: Economic Inflation Flares 1926 – 2007[1]; The strongly fluctuating dark line represents the Livermore CPI Indicator and the grey line is the Dow Jones Index. Source: More detailed figures for individual decades can be found here for downloading: www.bankingcockpit.com[2]

4. Endonomics

Cybernetics does not interpret why something happens but rather gives a detached descriptive view of what happens. Price fluctuations are always relative to each other which means that absolute numbers say nothing about the stability of the direction an economy is taking. Price fluctuations are therefore all an economy is about, no matter whether we speak of growth, stagnation, recession or depression. A depression like that of the 1930s is therefore nothing else than an exceedingly large hyper flare - a flare that overshoots the normal maximum over a long period (see Figure 2).

Considering these results, we can say that the relativity of prices explains the relativity of purchasing power.[3] If purchasing power in relative terms gets weaker, economic downturns are unavoidable. Because the economy consists of many cybernetic feedback loops between price differences, we can now see that, surprisingly, the use of just two economic indicators can make the hidden structure of the market transparent. The software I-Matrix makes the holistic character of the Consumer Price Index-Flare visible to the eye. It opens a way to a new economic understanding. This relativity theory of prices can be called endonomics.

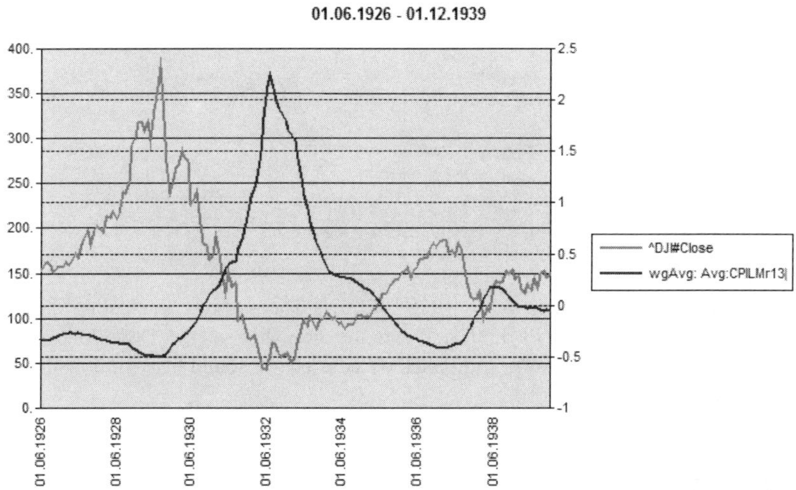

Figure 2: Great Depression Flare 1926 – 1939[1]; The strongly fluctuating dark line represents the Livermore CPI Indicator and the grey line is the Dow Jones Index. Source: More detailed figures for individual decades can be found here for downloading: www.bankingcockpit.com[2]

5. Sum-up

Human action tries to make a difference. Markets handle these differences through price fluctuations. The stronger the feedback loops in the economy become, the faster will be the price changes. The cybernetic consequences of any action have the potential to make a huge difference. An increasing number of feedback loops can accelerate the change processes in an economy even to the point of domino effects. Our cybernetic analysis suggests that the method may work under all market conditions, even under the worst market psychology, like the depression of the 1930s, the oil crisis of the 1970s, the Dow Jones Crash of 1987 or the internet depression of 2000 - 2003.

A kind of memory of the overall market development is hidden in every relative fluctuation of the market. Therefore, cybernetics is an indispensable tool to bring the hidden patterns behind market fluctuations to the fore. With the new interface "I-Matrix", we tried to demonstrate that apparently any boom-and-bust cycle can be identified if one uses the spectacles of a detached observer in the approach we have called endonomics. Einstein would not object if anyone would say that even the economy follows a relativistic trajectory.

References

1. For those who are interested in a more detailed view, here are the hyperlinks to the detailed analysis:
   www.bankingcockpit.com/images/DOWJONESLIVERMORECPI19262007.gif
   www.bankingcockpit.com/images/DOWJONESLIVERMORECPI19261930.gif
   www.bankingcockpit.com/images/DOWJONESLIVERMORECPI19301940.gif
   www.bankingcockpit.com/images/DOWJONESLIVERMORECPI19401950.gif
   www.bankingcockpit.com/images/DOWJONESLIVERMORECPI19501960.gif
   www.bankingcockpit.com/images/DOWJONESLIVERMORECPI19601970.gif
   www.bankingcockpit.com/images/DOWJONESLIVERMORECPI19701980.gif
   www.bankingcockpit.com/images/DOWJONESLIVERMORECPI19801990.gif
   www.bankingcockpit.com/images/DOWJONESLIVERMORECPI19902000.gif
   www.bankingcockpit.com/images/DOWJONESLIVERMORECPI20002007.gif
2. Artur P. Schmidt, www.bankingcockpit.com
3. L. von Mises, Human Action. A Treatise on Economics. London (1949).

CHAPTER 27

POSSIBLE QUANTUM ABSORBER EFFECTS IN CORTICAL SYNCHRONIZATION

Uwe Kämpf

Department of Psychology, Dresden University of Technology
Mommsenstr. 13, D-01062 Dresden
E-mail: uwe.kaempf@psychologie.tu-dresden.de

The Wheeler-Feynman transactional "absorber" approach was proposed originally to account for anomalous resonance coupling between spatio-temporally distant measurement partners in entangled quantum states of so-called Einstein-Podolsky-Rosen paradoxes, e.g. of spatio-temporal non-locality, quantum teleportation, etc. Applied to quantum brain dynamics, however, this view provides an anticipative resonance coupling model for aspects of cortical synchronization and recurrent visual action control. It is proposed to consider the registered activation patterns of neuronal loops in so-called synfire chains not as a result of retarded brain communication processes, but rather as surface effects of a system of standing waves generated in the depth of visual processing. According to this view, they arise from a counterbalance between the actual input's delayed bottom-up data streams and top-down recurrent information-processing of advanced anticipative signals in a Wheeler-Feynman-type absorber mode. In the framework of a "time-loop" model, findings about mirror neurons in the brain cortex are suggested to be at least partially associated with temporal rather than spatial mirror functions of visual processing, similar to phase conjugate adaptive resonance-coupling in nonlinear optics.

1. Introduction

Imagine you were the creator, and you had no space at all, only time and no more than just one single point to manipulate – what would you do?

In other words: How would you manage to transform time into space? And why should you do so in the first place? And what does this space-time transform have to do with the topic of this chapter, namely, the hypothesis of quantum absorber effects in the brain?

Below, we try to answer these questions consecutively, one by one. In fact, they were inspired by a 200-year-old challenge by the German transcendental philosopher Kant, who argued that the spatio-temporal appearance of our perceptual outlook on the world outside us is a purely subjective construction. If we took all material objects away from the world, leaving empty space and time, then, according to Kant, these two categories cannot be understood as basic properties of an objective outside reality. Rather they should be understood as internal sensory "qualia" of our "interface" to the world, hiding the "thing in itself" (*Ding an sich*), whose intrinsic properties, which exist behind the perceptual surface, cannot be perceived[a].

Kant's position seems to be supported by recent striking results in sub-atomic physics on the entanglement of distant elementary particles, including non-locality effects, etc., which have initially been obtained by Alain Aspect and his colleagues[1]. These instantaneous quantum correlations, also called the Einstein-Podolsky-Rosen (EPR)[b] paradox[2],

---

[a] Over the centuries, outstanding scientists and philosophers, following Karl Marx, have counter-argued against Kant's ideas from the rather pragmatic position that everyday practice shows – in spite of all philosophical reasoning – space and time plus the included objects to really behave as if they perceptually appear to us, namely, located outside of us, being thus part of an observer-independent reality. And as long as there were no practical reason for another view, runs the argument against Kant, there seems to be no need to complicate our lives by distinguishing the inner quality of the perceived reality from the objective one of the outside world. However, the experimental practice of quantum theory has confronted us, anew, with Kant's original formulation of a pure subjectivity of time and space as perceptual categories. In fact, experimental results, which will be referred to below, provide strong evidence against the above-cited pragmatic position from a hitherto unknown experimental scientific "practice" of quantum non-locality which is rather different from common sense.

[b] Strangely, this paradox has been named after the gedanken experiment of a man who did not even believe in the reality of this phenomenon. Einstein initially proposed it as a striking argument against the absurdity of the consequences of quantum mechanics. Thus, he intended to rule out "spooky" instantaneous action at distance, resulting in a "collapsing" wave function of a particle in a certain place, whose quantum state is correlated (entangled) to the state of another one being subject to a measurement in a different place at the same time.

challenge not only the relativist approach to the physics of time and space, but, moreover, the whole traditional realistic and "local" worldview of science. From special relativity, instantaneous – superluminal – "communication at a distance" between particles in quantum entangled states, due to the measurement of only one of the separated "twins", seems to imply one or both of the following conclusions: Either that spatial extension, the main property of space as such, does not exist, at least as an objective feature of the outside world, but only as a subjective one. Alternatively, or complementarily, this could mean – according to the Wheeler-Feynman absorber theory, which will be sketched below – that some kind of a "back in time" signaling occurs, contrary to the flow of causality. Thus, it appears that something might be wrong with the assumption of an isomorphism of space and time as we ordinarily perceive them in our everyday life, as corresponding to a physical reality "out there".

Speaking anachronistically, this supports Kant's idea that even "hard" physics may be merely understood as a kind of psychophysics, as far as it is concerned with no other reality than the "interfacial screen" of our subjective "quantum computer display": Even if we studied all the screen's spatio-temporal properties carefully, we would still not know anything about the computer's "non-local" reality itself[c].

Starting from Kant, we attempt to discuss in this chapter envisaged consequences of quantum non-locality for an internal psychophysics of visual time-and-space representation and cortical neuronal network synchronization. From our psychophysical view, i.e., as introspective observers, we have a "cockpit perspective" into a spatially extended, holistic "visual world" representation with simultaneous awareness of all perceived parts of the environment. Neuronal processing retransforms spatial extension into temporally extended codes, being registered, e.g., as an EEG pattern of successive phases or stages of cortical oscillations.

---

[c] To illustrate that, think of an "ego-shooter" computer game, containing not only the outside world's environment, but also your hands with the tool and your whole body within the frames of the screenplay. Studying, thus, the architecture of your body and brain, down to the nerve cells, atoms and electrons given with no other reality than the screen would always result in still "another brick in the wall" of the pixel arrangement on the display, but not in the processor's layout.

But how are these constituents of visual space-time psychophysically and neuro-cortically interconnected?

It will be shown below that this temporal extension of perceptual processing is in a fundamental conflict with the internal psychophysics of real-time comprehension of ongoing events. In fact, the problem of simultaneity and succession in vision, i.e. of the perceptual processing modes of spatial and temporal integration, is as yet unsolved, in spite of the many approaches addressing that issue[d]. In a comparatively unusual proposal, brought forward in the context of the so-called feature-binding problem[3], oscillatory synchronization of different brain loci has been applied to the psychophysics of spatio-temporal integration. However, the mechanism of this process has not been revealed. Is it possible, then, that the entanglement of distant particles may provide us with a "Kantian" approach to solving the communication problem between the spatially separate neuronal circuits?

Left with both space and time as pure subjectivity, as has been proposed by Kant, it would be hard to find a way out. From the perspective of quantum brain dynamics, the EPR effects rather show space to be the more obsolete of the two categories, raising doubts about its most evident feature, namely the reality of spatial extension. Time appears to be the more veridical category, and only some of its more unusual features seem to be in need of reform. Thus, the quantum non-locality results suggest space to be the best candidate for the subjective category, and so time is preserved as the more objective one. So what if we assigned – as a working hypothesis and in contrast to Kant – only one of the two categories, namely, space, to the subjective sphere, and left the other one, i.e., time, anchored in the objective domain?

In the following, a quantum non-local absorber model is proposed for handling psychophysical space with the hypothesis of a "time-loop" in

---

[d] The reader may object: The retina consists of a spatially distributed array of receptive fields, the activation patterns of which are transferred into cortical receptive fields by thick bundles of axons in the optical nerves in parallel, so what is the problem? The problem is that these "pixels" of information should be serially interconnected and this takes time. Alternatively, a "field-like" structure of the psychophysical correlate of visual consciousness must be *a priori* excluded now – at least, in the form of a classical electromagnetic field – since our head (and the whole body as well) can be introduced into a mega-tesla coil of fMRI scanners with no damage to ongoing cortical processing.

recurrent visual processing – inspired by the initial question presented above: What could we do, from the creators view, if we were left without space, having at our disposal only time and one single point?

2. A "Time-Loop" Model for Absorber-Type Cortical Representation of Visual Space

What we could do is to remember childhood: As small children, we used to meet on New Year's Eve to light sparklers. We would move them around repeatedly, very fast, to and fro, over one and the same area of our visual field, oscillating them fast enough to reach a speed above the critical flicker fusion frequency (CFF). If successful, we managed the transition from a sparkling moving point to an apparently stationary bright line. Extending this childhood experience into a new gedanken experiment, in which we keep moving the resulting one-dimensional line orthogonally to the initial direction up and down along a second dimension, again at a speed high enough to overcome flicker fusion, we obtain a bright two-dimensional square. If, in addition, we modulated brightness and color according to a certain highly synchronized repetitive code, we would obtain a stationary TV-picture. If, finally, we moved this arrangement along a third dimension, say, around us like a cocoon, we would obtain a three-dimensional spatial environment.

But stop: How could we manage to generate, by repetitive movement, a "to and fro" with our sparkler in the first place? Were we already "cheating" in the initial step of our reasoning? As a prerequisite for that movement, do we not need exactly that spatial extension we want to "create"?

No. We would not need it, if we were moving the spot forward and backward in time rather than to and fro in space. Thus, our linking hypothesis to the issue of how to generate space from time is, essentially, to include a "time loop" into the phase structure of visual processing in order to create a spatial dimension. Since this is conceived of as an oscillatory process, we obtain a kind of standing wave in time as the psychophysical basis of space perception. This idea seems to be applicable not only to the first dimension, but to the second and third one too, since the orthogonal oscillatory components can be conceived to constitute several levels of "sub-loop" structures in time, being nested in one another similar to the relation between the rows' frequency and the picture-frames' frequency in an ordinary TV screen, that is, one with a cathode ray tube rather than a liquid crystal display[e]. Time, thus, is conceived of as possessing an internally fractal substructure, which is the basis of our ability to perceive three-dimensional space.

Note that such a model proposes an elegant explanation of the origin of dimensions, which is not accounted for in most approaches, neither in psychophysics nor in "hard" physics. Ordinary Euclidean axiomatics as well as multidimensional models – e.g., the 11-dimensional structure of superstring theory etc. – cannot really explain what a dimension is, i.e. "what it is made of", where it comes from; so they take their extra dimensions from a nowhere like a *deus ex machina*. By contrast, the time-loop approach proposes dimensions of spatial extensions to be constituted by a standing wave pattern of nested oscillations in time, every additional sub-component of which can be generated in the form of another additional extension of a set of fractal dimensions.

Thus, the view presented here may imply the additional potential to propose more than simply an *ad hoc* speculation in order to get a handsome psychophysical model for a perceptual transformation of time into space. In fact, its basic approach may be proposed to correspond – as a psychophysical counterpart – to the core of the Wheeler-Feynman absorber theory[3] of quantum mechanics which has been proposed as an alternative explanation to the traditional Copenhagen interpretation. Following a proposal by John Archibald Wheeler, Richard Feynman developed this idea on the basis of his contributions to the theory of quantum electrodynamics (QED), as he tells us in an excerpt[5] from his Nobel Prize speech[f].

---

[e] Of course, this is a rather oversimplified picture, serving only for illustrative purposes. With respect to neuronal coding, as will be explained later on, the idea of a fractally organized hierarchy of spatial frequencies will be proposed as a more realistic idea.

[f] The cited excerpt of Feynman's Nobel Prize speech in Stockholm has been adapted from "Gehrtsen Physik", which is, so to speak, the bible of German university physics:
"One day my professor of physics, John Archibald Wheeler, called me by phone: 'Hey, Feynman, I know why all these electrons appear to be so identically exactly the same ones!' – 'Namely, why...?' – 'It's all the time just one and the same electron! There exists only a single one!' – 'And why do we think there are 10^{80} of them or so?' – 'Very

In fact, the absorber theory, which was developed later on by Cramer[6,7] into a transactional interpretation of quantum mechanics, attempted to explain the EPR paradox in a straightforward way. It rests on the idea that a "handshake" between the ordinarily transmitted signal and an anticipative effect deeply rooted in the quantum world is fed back from the "absorber" to the "emitter". Coupled in-between the "retarded" and "advanced" components in the collapsing wave function of the quantum event under measurement, this process appears to an external observer as a seemingly[g] time-reversed transaction. The mathematical roots of this may already be found in the standard quantum-mechanical procedure for calculating the so-called "collapse" of the wave function, by computing the square of the probability amplitude. This is done by multiplying a complex number (cos t + i sin t), representing a quantum state vector, by its conjugate (cos t – i sin t). Since the angle t represents time[h], a change of the sign of the imaginary part in the complex conjugate number mirrors the angle against the real axis, with the result of changing the angular rotation direction of time flow into its opposite. In a similar way, as discussed below, this argument has been adopted in the framework of phase conjugate optics by Marcer and Schempp[8,9,10] in order to give an account of quantum holography.

3. Oscillating Cycles of Synfire Chains as the Representational Basis for Spatio-Temporal Interaction in the Brain

Returning to the last of our initially posed questions, we can now try to sketch the connection of the "time-loop" model introduced above, which

---

easy: It runs very often zigzag to and fro along time. Anytime it comes from the past through the present we do think it is a new electron…' – '… and then it comes from the future, we think it's a positron?' – 'Bravo, Feynman, make a theory of that!'"

[g] In fact, the "back-in-time effect" is understood by the absorber model to result from an illusion from the point of view of the external observer since, in the relativist framework, the light beam from its inner aspect has zero time flow and no spatial extension. Thus, the absorber theory is an attempt to unify certain contradictions of the relativist and the quantum mechanical paradigms.

[h] More precisely, time is represented here by omega t, with omega as a constant 2Pi/T (with T as the time period measured, e.g. in msec) which transforms the temporal periodicity into the phase of an oscillatory "angular time" corresponding to circular revolutions in radiants.

is based on a quantum-absorber approach, with certain aspects particular to cortical synchronization and the representational psychophysics of space and time within the brain.

According to this hypothesis, representational simultaneity, as a brain process spread over spatially distant loci, is achieved by temporally bidirectional interactions of neuronal processing on a quantum scale. Absorber effects between the presumed "advanced" and "retarded" signal components are supposed to generate a standing waves pattern which might be speculatively assigned to the carrier process of an internal psychophysics of the representation of visual space. Evidence in support of this idea, as well as the one re-interpreted in the framework of the presented view, will be reported below (see sections 5 and 6 of this Chapter).

However, beyond these representational aspects, it is argued that the absorber model implies essential conclusions for the cortical synchronization process *per se*, i.e. on a neurological scale. Absorber effects appear, on the operational side, as anticipations of future states of the system. In the following, we present arguments for the significance of advanced signals as an anticipative feedback for the synchronization of spatially distant retarded processes. Our argumentation will be derived from an analogy between the synchronization of neuronal activity and that of chains of coupled oscillators on different scales (including a cosmological one).

This analogy holds true because the brain's neuronal feedback loops consist, in principle, of an assembly of hierarchically intertwined coupled oscillators. We assume, as a (meanwhile) well-established fact, that the functional architecture of recurrent visual processing is organized as a hierarchy of interdependent levels of loop structures, the so-called synfire chains[11,12]. Their timing is known to be coordinated by neuronal cycles of cascade-like activation lasting up to more than 500 msec of temporal delay. In spite of the considerable duration of the reverberations of these neuronal loops, their timing accuracy is synchronized with a precision of less than 1 msec, relative to the repetitive reoccurrence of activating one and the same neuron in the chain. Obviously, if an actual visual input is fed into such a synfire network, its processing results are retarded with respect to the changing environmental situation. Therefore, it is argued the visual signal's recurrent visual processing is critically dependent on advanced predictions which anticipate the next input. As will be shown below, these anticipations are needed in order to adjust the timing aspects of the coupled oscillators' synchronism in a way which stabilizes the equilibration of non-linear brain dynamics.

As has been shown for any kind of a coupled-oscillator system by the theory of non-linear dynamics, delayed feedback loops can trigger the emergence of chaotic attractors. One may get a "sensory" illustration of this type of regulation problem with the help of a coupled-oscillators system sold in gimmick shops, the so-called "chaos pendulum". On the one hand, such a device usually consists of a regularly oscillating primary pendulum (whose cosmological analogue might be, e.g., the sun-earth system). On the other hand, the construction involves a secondary sub-pendulum (comparable, e.g., to the earth-moon rotary system), whose oscillations are, in contrast, totally chaotic. Fortunately, however, chaotic pendulum dynamics has never been observed in the interactions of our sun, its planets and moons, although it can be observed in the case of brain dysfunction, as will be shown later.

Beyond the brain's and neuronal system's dynamics, the same architecture of oscillators in hierarchy – which is sensitive to chaos induction by retarded feedback – also underlies other bio-regulatory processes, such as the enzymatic metabolism, the control of the heart rate and blood circulation, as well as many of the organic and non-organic chemical and even pure mechanical oscillatory regulation processes. Helio-geocyclic processes might serve as an example of the latter, as do, last but not least, the loop-synchronization of the solar system's reverberations and those of the whole universe. Since these have been extensively studied, starting from Newton's work on the planets' revolutions around the sun up to the discovery of chaos in the three-planets problem by Poincaré, they may serve in the following as a fitting analogy for processes taking place in the brain.

4. Anticipatory Synchronization of Coupled Oscillators

According to a framework of anticipatory regulation as developed by Daniel Dubois[13], our universe – as well as our brain – has to cope with a synchronization problem of regulatory stability which has hitherto been ignored[i]. The reason for this is apparently hidden in Isaac Newton's differential equations describing the planets' periodical revolving movements. Since the underlying differentiation process dx/dt assumes a temporal delay approximately equal to zero, Newton's equations are implicitly based on the assumption of an infinite speed of gravitational signal transmission. However, in the light of Einstein's relativistic physics, gravitational distortions should have a finite transmission speed too, less than or equal to the velocity of light c[j]. If this were so, the gravitational signals would travel at least about one second from earth to moon and about eight minutes from sun to earth; on their way back, again, the same temporal delays should accumulate. Instead of Newton's infinitesimally small dt, therefore, a considerable amount of delta t has to be taken into account in the solar system's regulatory feedback loops, not to speak of the galactic and extra-galactic measures of spatial and temporal distance.

With respect to the topic of the present chapter, it should be admitted that this problem is neither an exotic issue concerning distant stars, nor is it just an issue of funny gimmick pendulum toys. Since, in fact, all bio-regulatory processes, e.g., human and animal metabolism, heart rhythm, blood circulation, and – last but not least – the brain's and the nervous system's exhibitory/inhibitory balance are based on an architecture of hierarchically coupled harmonic oscillators with non-linear dynamics, the delayed feedback problem and its consequence of the system slipping into chaos appears to be almost the same in all these cases.

Daniel Dubois gives an impressive example of the regulatory problem with delayed feedback loops in coupled harmonic oscillatory systems,

---

[i] The author of the present paper adopted selected issues from the work of Daniel Dubois in order to discuss corresponding topics of the view presented here. Being intentionally speculative about interrelations between brain synchronization, the theory of visual representations of spatial extension by a "time-loop" model is inspired by the absorber theory and, as will be seen later, the cosmological basis of the Kosyrev effect. The full responsibility for the presented view rests – here and below – with the chapter's author.

[j] In Einstein's general relativity, gravitation is due to the curvature of space, induced by the masses of the stars, planets, etc. However, distortions of the space curvature should be transmitted by a finite speed. Therefore, Dubois argues, space curvature should anticipatively adapt to gravity distortions in the way an absorber system does.

which is especially appropriate within the context of the visually guided behavioral control by oscillatory synchronized brain processes[14]. Considering the visuo-motor synchronization of a robot's arm trying to grasp an object in rotary motion, he argues that the cycle of registering and processing the input, generating and executing the motor commands etc., takes a certain amount of time. So it may easily turn out that this delay of minus delta t is in harmonic relation to the temporal cycle of the object's revolutions. In this case, the robot's attempt to grasp the object may become systematically erroneous or even chaotic according to the standard Feigenbaum[15] scenario. The same problem, however, should apply to the Sun, when it "attempts to grasp" the revolving Earth in its delayed position and, similarly, to Earth itself, with respect to the moon's delayed position. So why then, one might ask, is the moon stable in its revolutions around the Earth, just like the Earth revolves around the Sun and the Sun around the galaxy? And why are the brain's processing circuits of synfire chains, as a rule, also stable?

If delayed feedback is a chaos-inducing condition, Dubois proposes, vice versa, that the anticipated future state after the delay, fed forward into the system's regulation, leads to its "dechaotization". In fact, the robot's arm's control processes become smooth and synchronized after introducing an anticipative (in Dubois' nomenclature: incursive or hyper-incursive) term into the discretized feedback equations. In light of this, these equations were reformulated by Dubois, by using, instead of the state parameters of the delayed position, (i.e. the state at "– delta t"), the parameters of an anticipated future state, corresponding to the absolute value of the delay time, but with an inverse sign (i.e., the state at time "+ delta t"). Therefore, the anticipating feed-forward neutralizes the delayed feedback, and the resulting dt equals ± zero.

In a anticipation-based calculation, Dubois was able to show for the planet Mercury that this idea seems to be applicable to cosmology too[16]. Using the discretized Newtonian equations, retarded in correspondence with a negative delta t based on the velocity of light as the speed of signal transmission (a delay of about –1,8 minutes), his calculations resulted in big rosettes of "epicyclical" cycloid revolutions of the planet's orbit which, in fact, have never been observed. However, if one inserted into these equations a set of advanced parameters corresponding to exactly the same interval but with an inverted temporally positive sign (delta t of about +1.8 minutes ahead), this feedback from the system's own anticipated future state was able to produce a nearly ideal trajectory, which only marginally deviated from the Newtonian one, whose ultra-small rosettes were consistent with the recorded astronomic observations of many decades. As has been mathematically proven by Dubois, anticipation appears to be a vital requirement for avoiding planetary resonance-driven coupling to a chaotic attractor. The same principle should hold true for the analogous problem in brain synchronization.

5. Absorber-Type Effects in Cortical Synchronization – Evidence from the Spatio-Temporal Integration of Visual Processing

In order to obtain a synchronous visual flow in both left and right brain hemispheres, which is needed to account for the coherence, for instance, of binocular stereo vision, the temporally delayed signals of both eyes' left and right visual hemi-fields (which are processed in different hemispheres), should be integrated with anticipated versions of their complements in order to close the time gap.

In the light of this, it should be admitted that the functional structure of stereo vision displays several features relevant to the context of the present discussion. Anatomically, the visual nerve, which transfers the optic signals into the brain, is only partially crossed over with respect to the addressing of brain hemispheres via the so-called chiasma opticum. The crossing is true only for the nasal (inner) sides, with respect to the focus of fixation of the visual field of each eye, which is connected to the contra-lateral brain hemisphere. By contrast, the temporal (outer) side of the visual field of both eyes is fed into the same-sided ipsi-lateral hemisphere. In other words, the visual signal about the right half of the visual world is transmitted, from both eyes, first via the left lateral geniculate body (LGN) into the visual cortex of the left brain hemisphere. At the same time, vice versa, the left half of the visual world is transmitted via the right LGN into the right visual cortex. Only afterwards the information about left-hemispheric and right-hemispheric hemi-fields' vision is re-united via contra-lateral connections through the corpus callosum (CC). That means the contra-lateral visual signals from the opposite hemisphere are, in the absorber terminology, "retarded" with respect to one another. Does the visual system perform a correction for this delay, sending an "advanced" copy of its own input from each hemisphere to the contra-lateral counterpart, much in the same way as discussed above in the context of the coupled oscillator problem?

There is a perceptual phenomenon in the binocular vision: the so-called Pulfrich effect[23], which may be interpreted in accordance with this idea. Occluding one eye with a dark sun-glass while looking at a pendulum swinging in line with a parallel trajectory in front of the observer, the following phenomenon might be perceived: Apparently, the pendulum is seen to deviate from the fronto-parallel plane, describing a rotation in three-dimensional space. The effect is based on the induction of delayed retino-cortical transmission rates by the unilaterally darkened visual input. Inducing, therefore, a spatio-temporal shift between the left and the right eyes' stimulus patterns, this is interpreted by the stereo vision system as a depth parallax giving rise to the three-dimensional rotary movement. Every time the pendulum crosses the fixation focus, it is turning from the left to the right brain hemisphere, or vice versa. The result is an additional retarding moment which should be anticipatively counterbalanced by an advanced signal component.

So our brain's inter-hemispheric communication, which is required to account for the micro-timing of stereo vision – as exemplified by the Pulfrich phenomenon –, might be based on an absorber-type "tuning into the future" mechanism[k] which is comparable to the one proposed for the quantum-cosmological scale. In order to obtain a synchronous visual flow in both the left and right hemispheres of the brain, thus integrating the temporally delayed information of the left and right eyes' hemi-fields, which are initially processed in different hemispheres, anticipation is needed in order to close the time gap.

[k] There exists, of course, a textbook-standard explanation of the Pulfrich effect, stating that the visual integration in depth is calculated from memory and retrieval procedures. These are even additionally more retarded in time, since the underlying calculation must wait for the inter-hemispheric transfer's backward loop. However, the present explanation has the interesting feature of producing "just in time" output, the anticipated effects of which were advanced via an absorber-type mechanism.

Another neuro-ophthalmologic disorder should be considered in the present context, namely, the perceptual deficiencies occurring so-called amblyopia[17]. Associated, as a rule, with strabismus (squinting), amblyopia means, as a rule, not only the loss of stereo vision, but also a unilateral decay of visual acuity due to deficits in central visual function rather than peripheral optical damage. As has been shown for the amblyopic cat by the Singer research group at the Max Planck Institute Frankfurt, this is not due to losses in the visual brain channels *per se* but rather to deficits in their co-operative processing. In fact, it is the synchronisation between neuronal processing at different loci (coherence) which is preserved for the good eye and lost for the lazy eye[18,19]. This is interesting with respect to the model of holographic processing in the brain, as has been proposed by Karl Pribram[20] and others, since coherence of the neuronal processing is a prerequisite for the presumed "holochoric" brain states. Standard neuroscience has decisively argued against holography as a brain process, based on the finding that the nervous system cannot guarantee a linear transmission of the phase since it has been shown to perform in a rather non-linear way. In other words, it was argued that different time courses of retarded neuronal transfer functions in different neuronal circuits destroy the required phase locking of the neuronal signals.

In contrast, it is argued from the view of the absorber theory that a transaction between advanced neuronal states might compensate for the retarded ones, resulting in a temporal zero shift of neuronal processing. Analogously, the neuronal transmission circuits may, accordingly, be conceived of as an arrangement of standing waves in time and space. In fact, such a transactional interpretation is incorporated in the concept of quantum holography, as has been proposed by Marcer and Schempp[8,9,10], which is based on a neuronal mechanism of phase-conjugate adaptive resonance. With respect to amblyopia, this could imply – as a hypothesis – that the losses of neuronal processing coherence shown in the physiological recordings of the strabismic cat may be due to deficits in the advanced (absorber-type) component of visual processing, leading to a breakdown in the coherence of the reverberating time loops of neuronal synfire chains. Here, an absorber-type behaviour of the so-called mirror neurons might be proposed as a hypothesis: Assuming an anticipatory non-locality mechanism in the nervous system[21], these units may be conceived of as constituting a kind of "time-mirror" function[22]. This is currently, of course, a speculative assumption, as is the whole absorber model of quantum brain dynamics, which needs further investigation in order to be proven.

6. Is the Outcome of the Libet Experiment Due to a Possible Absorber Effect in Sensory-Motor Coordination?

If one accepts, as a hypothesis, the assumptions of the absorber theory, one might ask for further neurophysiological evidence in favour of a quantum-transactional analogue,[l] corresponding to an anticipatively timed architecture of sensory motor control and cortical synchronisation of the underlying brain processes. In such a context, the frequently discussed Libet[24] experiments might be considered. These experiments show an apparently paradoxical time difference between an action which is already under way, and the time when the decision for that action was made. Apparently the decision took place later[m] than the onset of the action. In the experimental setting, an observer is instructed to decide "by free will" within a certain interval the actual moment to press a button and to fix the time of decision making (and, in a control condition, also the time of executing the action) by observing a clock. Certain physiological data (EEG, EMG) are recorded, resulting in the paradox that these parameters give an earlier response (up to 500 msec) than the estimated time of the decision-making itself. This is especially curious in the context of subsequently collected data from similar experiments with so-called trans-cranial magnetic stimulation (TMS) of the motor cortex,

---

[l] At present, it is far from clear whether this analogy has to be conceived of only as a useful heuristics or whether there is something in common in the presumed brain's and the quantum mechanical "absorber" behaviour. However, the present chapter discusses this idea from an intentionally speculative view, looking for certain conclusions which could be drawn, giving the transactional explanation the preliminary credit of a working hypothesis.

[m] At least this might be concluded under acceptance of the chronometric conditions of these experiments, involving the procedure of judgment-based assignment of the subjective decision time – as is shown on the observed clockwork – to the reaction time, the time course of the EEG, EMG and other parameters.

evoking unwillingly forced finger-movement responses, the estimated timing of which did not show the typical temporal reversal[25,26]. The Libet experiments are very popular nowadays, since their interpretation in the framework of standard theory, according to some theorists, seems to give rise to the conclusion about the non-existence of free will. It is argued by certain leading neuroscientists, that the neuronal circuitry autonomously does the whole job for itself, whereas the mentally-timed feeling of "having made the decision made by myself" comes later, due to the compartmental architecture of the brain, rather than being induced in advance by an act of "free will" [n].

Alternatively, based on Cramer's transactional interpretation[6,7], Fred Alan Wolf has argued that the outcome of Libet's experiments is the result of a quantum mechanical absorber effect[27] which can be accounted for in the terms of the Wheeler-Feynman theory. Indeed, visual processing, on the one hand, takes up to 500 msec of time. However, the perceptual locomotion control requires a temporally much closer sensory-motor interfacing loop (preferably "just in time", i.e., as close as possible to dt = 0). Therefore, a straightforward idea has been proposed as a solution for this conflict. It has been argued that the interrelations between certain brain states associated with the sensory-motor processing control and the underlying external events are entangled in much the same way as the absorber and the emitter are entangled in a quantum-correlated transactional state. Thus, the retarded brain processes might serve as an internal absorber for an advanced (up to 500 msec) component of the sensory signals. Accordingly, the transactional results of sensory processing are – in analogy to non-local quantum absorber states – supposed to be dated back then, into their "entangled" temporal context of the ongoing external events.

---

[n] As has been frequently discussed, the whole debate is, of course, probably obsolete in several aspects, as far as it is applied to the topic of free will, as the experimental procedure is not able to really give a decisive answer to the controversy. One of the problems is that the conclusions drawn are anchored in the chronometric arguments of a classical physics and psychophysics of the 19th century which has been made obsolete by the quantum-mechanical revolution, as is also argued in the present chapter.

So the question arises: Are quantum absorber effects, such as the ones sketched out above, likely to occur within our biological tissue, "inside" the neuronal system? Nowadays, such effects are preferably searched for at the "microscopic" scale of sub-cellular nano-structures. This has been proposed by several authors associated with the ideas of, for example, micro black holes in the neuronal tissue[28], quantum computations within the microtubuli skeleton[29] interpreted as a non-classic wave guiding system, the genome apparatus as a biophotonic quantum holographic non-local processing system[30], etc.

Complementary to the rethinking of brain processes at the sub-cellular nano-scale, the present chapter additionally proposes to take the risk of having a look at some preliminary evidence about rather long-range non-locality effects. These can be found, if the reported evidence is valid, even on a light-year-wide cosmological scale. The cosmological scale is argued to be of relevance not only in the light of the regulatory problems of cortical synchronization which have been discussed above (see sections 3 and 4 of this chapter). Beyond that, it is also relevant with respect to the issue of visual consciousness in view of the presumed non-local aspects of brain processing ("cosmic consciousness"). If you look at the stars, consider, for example, an astrophysical experiment on the observer-dependent alternating registration of wave vs. particle aspects of transmitted radiation, the source of which is hidden light years away behind a gravitational lens (a so-called Einstein-cross), as a cosmological variant of the EPR paradox. Is this an observation-induced absorber effect, i.e. a transaction to an event, the origin of which is dated "back in time" some millions of light years ago?

Thus, an intentionally speculative look "outside" the nervous system – namely, into outer space – is appended to the context of the present chapter as a "side step" in order to provide us with an example of a rather "macroscopic" absorber-type quantum event, which is hidden behind a rather striking result of Soviet astronomy: the so-called Kosyrev effect.

7. The Kosyrev Effect

The Andromeda nebula is the centre of our neighbouring galaxy, which is located at a distance of about 2.5 million light years away. Its spiral arms cannot be seen without a radio telescope. However, in projective size, they form the apparently largest stellar object in the firmament. Altogether, it appears to be about six lunar diameters wide and cannot be mistaken for any other stellar object. At the end of the 1970s, the Soviet astronomers Nicolai Kosyrev and Victor Nasonov performed a series of observations, mainly of the Andromeda galaxy, using the facilities of a great reflector telescope[31]. However, instead of the usual ocular lens system, they inserted into the focus of the reflector a newly developed special CCD-type sensor. This detector consisted of a matrix of ultra sensitive piezo-electric crystals. Convinced that their device was highly sensitive to ultra-weak gravitational fluctuations[o], Kosyrev and Nasonov used it to scan a spatial profile of the gravitational signals of stellar objects, notably those of the Andromeda galaxy.

To their astonishment, the two astronomers reported registering the spiral profile of the galaxy not only at Andromeda's observed proper location, i.e., the place of the source of electromagnetic radiation, which in fact was sent out 2.5 million years ago. Additionally, they reported observing the same profile also at two distinct subsequent locations. One of them was the position which the Andromeda galaxy, according to calculations of its displacement's direction and velocity parameters, should have reached nowadays, "here and now", so to speak, at it's apparently instantaneously transmitted actual position. This was a quite unexpected observation, since it clearly seemed to be in conflict with the basics of Einstein's relativistic cosmology. According to Einstein, the velocity of light sets an upper limit to any kind of signal or matter transmission in space. However, even more astonishingly, Kosyrev and Nasonov registered one more of Andromeda's profiles at a third location, which the galaxy will have reached, based on today's predictions, in a far away future. This third piezo-detected profile was spatially located just the same 2.5 million light years ahead as the actually observable conventional radiation is temporally delayed. Thus, as compared to the ordinarily observable position, the future one appears to be in a "space-

---

[o] In Kosyrev's own theoretical thinking, the details of which are beyond the scope of this chapter, although, maybe, of interest in the context of the present book, gravitation is intimately linked with hypothetical physical properties of time, such as its "temporal density" aspect.

time mirrored" state, since it corresponded reciprocally – i.e., with an "opposite sign" – to the temporally delayed spatial position of the actually observable light signal. Apparently, the Kosyrev effect seemed not to depend on conventional electromagnetic radiation, since it persisted even after shielding the telescope with a metallic cover.

All in all, the Kosyrev effect is one of the strangest stellar phenomena which have been ever reported and is far from being accepted by the majority of astrophysicists. The reported evidence is nowadays far from established[p] and only a few attempts at replication have been made by other observers[32,33]. Is this phenomenon only a marginal curiosity in the history of cosmological observation? Or does it provide, alternatively, a major hint at the way in which the dynamic revolving equilibrium states of our universe are synchronized in a similar way, as is argued in the present chapter[q] for the dynamic equilibrium states of oscillation-based neuronal communication processes in the brain too? And does it provide, last but not least, an argument for an alternative hypothesis about the regulation of resonant tuning-based anticipatory visually-guided action control? Is the neuronal "time-loop" architecture in visual space representation, which is proposed as a hypothesis here, of the same origin as the astrophysical basis of the Kosyrev effect and the absorber-type quantum entanglement effects on the nanoscale?

Since the reported evidence of the Kosyrev effect is not sufficient for decisive conclusions at present, its significance as an argument in the context of the discussed topics has to be regarded as a speculation, to be verified or falsified by future investigations. These are continued, for example, on the lines of the work of Korotaev et al.[32,33] who established

[p] The relevance of the Kosyrev effect in the context of visuo-motor action control and neuronal communication and synchronisation processes depends, among other factors, on the verification or falsification of the underlying astronomical findings. Accordingly, the conclusion drawn is at present a rather preliminary one, which might be better understood as a heuristics to be discussed, rather than providing the basis of a comprehensive theory.

[q] In the context of this chapter, we discuss only the consequences of Kosyrev's findings in a way in which they appear "tailored" to fit as complementarily as possible to the standard relativistic and/or quantum cosmology. Kosyrev's own theoretical ideas as an issue of discussion about the physical properties of time – e.g., as a source of "anti-entropic" radiation, which is in intimate relationship with gravitational forces of different temporal density coming out of stellar objects due to their angular momentum of rotation – are beyond the scope of this paper.

anticipatory correlations between the dissipative components of helio- and geomagnetic cycles. As a hypothesis, however, the possible role of a common basic anticipatory mechanism in the framework of a quantum-transactional absorber-type model of cortical synchronisation and space-time representation might be discussed, whose effects can be observed at different (macroscopic, mesoscopic, microscopic) levels of the universe. This anticipatory mechanism might be argued to provide a hyper-incursive (in Dubois'[13,14] terms) solution, e.g., for tracking a circularly moving object as can be shown for the visuo-motor control of the above cited robot's arm[14] and, maybe, for the synchronization of the neuronal system's synfire loops of space-time representational control[r].

8. Concluding Remarks

Taken together as a hypothesis, the reported findings seem to support the idea of the existence of macroscopic, mesoscopic and microscopic "time loop" absorber effects. It is suggested that these provide us with a heuristics for understanding the relation between resonant transactional control of cortical synchronization, the internal psychophysics of visual space-time representation and spatio-temporal non-locality effects on a neurophysiological scale in analogy to those on a quantum mechanical scale as well as, possibly, even on a cosmological scale.

References

1. A. Aspect & P. Grangier, Experiments on Einstein-Podolsky-Rosen-type correlations with pairs of visible photons. In: Penrose, R. & Isham, C.J. (eds.) *Quantum concepts in space and time*. Oxford: Oxford University Press (1986).

---

[r] Note that the Kosyrev effect shows exactly the properties required for transforming succession into simultaneity: Assuming the velocity of light c as a transmission speed, with a convergence on a temporal interpolation of dt = ±0, the anticipatory spin off into the future (+ delta t), according to the soviet astrophysicists' calculations, appears to be advanced by exactly the same amount (2.5 million light years) ahead, as the ordinary electromagnetic signal (– delta t) is retarded. Applied to the nervous system, however, the anticipation/retardation parameters should be adjusted to about ±500 msec and the propagation velocity to approximately the range of the speed of sound.

2. A. Einstein, B. Podolsky, N. Rosen, Can Quantum-Mechanical Description of Physical Reality be Considered Complete? *Phys. Rev.*, 47, 777-780 (1935).
3. C. von der Malsburg & W. Schneider, A neural cocktail-party processor. *Biol Cybernetics*, 54, 29-40 (1986).
4. J.A. Wheeler & R.P. Feynman, Interaction with the absorber as the mechanism of radiation. *Rev. Mod. Phys.*, 17, 157-181 (1945).
5. D. Meschede, (ed.) *Gehrtsen Physik*, 21st edition, p. 892 (2002).
6. J.G. Cramer, The transactional interpretation of quantum physics. *Reviews of Modern Physics*, 58, 647-688 (1986).
7. J.G. Cramer, The transactional interpretation of quantum mechanics. In D.M. Dubois (ed.) Proceedings of the Fourth International Conference on Computing Anticipatory Systems, *AIP Conference Proceedings*, 573, 61-76 (2001).
8. P. Marcer & W. Schempp, Model of the Neuron working by Quantum Holography, *Informatica*, 21, 5 19-534 (1997).
9. P. Marcer & W. Schempp, The Model of the Prokaryote cell as an Anticipatory System Working by Quantum Holography, Proceedings of CASYS 97, *International Journal of Computing Anticipatory Systems*, 2, 307-315 (1997).
10. P. Marcer & W. Schempp, The brain as a conscious system, *International Journal of General Systems*, 27, 1/3, 231-248 (1998).
11. M. Abeles, *Corticonics*. Cambridge, UK: Cambridge Univ. Press (1991).
12. M. Abeles, H. Bergman, E. Margalit, E. Vaadia, Spatiotemporal firing patterns in the frontal cortex of behaving monkeys. *J. Neurophysiol.*, 70, 1629–38 (1993).
13. D.M. Dubois & G. Nibart, Towards a Computational Derivation of a Dual Relativity with Forward-Backward Space-Time Shifts, *International Journal of Computing Anticipatory Systems*, 5, 25-37 (2000).
14. D.M. Dubois, Incursive Anticipatory Control of a Chaotic Robot Arm, Computing Anticipatory Systems: CASYS - First International Conference, edited by D. M. Dubois. Published by The American Institute of Physics, *AIP Conference Proceedings*, 437, 406-417 (1998).
15. M.J. Feigenbaum, Quantitative universality for a class of nonlinear Transformations. *J. Statstical. Phys.*, 19, 25-52 (1978).
16. D.M. Dubois, Anticipative effect in relativistic physical systems, exemplified by the perihelion of the Mercury planet. Vigier IV Symposium, September 15th to September 19th (2004).
17. U. Kämpf, F. Muchamedjarow & T. Seiler, Unterstützende Amblyopiebehandlung durch Computerspiele mit Hintergrundstimulation: Eine placebokontrollierte Studie. *Klinische Monatsblätter für Augenheilkunde*, 218, 243 –250 (2001).
18. P. König, A.K. Engel, S. Löwel & W. Singer, Squint affects the synchronzation of oscillatory responses in cat visual cortex. *Eur. J. Neurosci.*, 5, 501-508 (1993).

19. P.R. Roelfsema, P. König, A.K. Engel, R. Sireteanu & W. Singer, Reduced synchronization in the visual cortex of cats with strabismic amblyopia. *Eur. J. Neurosci.*, 6, 1645-1655 (1994).
20. K.H. Pribram, *Brain and Perception: Holonomy and Structure in Figural Processing*, Lawrence Eribaum Associates, New Jersey (1991).
21. D.M. Dubois, Non-locality Property of Neural Systems Based on Incursive Discrete Parabolic Equation. *International Journal of Computing Anticipatory Systems*, 7, 233-244 (2000).
22. U. Kämpf, W. Mascolus, F. Muchamedjarow, A. Shamshinova & T. Kaschenko, Coherence induction via phase conjugate adaptive resonance coupling: computer generated sinusoidal grating patterns as a background stimulus in amblyopia treatment. *Journal of Psychophysiology*, 20 (2), 149 (2006).
23. C. Pulfrich, *Stereoskopisches Sehen und Messen*. Jena, Fischer Verlag (1911).
24. B. Libet, E.W. Wright, B. Feinstein, & D.K. Pearl, Subjective referral of the timing for a conscious sensory experience. *Brain*, 102, 193-224 (1979).
25. P. Haggard & E. Magno, Localising awareness of action with transcranial magnetic stimulation. *Exp. Brain. Res.*, 127, 102–107 (1999).
26. P. Haggard, S. Clark, & J. Kalogeras, Voluntary action and conscious awareness. *Nature Neuroscience*, 5, 382-385 (2002).
27. F.A. Wolf, *The timing of conscious experience: A causality-violating two-valued transactional interpretation of subjective antedating and spatial-temporal projection.* (1989).
28. M. Pitkänen, Biosystems as macroscopic quantum systems. In: Dubois, D.M. (ed.) CASYS – *International Journal of Computing Anticipatory Systems*, Liege: CHAOS ASBL, 10, 368-383 (2001).
29. S. Hameroff, Funda-mentality: Is the conscious mind subtly linked to a basic level of the universe? *Trends in Cognitive Sciences*, 2(4), 119-127 (1998).
30. P.P. Gariaev, B.I. Birshtein, A.I. Iarochenko, P.J. Marcer, G.G. Tertishny, K.A. Leonova & U. Kaempf, The DNA-wave Biocomputer. In: Dubois, D.M. (ed.) CASYS – *International Journal of Computing Anticipatory Systems*, Liege: CHAOS ASBL, Vol. 10, 290-310 (2001).
31. N.A. Kosyrev & V.V. Nasonov, Properties of time found by astronomical observations. Academy of Sciences of the USSR, *Problems of cosmological research*, 9 (1980).
32. S.M. Korotaev, V.O. Serdyuk & M.O. Sorokin, Effect of macroscopic nonlocality on geomagnetic and solar-ionospheric processes. *Geomagn. Aeron.*, 40, 323-330 (2000).
33. S.M. Korotaev, V.O. Serdyuk, V.I. Nalivayko, A.V. Novysh, S.P. Gaidash & Yu.V. Gorokhov, Experimental estimation of macroscopic nonlocality effect in solar and geomagnetic activity. *Phys. Wave Phenomena*, 11 (1), 46-54 (2003).

CHAPTER 28

TIME AND TIMING IN OUR BODY AND LIFE

Otto van Nieuwenhuijze

Independent Research Scientist, MSc, MD
17-II Gerard Doustraat, Amsterdam, NL-1072 VJ
E-mail: research@provisions.nl

The cells in our body all synchronise their action 1) on the basis of their common origin and 2) in correspondence with changes in our context; of which we form an integral part. The Time Fractal helps us to better understand this interaction, and helps understand Time Base diseases.

1. Introduction

One day I heard Susie Vrobel[1] talk about the Time Fractal (2005). This chapter describes its importance for understanding Life. For me, the Time Fractal offers a direct handle for my work on the 4D nature of time[2]. This is of direct importance to understanding not just how something can exist, but also how it comes into being and ceases to exist.

The following story describes how the Time Fractal can be used to understand the way the cells in our body synchronise what they do in their context. The Time Fractal also helps us to understand how the body organs synchronise their function, in our changing context. It thereby offers the basis for a model of the explicit formulation and understanding of Health. Not just the health of the *cells* in our body, the *organs*, or the health of our whole *body*, but also the health of *society*. Humanity as a whole functions in the same manner as our body. Just as our body is composed of cells, so is humanity composed of individual humans. The Time Fractal is of use also in enhancing social interactions.

(This text makes use of layout-coding. Some of the sentences in the text are the equivalent to mathematical operators. **Bold** is used to group matrix components; *italic* is used for indicating variables in a set and underlining as vectors to link to other operators (paragraphs) elsewhere. They are used for emphasis also.)

This chapter presents some fundamental ideas about the origin of the universe. This is needed to point out the relationship of time as part of the 4D complex it forms part of: **Consciousness, Energy, *Time* and Space**. The story then continues with a description of the way the Time Fractal helps us to understand the complex organisation of our body. The body anatomy is based not on energy and matter, but on the synchronisation of living cells. **Anatomy, physiology, neurocrine regulation and information integration** form one integral unity. The third part of the story makes explicit how all the cells of our body are part of one integral unfolding dynamic. **Cells, organs, our body and humanity are one integral system**. This is the basis of health. This extends the interpretation and, therefore, the meaning, of the Time Fractal in our body. A *Virtual* Time component needs to be included (This is also the basis for understanding the meridian and chackra system in our body). Finally the story takes us to a pragmatic interpretation of the meaning of the Time Fractal in our body: the existence of Time Base Diseases. Changes in cell rhythms, in organ rhythms, and in circadian body rhythms, and the time of life of a person are therein all interrelated.

The conclusion will be simple: The Time Fractal[1], of which I heard Susie Vrobel speak on that day, helps us understand the complex 4D time processes in our body in a very straightforward manner. It thereby offers a handle for resolving a complex group of diseases: diseases of disturbed Time Base. This can be succinctly described in a simple model: the relationship in the time dynamics between one cell and another. In realising that, in our body, all cells therein have a common basis, we also have a direct understanding of the nature of health in our body. Susie Vrobel's Time Fractal thereby helps bring our understanding of medicine a vital step forward. We can understand the basis of the maintenance of integrity of our living body, in a changing context.

2. Space, Time, Energy & Consciousness

Time does not have a reality unto itself. It is part of a 4D Unity: $STEC^3$ (Space-Time–Energy-Consciousness), in which our involvement matters.

Time is measured by cycles of repetition: clocks, planetary revolutions, atomic cycles. This means that time is connected to the continuity of cycles. The perception of Time depends upon the presence of an observer. Without an observer and repetition (continuity), we do not experience time as we customarily know it. There are other forms of time, such as of a system in *creation*, or in system *decay*. We know this as life and death. In these cases, the measure of time is chronically changing. It means we cannot gauge time without defining a **time base**. Figure 1 shows a schematic representation of a time object: it has 1) a Time Base, with 2) a defined variational range, which 3) has a Critical Time border, where 4) it is separated from other Time Objects by a Time Leap/Gap.

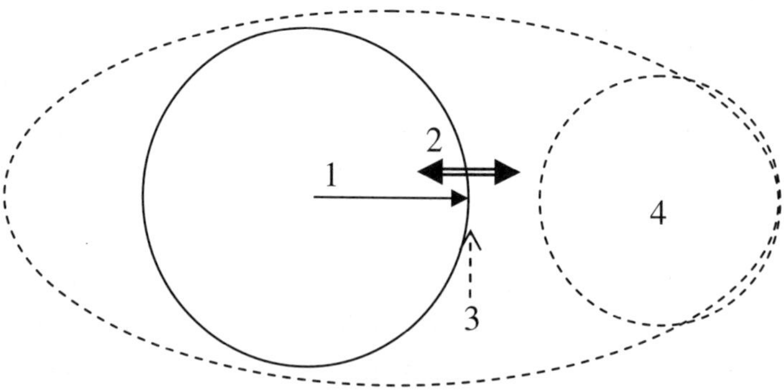

Fig. 1. The Definition of an Object in Time.
1) Time Base
2) Variational Vital Range
3) Critical Timing Leap - Gap phase
4) Environment Time Base

In the following story, we need to understand this in a general manner: *the time of our life, and our life time, is not regulated by the dynamic of clocks*. It is based on the *living* cell dynamics of intelligent cells. In understanding the measure of time in our body, we must take into account how our body was born, and can die.

This turns out to connect the time *in* our body to the materials *of* our body, but also to the exchange of processes and their rhythms by which *we are connected into our context.*

We need to understand the *time* in our body, with respect to the *energy* on which the time dynamics is based. This, again, we need to relate to the *phase space* synchronisation of our body *as part of our context.* It means that time is but one aspect of the description. We need to understand time in its origins of *energy transformation*, as part of the *information integration* as our body is always part of our context. Therefore, Time as such does not exist on its own. It is always related to the process dynamics of the interaction of the Living participant Observer. This is seen in its most explicit form by looking at the role of Time in our living body: in the synchronisation of living cells. Space-Time–Energy-Consciousness is an integral 4D Unit(y), STEC^3. This relationship needs to be made explicit, because each of the components reflects on our interpretation/experience of the Time domain.

There is no easy version of the next part of the story. We have been taught to think about the universe the wrong way around. The current customary story is that we live in a *material* universe. Science refers to *matter* in describing what is held to be real. That is *not* correct. **Matter** is composed of **Molecules**, composed from **Atoms**, which are composed by subatomic fields of **Vibration**. *Matter is immaterial.* There is no physical basis for what we know. All is *Phasical*: *Phase Space* is the only basis for the reality that we are part of. As Quantum theory put it: Reality is a Realisation. Our involvement determines our perception ("Collapse of the vector of State"). It means that *we cannot regard reality as if material.* Matter is but one of the forms of manifestation of phase in formation: *information.*

We were told every*thing* backwards; we need to turn it all around.

- Matter is made of Molecules, made of Atoms, made of Information (Phase space).

- Structure is thus the result of Process, which is a result of Transformation, which is a form of Integration.

Space is form of Time, which is a form of Energy, which is a form of Information.

This can be condensed into the following understanding:

- Consciousness is the basis of Energy, which is the basis for Time, which is the basis of Space.

This means that *all* our descriptions and formulations in science and all that is based on science need to be rewritten. In fact, it means that science has now come to the place where it originally started, in opposing the approach of religion. Science has now come to conclude that *information* (spirit) is the basis of *manifestation*. Cosmology is the contemporary version of the ongoing story of creation: out of the Information (*phase*) space, there was a shock wave (*energy*) which created Cosmic Gas (ionised matter), which precipitated into recursive patterns (cyclicity, a.k.a. Time) which we now know as the Stellar Process; which then again led to a next phase of condensation of Planet formation. Space, Time and Energy are different modes of aggregation of Phase.

- Consciousness (Phase Space information), Energy, Time and Space are all interrelated aspects of this process of manifestation. It is a dynamic of Phase (re)organisation.

As is the case for the Four Fundamental Forces of Physics: they are merely different aspects of the same essence. The difference in perspective (involvement) is the basis for the differences in manifestation (realisation). As has been mentioned, there is no simple way to describe it: we are part of the universal process of creation. And our involvement matters. It also means that we can never describe the actual ongoing process, because of limitations of language and communication. Due to the 4D nature of the interwovenness of consciousness, energy, time and space, we need to describe our own involvement (realisation) in any (reality) description[4]. As this is not currently customarily the case ... there is no simple way to describe this. However: we do *experience* it; we are part of it. Thus, instead, we can focus our description on our *experience* of creation.

The next section describes how we experienced it in the way our body came into existence; the 3rd section thereafter describes how we experience it in the way we interact with our body. The final 4th section describes what goes wrong when we make mistakes in these interactions.

The conclusion of this 1st section is important: in what follows whatever is described in terms of Time (as the Time Domain), is actually a formulation of a particular resonant mode of *energy* transformation, which is but one of the modes of expression in the emergence of *information* (change of coherence of phase). In other words:

- Time is a form of *Energy*, which in itself is but a mode of *Information* in formation.

3. Body Time – Cell Synchronisation

The metric of time in our body is based on our cell dynamics. *Everything* in our body is based on cell dynamics. The following describes this as the interrelationship between body, mind, soul and spirit:

Body: The physical body *structure* which we are told 'is' our body (Anatomy) is created and secreted by living body cells. (It is a consequence, not a cause.)
Mind: The *process* of our body cell dynamics composes what we know as our mental process: the exchange of materials and information between our living cells, and their context (Physiology).
Soul: These process dynamics are variable but stable. The internal cell exchange and dynamics is always in balance with the context. This is known as the Soul body (Regulatory System); that which calibrates and maintains the cell process dynamic in relationship to our context.
Spirit: The most important level, however, is that of preservation of information. The dynamics of our body *must* match the dynamics of the

context (Information Integration System). If not, we become sick, and die.

Body, Mind, Soul and Spirit are all interconnected. Yet they are all merely aspects of the functioning of the cells in our body.

What we know as our Anatomy, Physiology, Regulatory system and Intelligence are simply the various system aspects of the functioning of our body: the **Structural** level, the level of **Time** processes, the level of **Energy** transformation, and the level of **Information** integration.

Although this is commonly understood – also, to some extent, in medical science - it is not realised that the integrity of this system is *not* based on the structure of the body, but on the *integrity of information integration* within, and between, living cells.

Our body functioning is based on the synchronisation between cells. Atomic resonance frequencies are phase-locked into molecular complexes, operating as vibrating cellular units in forming the whole body dynamics. This dynamic coherence operates not just at the level of the momentary body functions, but also in the way the body developed out of our ancestors and previous life forms. We need to understand *how* our body was formed.

There are two parts to that story, and both are important for the understanding of time in our body:

- The first part of the story is the long-term perspective: it is about the way our body was created as part of the manifestation of the cosmos (see above) and the emergence of all life forms.
- The second part of the story involves the formation of your specific unique physical body.

The following summarises some aspects of the history of your body to help appreciate the relevance and role of the Time Fractal to describe this.

Please bear in mind that we always need to understand the time and timing in the living body cell, *in its context*. The development of the human species is part of the development of *all* life forms, the planet and the cosmos.

Emergence of molecules of Life

We can see that the time processes of our body can be traced back from DNA from proteins from amino acids that formed out of primordial gasses on primordial Earth. The process dynamics of Earth is thus in-built in our body.

DNA memory record

Bruce Lipton[5] described how the experience of every person is recorded into the DNA of our cells. Covering our DNA, there is a layer of Mantle protein which transcribes (onto your DNA) the dynamics of the Linking protein which co-ordinates the activities of your *sensor* proteins and *portal* proteins on your cell membranes. DNA is thus an archive record of experience of life on Earth, and (in rare cases) a back-up record for restoring cell function.

Symbiosis of life forms

Lynn Margulis[6] has described how we can understand the origin of our human Eukaryote cell from the symbiosis of four different types of microbes.

Cell Fusion

At each step of your history, you had two ancestors who combined egg and sperm. Millions of people, ancestors, are involved in (y)our creation.

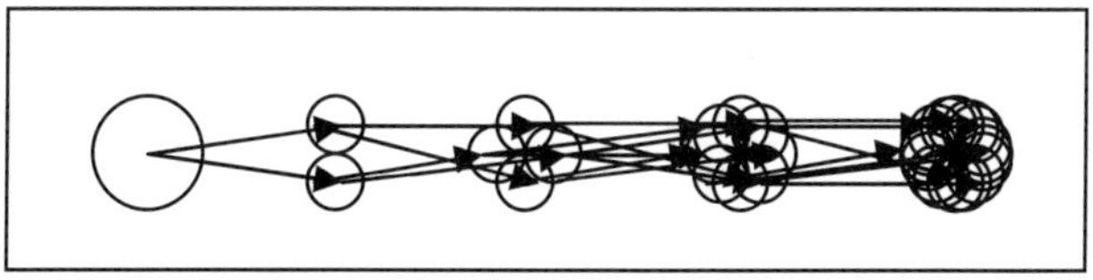

Fig. 2. The Time Fractal as Trace of Zygote Cell Division.
(In inverse this can be used to describe the process of ancestral cell fusion.)

In each stage, the memory of the previous stage still exists as a memory (an internal relation). This makes the mechanism of memory explicit. Also, it shows that Real and Imaginary aspects both play a role: the Time Fractal is a *Complex* Vector Matrix.

Zygote cell division

Your whole body is based on a sequence of divisions of your first body cell, the Zygote. Every body cell, every organ, and your whole body unfolded from that initial cell. It is the pattern of cell division, from the first cell onward. This specifies the position and characteristics of every body cell[7].

The Time Fractal in our Body

Every cell in our body can be traced back from its unfoldment of the Zygote. The origin of the Zygote can be traced back from the fusion of cells; from your ancestors and previous life forms.

The Time Fractal offers a means to – conceptually – graphically represent this. It means that the Time Fractal serves a tool for research on the integrity of your cell in its context, throughout the various phases of its formation and creation.

Technically the Time Fractal is a scalar, unfolding to a complex domain vector, thence unfolding again as a complex space matrix (in 4D Dual Minkovsky space[3]; 4 components in Space, and 4 in Time. Therein 3 Time components are Imaginary, (as is 1 Space component.)

Time Architecture

This brings out the full richness and meaning of the concept of the Time Fractal[1]. The concept of **the Time Fractal makes it possible to** trace the origin of *every* cell of our body back to its emergence from the Zygote, *and* understand how the Zygote itself was formed out of earlier cell fusions.

In this approach, we need to understand the process of cellular division as a complexed **branched chain reaction**. (This pertains to the development of a wide range of interrelated body cells, instead of the more usual limited interpretation as a reaction cascade of chemical products.)

Bio-Holograms

The Time Fractal helps understand the properties of unique living cells. All cells have their own specific Time Base. From their development, from the Zygote, we can infer that all cell Time Bases are interrelated. The body as a whole is a Symphony of cell dynamics. It is a coherent phase organisation (Bio-Hologram[8]): all vibrational patterns of all cells are interrelated. The body as a whole is a result (and manifestation) of their common vibration[9].

The Time Fractal offers a basic method to understand the complexity of the Bio-Hologram in a simpler manner: as the unfolding of a Time Fractal[7].

Cell Synchronisation – the Gap Phase

This we can see in the functioning of the individual cells: in the cell cycle, we see the Gap Phase. This is a pause in the time cycle of cell division. It is a requirement for cell synchronisation in a multi-cell complex (organs, the body). Therein all cell dynamics are interrelated. The first cell, the *Zygote*, has its own particular time base. In dividing, the Zygote also divides its Time Base amongst more cells. This sets the Time Architecture of the body as a whole; each stage of division maps the temporal (cell) dynamics onto the spatial of the body.

Time Fields in our Body

The Time Base of each ensuing cell differs from the cells it derived from. This defines the Time Fractal of cellular division. In the body as a whole, **all cells are time-base connected**. The Gap Phase in cell division serves for cell synchronisation for the continuation of the cycle. In this moment, cell dynamics await a synchronisation pulse from their context. Oriental Medicine makes use of body cell synchronisation (the Meridian System) for the preservation of the integrity of the dynamics of the body.

Body Materials – liquid crystals

That the Time Fractal is important in our body is seen also in the plasticity of the body materials. Nearly all are Liquid Crystals. Although Western medicine has taken the perspective of regarding the *physical* structure as the basis for describing the body, from the body itself, it can be seen that all body materials are secreted by living cells, and are the *consequence* of living cell dynamics. The interchange between material phases in our body demonstrates that the most important level of our body functioning is not matter but *information integration.* (See above).

Synchronisation for Synergy for Syngnosis
All body materials are responsive to our processing of information. The body time process dynamics are relevant: they determine the speed by which different signals can be recorded and processed. This is the basis of the *memory* capacity of our living body. The body time processes thus must be understood as subsidiary to the more important level of energy transformation in the body. This, however, merely serves to substantiate the information integration between the living cells in our body, and our context (*Syngnosis*).

Time, and timing in our body, is but a means for the preservation of bodily health.

4. Health & Timing

The unfolding of our whole body from one single cell is possible only because they remain *one functional unit.* The body is the unfolded Zygote. The Time Fractal makes this explicit. This makes the Time Fractal a powerful tool for research. It makes it possible to obtain an integral understanding of Health[7].

The multiplicity of the complex temporal body dynamics can all be traced back to *the unfolding from one single cell.* The body is a temporal hologram (phase space organisation) constructed from the unfolding of *one time fractal.* (The double-shafted arrows in Figure 3.)

The interplay between all body *cells* is an explication of the interplay that pre-existed in the Zygote. In this connotation, it is clear that this also determines the interplay between all our body *organs.* Rather than regarding our organs as separate - as is done by western somatic medicine - it can be seen that they all form aspects of one ongoing temporal cycle - as described by oriental medicine. Every organ therein has a place, and a musical note (time base frequency), in an ongoing transformative cycle in which the exchange between the body and the context is maintained.

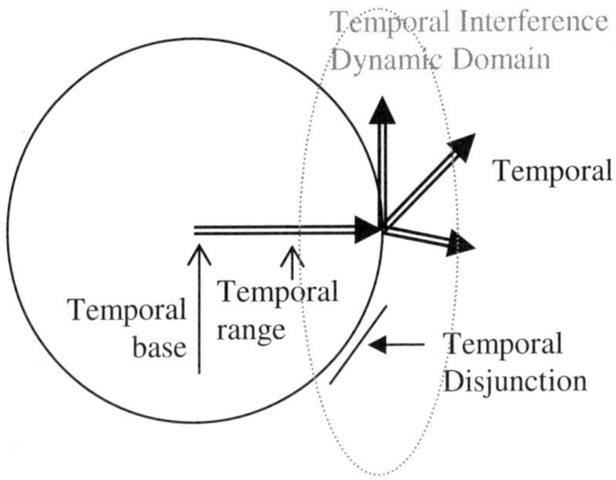

Fig. 3. The system definition is physical, as a result of an underlying temporal separation. In the physical domain, every boundary is a separator and connector. In the temporal domain, every boundary is a filter for linking/separating internal/external processes of/for the system.

The interplay between *people* also can be described on the same basis. All our interactions take place on the basis of (temporal) processes in our body. Synchronisation (synergy and *syngnosis*) between people is thus based on the same dynamics as that of the unfoldment of the Zygote.

This adds a different level of understanding of the Time Fractal in the body: there are levels in which the unfolding zygote interacts with its origin in the form of time cycle harmonics. The organ cycle, and the *metabolomes* (molecular exchanges between related cell groups) are examples. The notation of the Time Fractal makes it possible to discern time loops and time cycles in our body. (In part 5, this is used to identify the existence of time-based diseases in the body) The living body is not determined by the *materials* in which it is seen, but by the stability of the *processes* by which those materials are secreted.

At a practical level, this means that we need to discern the time architecture of the living body. This encompasses the resonant **time base** of the local system, its **variational set** of vibrations in its stability range, the **critical frequencies** of bandwidths by which it interlocks with/in, or unlocks from its changing context, and the **potential set** of frequencies from which new wave forms can emerge.

Especially with the new understanding of physics, it makes no sense to regard the living body as a material object. We know that all materials are secreted by living cells. What matters is the manner in which these cells are all interrelated, and together define the dynamic integrity of the whole body. We thus need to regard the body as one integral whole, and a flexible resonant field of vibrations.

This resonant phase field must match the frequency patterns and characteristics of the field from which it emerged. If this is not the case, the wave integration will collapse and the living body will disintegrate within its context. (This is a complex form of "the collapse of the vector of state" in Quantum Physics. Life is based on operating this "collapse".)

Life and death depend on the fractal of time and timing. We must thus not only regard the body cells, but also the way the cell processes are shared between cells, organs, the whole body, people and our context.

The Time Fractal will need to become a major field of research in medicine.

5. Time Base Diseases

Part 1 showed that Time is a form of energy; which helps define the organisation of space. Consciousness (information) is the basis of energy[3]. What we are dealing with are different modes of phase information; in formation. This is seen explicitly in the interaction between living cells.

Part 2 made explicit that all living body cells emerged out of division of the Zygote.

Part 3 then spelled out that these cells all maintain their connections, in sharing not only their same starting point, but also a shared time base.

This 4th part spells out that this can lead to diseases of Time Base. Of this there are many examples, such as circulatory problems, gall stones or kidney stones, memory problems, appendicitis and cancer. They are all have a common basis: two cells in interaction together create a shared Time Base. This we can represent by a time fractal (Figure 3).

When there is a disruption in the unfolding of the Time Fractal, the interrelated harmonics are replaced by disharmonics. We can see that as a shift in the temporal boundary definitions (Figure 4).

The temporal boundary is the process interface in which processes of the system and the context are yes/no connected. (This is part of the System Singularity set. This System Singularity Set has a very specific temporal component. It determines trans-cellular temporal continuity, by synchronisation.)

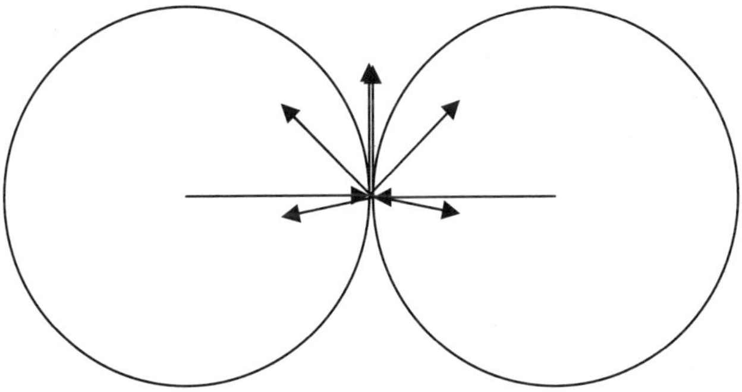

Fig. 4. Cells are temporally linked to their origin; and to their present. The reciprocity of the forward and backward timeline must be in balance with the cell dynamic balance in the present; between all cells and their context. (Space - Time Duality).

Any temporal object has a system definition in which the bandwidth for connecting processes is limited. This means that the system is systematically separated from its surrounding. This is a dynamic interference pattern, which determines the boundary of the system.

When this separation is lost, the internal processes and external processes start to con-fuse. The Regulatory System of the body is very specific in the way it accounts for this. System Regulation is based on the dissolution and resolution of the boundary of the system[10].

Instead of describing this as the re-setting of the boundary of the system (i.e. the (un)linking of processes that connect the body and its context), this can be described also in terms of shifts in the Time Fractal of the unfolding of our body.

This is the basis for the definition of Time base diseases in our body.

Every cell has its own eigen-time-base. Its rhythms determine its proper function. This function is not defined by the cell by itself. It is based by the node and branch and circuit of the Time Fractal of the Bio-Hologram of the body[8]. All intercellular interactions are part of this dynamic. The Time Fractal defines this dynamic at the local level also: within the cell interactions.

Drawing a line, from the core of the cell to the cell membrane, brings the time vector of that cell time base to a moment of bifurcation: the inner rhythms of the cell are interconnected with the rhythms of its context. For the survival of the cell, it is important that these 'separate' sets of rhythms remain connected. Every cell membrane is separator AND connector. It means that the integrity of the cell as a dynamic function is seen by 'reading the time fractal in reverse'. It spells the fusion of different rhythms, onto one common integral coherent time (phase) base.

Health and healing thus have to do with process coherence; thus, with time base integration. Each instant, the process dynamics must match the history of the dynamics of the time line of system unfoldment. (Time Fractal inversion). This can be understood as the need for each individual cell to integrate and unify the internal time rates and rhythms with its context. This calls for the understanding of each living cell as a symphonic harmonic. Like a hologram, it must be a coherent base of modulations of phase. It is a 'standing wave' of phase space vibrations.

Bearing this in mind, the body as a whole is one whole field of vibrations, in which every cell spells a node. Every cell is thus part of the overall field of vibrations, and a vibrational field in itself. They are connected by their shared time base. The Time Fractal[1] is a crucial tool for describing the integrity of this system's time dynamic.

This being the case, the same formulation of the Time Fractal can help us to understand what happens when the symphony of harmonising vibrations in the wave field of living cells gets disturbed. The symphony becomes cacophony, and the time base field becomes incoherent. The natural *constructive* interference pattern becomes a *destructive* interference pattern. Vibrations no longer synchronise and synergise, but de-energise: they drain synchrony, thus energy, from the whole and the local system. This is due to loss of integrity of the time base.

Some examples can be given. Medical lore describes gall stones as "the result of a small stone that blocks the outflow of the gall bladder. The stagnated flow causes massive bile stone formation". Invert this line of thinking: if, for whatever reason, the outflow of the bile duct is stagnated, the gall stones can form. (Bile is a liquid crystal.) The same is the case for kidney stones. Likewise, appendicitis can result from stagnation in the natural peristaltic rhythms of the appendix. (Microbes are sealed off and cause inflammation.) Diarrhoea and constipation are the result of a change of rhythm of peristalsis. Menstruation can have irregularities in flow and discharge when the natural monthly cycle of peristalsis is disturbed. Cold limbs and hot flushes are changes in the rate of circulatory rhythm in the body. Migraines and other symptoms of accumulated tension likewise are expressions of a change in the balance of the natural rhythms. Cancers are based on cells functioning out of rhythm with their context. There are many other diseases are based on the loss of shared rhythms of the cells in their context.

The Time Fractal helps to make explicit that ultimately the basis of the whole body organisation is the synchronisation of cell rhythms, in an integral system in which all cells originate from one common cell, and share one basic cell rhythm, as part of an integral resonant system.

The understanding of the importance of Timing makes it possible to identify and rectify time base diseases in our body. In traditional forms of healing, this is known and used. The meridian system and chackra system are examples of this understanding. Acupuncture points and Chakra points are locations for inducing a phase shift in the timing and assembly of body rhythms. Acupuncture uses an acupuncture point to resolve the spasm of the appendix. The inflamed appendix then can discharge its stagnated bacterial content into the gut lumen to resolve the problem. Likewise, the resynchronisation of cancer has been resolved by inverting the polarity of the surrounding electrical charge field[11]. This is why we need to understand the time rhythms of our body in relationship to the *synergy* (not energy) between living body cells. Local cell function needs to be understood as part of the integral vibration field that all the cells form part of. (All cells originated from the same original cell, the zygote, as part of the same Time Fractal.)

6. Conclusion

It is immaterial to regard our living body as a material object. It is a symphony of living cells, which together form the integrity of our body. That integrity disintegrates[12] when the rhythms of the body no longer match the dynamics of the environment that we live in. This cannot be understood from the perspective of the physical structure of the body (which is a 'deposit' of materials from living vibrating cells). It needs to be understood on the basis of the ongoing cell dynamic. All cells are interconnected, by their common origin from the Zygote. The Time Fractal offers a tool to see how the part relates to the whole.

From the perspective of the Zygote, the Time Fractal traces the rhythms of the cell unfolding to form the individual rhythms of each cell in our body.

From the perspective of any individual cells, the Time Fractal specifies how its eigen time base rhythms relate to its dynamics in its context. Inversely, we can use the Time Fractal to see and understand how vibrations from our context are attuned, assimilated, and integrated into the vibrational pattern (coherent system time base) of our integral body. If this does not happen, the temporal conjunction at the temporal interface causes a temporal disjunction. The body disintegrates, locally, along specific time fractals, from there in local or regional time circuits, and ultimately the whole body time base network collapses. This is a formal definition of disease and death.

The Time Fractal thus helps us to understand our body at a much more basic level of its existence and dynamics. This is of fundamental importance for our understanding of health and healing[7].

References

1. S. Vrobel, A Nested Detector with a Fractal Temporal Interface, Talk held at CASYS, Liège (2005).
2. O. van Nieuwenhuijze, 4-Dimensional Time (Modelling beyond Eternity; The Paradox of Time); Talk held at the Symposium Modelling Anticipative Systems, Baden-Baden (1996).
3. O. van Nieuwenhuijze, STEC: Space-Time_Energy-Consciousness (The Paradox in Model Making), Talk held at CASYS, Liège (1999).
4. O. van Nieuwenhuijze, The Blind Spot of Science (Life, Love, Consciousness and Health), Talk held at CASYS, Liège (2005).
5. B. Lipton, *The Wisdom of Your Cells: How Your Beliefs Control Your Biology*, Sounds True, (1980).
6. L. Margulis & D .Sagan, What is Life, University of California, (2000).
7. O. van Nieuwenhuijze, The Equation of Health (Our Body Map of Health and Knowledge), Talk held at CASYS, Liège (2007).
8. O. van Nieuwenhuijze, BioHolograms, (Phase Coherence of Life. Talk held at the Int. Congress on Cybernetics, Namur (1998).
9. R. Gerber, *Vibrational Medicine: The #1 Handbook of Subtle-Energy Therapies,* Bear & Company, (2001)
10. H. Selye, The Stress of Life, McGraw-Hill, NY, (1956).
11. B.E.W. Nordenström, Biologically Closed Electric Circuits: Clinical, experimental and theoretical evidence for an additional circulatory system, Nordic Medical Publications, Stockholm, (1983).
12. O. van Nieuwenhuijze, (Dis)Integrity; The one source of health, the one cause of disease, Talk held at SRIC, Baden-Baden (1998).

AUTHOR INDEX

SUBJECT INDEX